Lecture Notes in Physics

244

Walter Dittrich
Martin Reuter

Selected Topics in Gauge Theories

Springer-Verlag
Berlin Heidelberg New York Tokyo

Authors

Walter Dittrich
Institut für Theoretische Physik der Universität Tübingen
D-7400 Tübingen, FRG

Martin Reuter
Deutsches Elektronen-Synchrotron DESY
Notkestieg 1, D-2000 Hamburg 52, FRG

ISBN 3-540-16064-7 Springer-Verlag Berlin Heidelberg New York Tokyo
ISBN 0-387-16064-7 Springer-Verlag New York Heidelberg Berlin Tokyo

Printed in Germany

Printing and binding: Beltz Offsetdruck, Hemsbach/Bergstr.
2153/3140-543210

Preface

This volume contains a set of lectures and seminar talks on some recent developments in field theory (with particular emphasis on topological aspects and chiral anomalies in gauge theories) given by the authors at Tübingen University and elsewhere. The presentation of the material is not intended to be a self-contained introduction to the subject, but a workbook for the student who has a basic knowledge of quantum field theory and who now wants to acquire some routine in doing practical calculations in a modern and still rapidly growing branch of particle physics. Therefore, the standard discussions and explanations contained in every modern textbook are omitted to a large extent, whereas the computational details are presented in much greater detail than in the usual monographs or original publications. Hence, to set the stage for the calculations presented here, the reader should first consult at least some of the papers or books cited at the end of each chapter.

We hope that due to this (admittedly unorthodox) manner of presentation, these lectures will prove useful in studying some topics currently being widely discussed, such as chiral anomalies and differential geometry, fermion fractionization, the Wess-Zumino effective action, etc.

The authors would like to thank Mrs. V. Dittrich for the typing of the manuscript.

Tübingen, September 1985

W. Dittrich
M. Reuter

TABLE OF CONTENTS

1. Classical Yang-Mills Theory

Before we start with non-abelian gauge theories, let us recall that an ordinary abelian gauge theory (like QED) with free Lagrangian

$$\mathcal{L} = \bar{\psi}(x)(i\gamma\cdot\partial - m)\psi(x) \tag{1.1}$$

is invariant under the global U(1) phase transformation

$$\begin{aligned} \psi(x) &\longrightarrow \psi'(x) = e^{i\Lambda}\psi(x) \\ \bar{\psi}(x) &\longrightarrow \bar{\psi}'(x) = \bar{\psi}(x)e^{-i\Lambda} \end{aligned} \tag{1.2}$$

with Λ = const., $\Lambda \in \mathbb{R}$. This is called a first-class gauge transformation or gauge transformation of the first kind.

Now we want to make this symmetry a local symmetry, i.e., we want to gauge the U(1) symmetry by replacing Λ by $\Lambda(x)$, where $\Lambda(x)$ is an arbitrary function of space-time position. In this way, (1.2) is turned into a local phase transformation

$$\begin{aligned} \psi(x) &\longrightarrow \psi'(x) = e^{i\Lambda(x)}\psi(x) \\ \bar{\psi}(x) &\longrightarrow \bar{\psi}'(x) = \bar{\psi}(x)e^{-i\Lambda(x)} \end{aligned} \tag{1.3}$$

This is also called a second-class gauge transformation or gauge transformation of the second kind.

Our goal is to construct a theory which will be invariant under the space-time dependent phase change (1.3). Obviously, $\mathcal{L}$ given in

(1.1) is not invariant under (1.3), since we pick up a $\partial_\mu \Lambda(x)$ term. As is familiar, we can obtain a total invariant Lagrangian if we introduce a coupling to a vector field $A_\mu(x)$ (gauge field) which is shifted by $\partial_\mu \Lambda(x)$, while $\psi(x)$ is transformed according to (1.3).

$$\mathcal{L} = \bar{\psi}\, i\gamma(\partial - ieA)\psi - \bar{\psi} m \psi - \frac{1}{4} F_{\mu\nu} F^{\mu\nu} \tag{1.4}$$

$$F_{\mu\nu} = \partial_\mu A_\nu - \partial_\nu A_\mu \tag{1.5}$$

The Lagrangian (1.4) is then locally invariant under the gauge transformation of the second kind, eqs. (1.3), when $A_\mu(x)$ undergoes the transformation

$$A_\mu(x) \longrightarrow A'_\mu(x) = A_\mu(x) + \partial_\mu \Lambda(x) \tag{1.6}$$

Let us prove that $\mathcal{L}$ given in (1.4) is indeed invariant under the local U(1) transformation:

$$\begin{aligned} \psi(x) &\longrightarrow \psi'(x) = e^{ie\Lambda(x)}\, \psi(x) \\ \bar{\psi}(x) &\longrightarrow \bar{\psi}'(x) = \bar{\psi}(x)\, e^{-ie\Lambda(x)} \end{aligned} \tag{1.7}$$

In the following proof, we only need to concentrate on the derivative term:

$$\mathcal{L}(\psi', \bar{\psi}', A'_\mu) = \bar{\psi}(x) e^{-ie\Lambda(x)}\, i\gamma^\mu \left[\partial_\mu - ie(A_\mu + \partial_\mu \Lambda)\right] \cdot e^{ie\Lambda(x)} \psi(x) - \bar{\psi} m \psi - \frac{1}{4} F_{\mu\nu} F^{\mu\nu}$$

$$= \bar{\psi}(x) e^{-ie\Lambda(x)} e^{ie\Lambda(x)}\, i\gamma^\mu \left[ie(\partial_\mu \Lambda) + \partial_\mu - ieA_\mu - ie\,\partial_\mu \Lambda\right] \psi(x)$$

$$-\bar{\psi} m \psi - \frac{1}{4} F_{\mu\nu} F^{\mu\nu}$$

$$= \bar{\psi}\, i\gamma(\partial - ieA)\psi - \bar{\psi} m \psi - \frac{1}{4} F_{\mu\nu} F^{\mu\nu}$$

$$= \mathcal{L}(\psi, \bar{\psi}, A_\mu)$$

The special combination

$$D_\mu := \partial_\mu - ie A_\mu \tag{1.8}$$

is called the gauge-covariant derivative. Note that if we have several fields with different charges e_j, we write

$$\psi_j \rightarrow e^{ie_j\Lambda} \psi_j$$

$$D^j_\mu = \partial_\mu - ie_j A_\mu$$

Taken together, the Lagrangian (1.4) is seen to be formed out of various fields ψ, their gauge-covariant derivatives and the gauge invariant kinetic term of the photon field, $-\frac{1}{4}F^2$, constructed from the gauge field derivative according to

$$F_{\mu\nu} = \partial_\mu A_\nu - \partial_\nu A_\mu$$

A photon mass term $A_\mu A^\mu$ is not permitted, since it is not gauge invariant.

Also note that under gauge transformation of the second kind, $D_\mu\psi$ behaves like ψ, eq. (1.7), namely:

$$D_\mu\psi \longrightarrow [D_\mu\psi(x)]' = e^{ie\Lambda(x)} D_\mu\psi(x) \tag{1.9}$$

By direct substitution of (1.6) into (1.5), it is easy to see that F is, in fact, gauge invariant. It can, however, also be seen as follows:

the antisymmetric field strength tensor $F_{\mu\nu}$ is related to the commutator of two covariant derivatives as

$$[D_\mu, D_\nu]\psi = -ie\,F_{\mu\nu}\,\psi \tag{1.10}$$

From (1.9) it follows that

$$\left([D_\mu, D_\nu]\psi\right)' = e^{ie\Lambda(x)}\left([D_\mu, D_\nu]\psi\right) \tag{1.11}$$

or
$$F'_{\mu\nu}\,\psi' = e^{ie\Lambda(x)}\,F_{\mu\nu}\psi = F_{\mu\nu}\psi' \tag{1.12}$$

which means $F'_{\mu\nu} = F_{\mu\nu}$.

Thus, any Lagrangian constructed solely from ψ, $D_\mu\psi$, and $F_{\mu\nu}$ is invariant under gauge transformations of the second kind if it is invariant under gauge transformations of the first kind.

$$\mathcal{L}(\psi, \partial_\mu\psi) \longrightarrow \mathcal{L}_{matter}(\psi, D_\mu\psi) - \tfrac{1}{4}F_{\mu\nu}F^{\mu\nu}$$

or

$$\bar{\psi}(i\gamma\cdot\partial - m)\psi \longrightarrow \bar{\psi}(i\gamma\cdot D - m)\psi - \tfrac{1}{4}F_{\mu\nu}F^{\mu\nu}$$

We could add a Pauli term, $-\mu'\bar{\psi}\sigma_{\mu\nu}\psi F^{\mu\nu}$, which is also gauge invariant. However, we know that the perturbative calculations based on QED with minimal coupling yield the observed properties of electron-photon physics to such a high degree of accuracy that there is no reason to keep the $\sigma_{\mu\nu}F^{\mu\nu}$ term; it is also ruled out by the requirement of renormalizability.

There is, however, still another Lorentz- and gauge invariant combination:

$$\mathcal{G} = -\tfrac{1}{4}F_{\mu\nu}{}^{*}F^{\mu\nu} = \vec{E}\cdot\vec{B} \tag{1.13}$$

where the dual field strength tensor is given by

$$^{*}F_{\mu\nu} = \frac{1}{2}\,\varepsilon_{\mu\nu\rho\sigma}\,F^{\rho\sigma} \quad . \tag{1.14}$$

The pseudoscalar quantity can be excluded for two reasons: first of all, electrodynamics is parity-conserving and, secondly, $\mathcal{G}$ can be rewritten as a total divergence and so it does not show up in the equation of motion for A_μ ,since it would only affect the action at its endpoints.

$$\begin{aligned}
\frac{1}{4}F_{\mu\nu}{}^{*}F^{\mu\nu} &= \frac{1}{4}F_{\mu\nu}\,\frac{1}{2}\varepsilon^{\mu\nu\rho\sigma}F_{\rho\sigma} \\
&= \frac{1}{8}\varepsilon^{\mu\nu\rho\sigma}(\partial_\mu A_\nu - \partial_\nu A_\mu)(\partial_\rho A_\sigma - \partial_\sigma A_\rho) \\
&= \frac{1}{8}\varepsilon^{\mu\nu\rho\sigma}(\partial_\mu A_\nu \partial_\rho A_\sigma - \partial_\mu A_\nu \partial_\sigma A_\rho \\
&\qquad - \partial_\nu A_\mu \partial_\rho A_\sigma + \partial_\nu A_\mu \partial_\sigma A_\rho) \\
&= \frac{1}{8}\varepsilon^{\mu\nu\rho\sigma}(4\,\partial_\mu A_\nu \partial_\rho A_\sigma) \\
&= \frac{1}{2}\varepsilon^{\mu\nu\rho\sigma}\,\partial_\rho(A_\sigma \partial_\mu A_\nu) \\
&= \partial_\mu\,\varepsilon^{\mu\nu\rho\sigma}\,\frac{1}{2}A_\nu \partial_\rho A_\sigma
\end{aligned}$$

or

$$F_{\mu\nu}{}^{*}F^{\mu\nu} = \partial_\mu X^\mu \,, \quad X^\mu = 2\varepsilon^{\mu\nu\rho\sigma}A_\nu \partial_\rho A_\sigma \tag{1.15}$$

and

$$\mathcal{G} = -\frac{1}{4}\partial_\mu X^\mu \tag{1.16}$$

Since currents are going to play an important role, let us devote a few remarks to the construction of currents.

As is well-known, conserved (and partially conserved) currents are related to symmetry properties. In order to see this relationship, consider a field theory described by the Lagrangian

$$\mathcal{L}(x) = \mathcal{L}\left(\varphi_i(x),\, \partial_\mu \varphi_i(x)\right), \quad i = 1, 2, \ldots, n \tag{1.17}$$

where the fields transform according to some symmetry group G whose Lie algebra is given by

$$[T^i, T^j] = i f_{ijk} T^k \tag{1.18}$$

i=1,2, ... ,(dim. of Lie algebra of G);

the f_{ijk} are the structure constants of the (compact) Lie group:

$f_{ijl}\, f_{lkm} + f_{jkl}\, f_{lim} + f_{kil}\, f_{ljm} = 0$.

Now let us perform an infinitesimal linear local phase transformation on the fields

$$\varphi_i(x) \to \varphi_i'(x) = \varphi_i(x) + i \Lambda^l(x)\, T^l_{ij}\, \varphi_j(x) \tag{1.19}$$

$$\partial_\mu \varphi_i(x) \to \partial_\mu \varphi_i'(x) = \partial_\mu \varphi_i(x) + i \left(\partial_\mu \Lambda^l(x)\right) T^l_{ij}\, \varphi_j(x) + i \Lambda^l(x)\, T^l_{ij} \left(\partial_\mu \varphi_j(x)\right) \tag{1.20}$$

where the T^l are the generators of the (unitary) group, eq. (1.18), and the $\Lambda^l(x)$ are infinitesimal space-time functions.

To first order in $\Lambda^l(x)$, the change in $\mathcal{L}$ is given by

$$\delta \mathcal{L} = \frac{\partial \mathcal{L}}{\partial \varphi_i} \delta \varphi_i + \frac{\partial \mathcal{L}}{\partial(\partial_\mu \varphi_i)} \delta(\partial_\mu \varphi_i)$$

$$= \frac{\partial \mathcal{L}}{\partial \varphi_i} i\Lambda^\ell T_{ij}^{\ell} \varphi_j + \frac{\partial \mathcal{L}}{\partial(\partial_\mu \varphi_i)} i[(\partial_\mu \Lambda^\ell) T_{ij}^{\ell} \varphi_j + \Lambda^\ell T_{ij}^{\ell} (\partial_\mu \varphi_j)]$$

$$= i\left[\frac{\partial \mathcal{L}}{\partial \varphi_i} T_{ij}^{\ell} \varphi_j + \frac{\partial \mathcal{L}}{\partial(\partial_\mu \varphi_i)} T_{ij}^{\ell} (\partial_\mu \varphi_j)\right]\Lambda^\ell + i \frac{\partial \mathcal{L}}{\partial(\partial_\mu \varphi_i)} T_{ij}^{\ell} \varphi_j (\partial_\mu \Lambda^\ell)$$

or

$$\delta \mathcal{L} = \frac{\delta \mathcal{L}}{\delta \Lambda^\ell} \Lambda^\ell(x) + \frac{\delta \mathcal{L}}{\delta(\partial_\mu \Lambda^\ell)} (\partial_\mu \Lambda^\ell(x)) \tag{1.21}$$

where

$$\frac{\delta \mathcal{L}}{\delta \Lambda^\ell} = i\left[\frac{\partial \mathcal{L}}{\partial \varphi_i} T_{ij}^{\ell} \varphi_j + \frac{\partial \mathcal{L}}{\partial(\partial_\mu \varphi_i)} T_{ij}^{\ell} (\partial_\mu \varphi_j)\right] \tag{1.22}$$

and

$$\frac{\delta \mathcal{L}}{\delta(\partial_\mu \Lambda^\ell)} = i \frac{\partial \mathcal{L}}{\partial(\partial_\mu \varphi_i)} T_{ij}^{\ell} \varphi_j \tag{1.23}$$

Now we take the derivative of (1.23):

$$\partial_\mu \frac{\delta \mathcal{L}}{\delta(\partial_\mu \Lambda)} = i\, \partial_\mu \left(\frac{\partial \mathcal{L}}{\partial(\partial_\mu \varphi_i)}\right) T_{ij}^{\ell} \varphi_j + i \frac{\partial \mathcal{L}}{\partial(\partial_\mu \varphi_i)} T_{ij}^{\ell} (\partial_\mu \varphi_j) \tag{1.24}$$

and employ the Euler-Lagrange equation:

$$\partial_\mu \left(\frac{\partial \mathcal{L}}{\partial(\partial_\mu \varphi_i)}\right) = \frac{\partial \mathcal{L}}{\partial \varphi_i} \tag{1.25}$$

to replace the first term on the right-hand side of (1.24) so that

$$\partial_\mu \frac{\delta \mathcal{L}}{\delta(\partial_\mu \Lambda^\ell)} = i \frac{\partial \mathcal{L}}{\partial \varphi_i} T_{ij}^{\ell} \varphi_j + i \frac{\partial \mathcal{L}}{\partial(\partial_\mu \varphi_i)} T_{ij}^{\ell} (\partial_\mu \varphi_j)$$

$$\underset{(1.22)}{=} \frac{\delta \mathcal{L}}{\delta \Lambda^\ell} \tag{1.26}$$

So we have found the (Gell-Mann-Levy) equation

$$\partial_\mu \left(\frac{\delta \mathcal{L}}{\delta(\partial_\mu \Lambda^\ell)}\right) = \frac{\delta \mathcal{L}}{\delta \Lambda^\ell} \tag{1.27}$$

This equation looks like an Euler-Lagrange equation which it is not, because $\Lambda^{l}(x)$ is a (gauge-) function and not a dynamical variable. If we define $J^{l}_{\mu}(x)$ to be the current associated with the phase transformation (1.19), i.e.,

$$J^{l}_{\mu}(x) = -\frac{\delta \mathcal{L}}{\delta(\partial^{\mu}\Lambda^{l})} \underset{(1.23)}{=} -i\frac{\partial \mathcal{L}}{\partial(\partial^{\mu}\psi_i)} T_{ij}{}^{l}\psi_j \tag{1.28}$$

then its divergence is, according to (1.27),

$$\partial^{\mu} J^{l}_{\mu}(x) = -\partial^{\mu}\frac{\delta \mathcal{L}}{\delta(\partial^{\mu}\Lambda^{l})} = -\frac{\delta \mathcal{L}}{\delta\Lambda^{l}}$$

or

$$\partial^{\mu} J^{l}_{\mu} = -\frac{\delta \mathcal{L}}{\delta\Lambda^{l}} \quad . \tag{1.29}$$

Therefore, we can compute the divergence of J^{l}_{μ} directly from $\mathcal{L}(\Lambda)$ without making use of the equations of motion. Moreover, if $\frac{\delta \mathcal{L}(\Lambda)}{\delta\Lambda^{l}} = 0$, then the current is conserved.

So far, we have found the response of $\mathcal{L}$ under infinitesimal phase transformation (cf. eq. (1.21)) to be

$$\delta \mathcal{L} = -\left(\partial_{\mu} J^{\mu l}(x)\right)\Lambda^{l}(x) - J^{\mu l}(x)\left(\partial_{\mu}\Lambda^{l}(x)\right) \tag{1.30}$$

$$= -\partial_{\mu}\left(J^{\mu l}(x)\,\Lambda^{l}(x)\right) \tag{1.31}$$

In particular, if $\Lambda^{l}(x) = \underset{x}{\text{const.}}$, i.e.,

$$\delta \mathcal{L} = -\left(\partial_{\mu} J^{\mu l}(x)\right)\Lambda^{l}(x) \tag{1.32}$$

we learn that for constant infinitesimal phase transformation, $\delta \mathcal{L} = 0$ implies $\partial_{\mu} J^{\mu l}(x) = 0$. Thus, with any constant infinitesimal phase transformation which leaves the Lagrangian invariant, there exists a conserved vector current (Noether-current).

Let us illustrate our rather formal procedure by looking at the simple QED-Lagrangian (1.1):

$$\mathcal{L} = \bar{\psi}(x)(i\gamma\cdot\partial - m)\psi(x) \tag{1.33}$$

Under an infinitesimal local phase transformation

$$\begin{aligned} \psi(x) &\to (1 + i\alpha(x))\psi(x) \\ \bar{\psi}(x) &\to \bar{\psi}(x)(1 - i\alpha(x)), \quad \alpha(x) \text{ inf.} \end{aligned} \tag{1.34}$$

we can generate the (Noether-) vector current without use of the equations of motion.

Hence, let us consider the variation of (1.33) under the phase change (1.34):

$$\begin{aligned} \mathcal{L} \to \mathcal{L}[\alpha] &= \bar{\psi}(x)(1-i\alpha(x))(i\gamma\partial - m)(1+i\alpha(x))\psi(x) + O(\alpha^2) \\ &= \mathcal{L} - (\partial^\mu\alpha(x))\bar{\psi}\gamma_\mu\psi + O(\alpha^2) \end{aligned}$$

which yields

$$\frac{\delta\mathcal{L}[\alpha]}{\delta\alpha} = 0 \tag{1.35}$$

and

$$\frac{\delta\mathcal{L}[\alpha]}{\delta(\partial^\mu\alpha)} = -\bar{\psi}\gamma_\mu\psi \tag{1.36}$$

From equation (1.28), we obtain

$$J_\mu = \bar{\psi}\gamma_\mu\psi \tag{1.37}$$

According to (1.35), the right-hand side of (1.27) vanishes, showing that J_μ is a conserved current:

$$\partial^\mu J_\mu = \partial^\mu(\bar{\psi}\gamma_\mu\psi) = 0 \tag{1.38}$$

(This result is usually obtained by using the Dirac equation and its adjoint:

$$\gamma^\mu \partial_\mu \psi = -im\psi ,$$
$$\partial_\mu \bar{\psi}\gamma^\mu = im\bar{\psi}$$

so that $\partial_\mu(\bar{\psi}\gamma^\mu\psi) = (\partial_\mu\bar{\psi}\gamma^\mu)\psi + \bar{\psi}(\gamma^\mu\partial_\mu\psi) = im\bar{\psi}\psi - im\bar{\psi}\psi = 0$)

At this stage, it is convenient to make contact with Schwinger's action principle applied to a phase change in the original fermionic Lagrangian:

$$\mathcal{L} = \bar{\psi}(i\gamma\cdot\partial - m)\psi =: \mathcal{L}[0] \tag{1.39}$$

$$\psi \rightarrow \psi' = e^{i\alpha(x)}\psi$$
$$\bar{\psi} \rightarrow \bar{\psi}' = \bar{\psi}\, e^{-i\alpha(x)} , \qquad \alpha(x) \text{ finite!} \tag{1.40}$$

Under this finite phase change, we obtain

$$\begin{aligned}\mathcal{L}[0] \longrightarrow \mathcal{L}[\alpha] &= \bar{\psi}e^{-i\alpha}(i\gamma\cdot\partial - m)e^{i\alpha}\psi \\ &= \bar{\psi}e^{-i\alpha}e^{+i\alpha}(-\gamma\cdot\partial\alpha + i\gamma\cdot\partial - m)\psi \\ &= \bar{\psi}(i\gamma\cdot\partial - m)\psi - (\bar{\psi}\gamma^\mu\psi)\partial_\mu\alpha\end{aligned}$$

$$\mathcal{L}[\alpha] = \mathcal{L}[0] - J^\mu\partial_\mu\alpha \tag{1.41}$$

where we use J_μ as a shorthand for the expression $\bar{\psi}\gamma_\mu\psi$. The equation (1.41) reveals the role of $\alpha(x)$ as an external source to which the current J^μ is coupled.

Now let us rewrite (1.41) in the form

$$\delta_\alpha \mathcal{L} := \mathcal{L}[\alpha] - \mathcal{L}[0] = -J^\mu \partial_\mu \alpha \tag{1.42}$$

and look at the response of $\alpha \to \alpha + \delta\alpha$, $\delta\alpha$ infinitesimal. The result is

$$\begin{aligned} \delta_{\alpha+\delta\alpha} \mathcal{L} &= - J^\mu (\partial_\mu \alpha + \partial_\mu \delta\alpha) \\ &= \delta_\alpha \mathcal{L} - J^\mu \partial_\mu \delta\alpha \end{aligned} \tag{1.43}$$

or

$$\delta \mathcal{L} := \delta_{\alpha+\delta\alpha}\mathcal{L} - \delta_\alpha \mathcal{L} = -J^\mu \partial_\mu \delta\alpha$$

which we could have obtained from (1.42) by replacing $\alpha \to \delta\alpha$. Hence, if $\mathcal{L}$ changes according to

$$\mathcal{L} \to \mathcal{L} + \delta\mathcal{L}$$

then the action responds likewise:

$$W_{12} \to W_{12} + \delta W_{12}$$

where

$$\delta W_{12} = \int_{\sigma_2}^{\sigma_1} d^4x \, \delta\mathcal{L} = -\int_{\sigma_2}^{\sigma_1} d^4x \, J^\mu (\partial_\mu \delta\alpha)$$

$\sigma_{1,2}$ space-like hypersurfaces

$$= -\int_{\sigma_2}^{\sigma_1} d^4x \, \partial_\mu (J^\mu \delta\alpha) + \int_{\sigma_2}^{\sigma_1} d^4x \, \delta\alpha \, \partial_\mu J^\mu$$

If W_{12} is stationary with respect to an infinitesimal change in α, then

$$\partial_\mu J^\mu = 0 \tag{1.44}$$

and δW_{12} depends only on the endpoints

$$\delta W_{12} = -\int_{\sigma_2}^{\sigma_1} d\sigma_\mu\, J^\mu \delta\alpha = G[\sigma_1] - G[\sigma_2] \tag{1.45}$$

where

$$G[\sigma] = -\int^{\sigma} d\sigma_\mu\, J^\mu \delta\alpha \tag{1.46}$$

or, in flat space, $d\sigma_0 = d^3x$,

$$G(t) = -\int d^3x\; J^0(\vec{x},t)\, \delta\alpha(\vec{x},t) \tag{1.47}$$

In particular, for constant $\delta\alpha$:

$$\begin{aligned} G(t) &= -\delta\alpha \int d^3x\; J^0(\vec{x},t) \\ &=: -\delta\alpha\; Q(t), \end{aligned}$$

$$Q = \text{total charge.} \tag{1.48}$$

$$\dot{Q}(t) = 0 .$$

Introducing unitary indices i again, we can now rewrite (1.19) as

$$\frac{1}{i}[\psi_i, G] = \delta\psi_i = i\,\delta\alpha^\ell(x)\, T^\ell_{ij}\, \psi_j(x) \tag{1.49}$$

$$\overset{(1.47)}{\|}\qquad\qquad\qquad\qquad \equiv$$

$$i\Big[\psi_i(x), \int d^3x'\, J^{0\ell}(\vec{x}', x^{0\prime})\,\delta\alpha^\ell(\vec{x}', x^{0\prime})\Big] \qquad i\int d^3x'\, \delta^{(3)}(\vec{x}-\vec{x}')\,\delta\alpha^\ell(\vec{x}', x^0)\cdot T_{ij}^{\ell}\, \psi_j(\vec{x}', x^0)$$

$$\equiv$$

$$i\int d^3x'\, \big[\psi_i(\vec{x}, x_0), J^{0\ell}(\vec{x}', x^{0\prime})\big]\,\delta\alpha^\ell(\vec{x}', x^{0\prime})$$

Here we obtain the important equal-time commutator

$$\big[J^{0\ell}(x), \psi_i(x')\big]_{x^{0\prime}=x^0} = -\delta^{(3)}(\vec{x}-\vec{x}')\, T_{ij}^{\ell}\, \psi_j(x) \tag{1.50}$$

For constant $\delta\alpha^\ell$, (1.49) reduces to

$$\frac{1}{i}[Q^\ell(t), \varphi_i(x)]\,\delta\alpha^\ell = i\,\delta\alpha^\ell\, T_{ij}^\ell\, \varphi_j(x)$$

or (1.51)

$$[Q^\ell(t), \varphi_i(x)]_{x^0=t} = -\,T_{ij}^\ell\, \varphi_j(x)$$

The infinitesimal version (1.47) suggests that the following unitary transformation U(t) generated by G generates rotations in the internal space, spanned by the φ_i's:

$$U(t) = e^{i\,G_\alpha(t)}\ , \quad U^+ = U^{-1} \tag{1.52}$$

where $\alpha(\vec{x},t)$ is now finite! Then the transformed field operator is $(J^0\alpha \equiv J^{0\ell}\alpha^\ell)$

$$\varphi_i' = U\varphi_i\, U^{-1}$$

$$= e^{-i\int d^3x'\, J^0(x')\alpha(x')}\, \varphi_i(\vec{x},x^0)\, e^{i\int d^3x'\, J^0(x')\alpha(x')} \tag{1.53}$$

where all quantities are at the same time, $x^0 = x^{0\prime}$.

To evaluate (1.53), consider

$$F_i(\xi) = e^{-i\,\xi\int J^0\alpha}\ \varphi_i\ e^{+i\,\xi\int J^0\alpha} \tag{1.54}$$

If we differentiate (1.54) with respect to the parameter ξ, we obtain the differential equation

$$F_i'(\xi) = -i\, e^{-i\xi\int J^0\alpha} \int J^0(x')\alpha(x')\, \varphi_i(x)\, e^{i\xi\int J^0\alpha}$$
$$+\, i\, e^{-i\xi\int J^0\alpha}\, \varphi_i(x) \int J^0(x')\alpha(x')\, e^{i\xi\int J^0\alpha}$$

$$= i\, e^{-i\xi \int J^0_\alpha} \int \alpha(x') \left[\varphi_i(x), J^0(x')\right]_{x^0 = x'^0} d^3x' \; e^{i\xi \int J^0_\alpha}$$

$$\underset{(1.50)}{=} i\, e^{-i\xi \int J^0_\alpha} \alpha^\ell(\vec{x},t)\, T_{ij}^\ell\, \varphi_j(\vec{x},t)\, e^{i\xi \int J^0_\alpha}$$

$$= i\, \alpha^\ell T_{ij}^\ell F_j(\xi) \equiv i\,\alpha T_{ij} F_j(\xi)$$

This differential equation has the solution

$$F_i(\xi) = \left(e^{i\xi\alpha^\ell T^\ell}\right)_{ij} F_j(0) = \left(e^{i\xi\alpha^\ell T^\ell}\right)_{ij} \varphi_j$$

We are interested in $\xi = 1$, cf. eq. (1.53):

$$F_i(1) = \varphi_i'(x) = \left(e^{i\alpha^\ell T^\ell}\right)_{ij} \varphi_j(\vec{x},t)$$

$$= \left(\delta_{ij} + i\alpha^\ell T_{ij}^\ell + \cdots\right) \varphi_j(\vec{x},t)$$

α^ℓ inf.:

$$= \varphi_i(\vec{x},t) + i\alpha^\ell(\vec{x},t)\, T_{ij}^\ell\, \varphi_j(\vec{x},t) + O(\alpha^2)$$

which brings us back to our original equation (1.19). This, then, concludes our proof that the unitary transformation (1.52), whose generator is G_α, generates rotations in the internal (charge) space.

Two special unitary groups are encountered rather frequently:

(1) G = SU(2) with generators in the fundamental representation

$$T^i = \tfrac{1}{2}\sigma^i, \quad \sigma^1 = \begin{pmatrix} 0 & 1 \\ 1 & 0 \end{pmatrix}, \quad \sigma^2 = \begin{pmatrix} 0 & -i \\ i & 0 \end{pmatrix}, \quad \sigma^3 = \begin{pmatrix} 1 & 0 \\ 0 & -1 \end{pmatrix}$$

$$\mathrm{tr}\, \sigma^i = 0, \quad [\sigma^i, \sigma^j] = 2i\varepsilon^{ijk}\sigma^k, \quad f^{SU(2)}_{ijk} = \varepsilon^{ijk}$$

(2) G = SU(3) with generators in the fundamental representation [3] :

$$T^i = \frac{1}{2}\lambda^i \quad , \quad \lambda^i = \text{Gell-Mann matrices} \; , \; \mathrm{tr}\,\lambda^i = 0 \, ,$$
$$i = 1, 2, 3, \ldots, 8$$
$$[\lambda^i, \lambda^j] = 2i\, f_{ijk}\, \lambda^k \quad , \quad f_{ijk} \text{ totally antisymmetric.}$$

Omitting internal indices, the fields transform according to

$$\phi' = \Omega\, \phi \qquad , \; \Omega = (n \times n)-\text{matrix} \tag{1.55}$$

dependent on x .

while the derivatives respond as

$$(\partial_\mu \phi)' \equiv \partial_\mu \phi' = (\partial_\mu \Omega)\phi + \Omega\, \partial_\mu \phi \tag{1.56}$$

Next, we look for a covariant derivative with

$$(D_\mu \phi)' = \Omega\, (D_\mu \phi) \tag{1.57}$$

or, since $(D_\mu \phi)' \equiv D'_\mu \phi' = \Omega\, D_\mu \phi$

$$\Rightarrow \quad D'_\mu \Omega \phi = \Omega D_\mu \phi$$

$$\Rightarrow \quad D'_\mu \Omega = \Omega D_\mu$$

we want

$$D'_\mu = \Omega D_\mu \Omega^{-1} \tag{1.58}$$

As usual, we introduce a matrix-valued vector field

$$A_\mu(x) = A^i_\mu T^i \tag{1.59}$$

and try the ansatz (ϕ = column vector)

$$D_\mu \phi = (\partial_\mu - ig A_\mu)\phi \tag{1.60}$$

If we then demand for $A_\mu(x)$ to transform as

$$A'_\mu = \Omega A_\mu \Omega^{-1} - \frac{i}{g}(\partial_\mu \Omega)\Omega^{-1} \tag{1.61}$$

then the ansatz (1.60) yields the required transformation property for $D_\mu\phi$, eq. (1.57):

$$\begin{aligned}
(D_\mu\phi)' &= D'_\mu \phi' = (\partial_\mu - ig A'_\mu)\Omega\phi \\
&= (\partial_\mu \Omega)\phi + \Omega\partial_\mu\phi - ig\left[\Omega A \Omega^{-1} - \frac{i}{g}(\partial_\mu\Omega)\Omega^{-1}\right]\Omega\phi \\
&= \Omega(\partial_\mu\phi - ig A_\mu\phi) \\
&= \Omega(D_\mu\phi)
\end{aligned}$$

For an infinitesimal change

$$\begin{aligned}
\Omega_{kl} &= \delta_{kl} - i T^i_{kl}\theta^i + O(\theta^2) \\
\Omega^{-1}_{kl} &= \delta_{kl} + i T^i_{kl}\theta^i + O(\theta^2)
\end{aligned} \tag{1.62}$$

equation (1.61) becomes

$$\begin{aligned}
T^i_{kl} A'^i_\mu &= T^i_{kl} A^i_\mu + i\theta^j [T^i, T^j]_{kl} A^i_\mu - \frac{1}{g} T^i_{kl}\partial_\mu\theta^i \\
&= T^i_{kl} A^i_\mu - f_{ijm}\theta^j T^m_{kl} A^i_\mu - \frac{1}{g} T^i_{kl}\partial_\mu\theta^i
\end{aligned}$$

or

$$\delta A^i_\mu = f_{ijk}\theta^j A^k_\mu - \frac{1}{g}\partial_\mu\theta^i \tag{1.63}$$

To obtain the gauge field tensor and its transformation behavior under (1.62), we go back to eq. (1.10) which now takes the form

$$[D_\mu , D_\nu]\phi = -ig(T^i F^i_{\mu\nu})\phi \tag{1.64}$$

with $$F^i_{\mu\nu} = \partial_\mu A^i_\nu - \partial_\nu A^i_\mu + g f_{ijk} A^j_\mu A^k_\nu \tag{1.65}$$

From the fact that

$$([D_\mu , D_\nu]\phi)' = \Omega(\Theta)([D_\mu , D_\nu]\phi) \tag{1.66}$$

we obtain from (1.64)

$$T^i F^{i\prime}_{\mu\nu}\Omega\phi = \Omega T^i F^i_{\mu\nu}\phi$$

or $$T^i F^{i\prime}_{\mu\nu} = \Omega(\Theta)(T^i F^i_{\mu\nu})\Omega(\Theta)^{-1} \tag{1.67}$$

For infinitesimal transformation, this expression turns into

$$\delta F^i_{\mu\nu} = f_{ijk}\Theta^j F^k_{\mu\nu} \tag{1.68}$$

Unlike the abelian case $f_{ijk} = 0$, $F^i_{\mu\nu}$ transforms nontrivially, like a triplet under SU(2) or an octet under SU(3). However, the trace is gauge invariant:

$$\mathrm{tr}\,[(T^i F^i_{\mu\nu})(T^j F^{\mu\nu j})] \sim F^i_{\mu\nu} F^{\mu\nu i} \tag{1.69}$$

$$\delta[F^i_{\mu\nu} F^{i\mu\nu}] = 2\, \delta F^i_{\mu\nu} F^{i\mu\nu}$$
$$= 2 f_{ijk}\, \Theta^j F^k_{\mu\nu} F^{i\mu\nu}$$
$$= 0 \qquad (f_{ijk} \text{ totally antisymmetric}) \tag{1.70}$$

and, therefore, $-\frac{1}{4}F^i_{\mu\nu}\, F^{i\mu\nu}$ can be used as the kinetic term for the gauge vector field $A^i_\mu(x)$.

We can now summarize the above discussion by presenting the following scheme for constructing a local gauge invariant theory. Let $\mathcal{L}(\phi_k, \partial_\mu\phi_k)$ be a globally gauge invariant Lagrangian, i.e.,

$$\mathcal{L}(\phi', \partial_\mu\phi') = \mathcal{L}(\phi, \partial_\mu\phi)$$

with $\phi' = \Omega\phi$, Ω = const.
The transformations form a non-abelian group G which is assumed to be semi-simple and compact, e.g., G = SU(N), etc.
The number of generators of G is taken to be n , e.g., $n = N^2-1$ for SU(N). Then we introduce n vector gauge fields A^i_μ and define a covariant derivative

$$D_\mu \phi = (\partial_\mu - i g A^i_\mu T^i)\, \phi \tag{1.71}$$

with the second-rank field strength tensor

$$F^i_{\mu\nu} = \partial_\mu A^i_\nu - \partial_\nu A^i_\mu + g f_{ijk} A^j_\mu A^k_\nu \tag{1.72}$$

The local transformation properties are contained in ($A_\mu = T^i A_\mu^i$)

$$
\begin{aligned}
\phi' &= \Omega \phi \\
A'_\mu &= \Omega A_\mu \Omega^{-1} - \frac{i}{g} (\partial_\mu \Omega) \Omega^{-1} \\
\mathcal{F}'_{\mu\nu} &= \Omega \mathcal{F}_{\mu\nu} \Omega
\end{aligned}
\tag{1.73}
$$

with

$$
\Omega = \exp\left\{ -i\, \Theta^i T^i \right\}
\tag{1.74}
$$

The complete gauge invariant Lagrangian which describes the interaction between the gauge fields A_μ^i and the fields ϕ_k is then given by

$$
\mathcal{L} = \mathcal{L}(\phi_k, D_\mu \phi_k) - \frac{1}{4} \mathcal{F}_{\mu\nu}^i \mathcal{F}^{i\mu\nu}
\tag{1.75}
$$

If the group happens to be simple, there exists only one coupling constant. If, however, the group is a product of simple groups such as $SU(2) \times SU(2)$, where the respective set of generators closes under commutation and commutes with other sets, there exist independent coupling constants for each factor group. Here are some well-known examples:

(1) $G = U(1)$, electrodynamics

$\mathcal{L} = \bar{\psi}(i\gamma\partial - m)\psi$ is invariant under $\psi' = \Omega\psi = e^{i\Lambda}\psi$, Λ = const.

Number of generators: one; therefore, we have one gauge field, A_μ. The covariant derivative is given by ($e \equiv g$)

$$
D_\mu = \partial_\mu - ie A_\mu
$$

Since $f_{ijk} = 0$, it follows that

$$
\mathcal{F}_{\mu\nu} = \partial_\mu A_\nu - \partial_\nu A_\mu
$$

Particles and gauge field are then coupled according to the locally gauge invariant Lagrangian

$$\mathcal{L} = \bar{\psi}(i\gamma\cdot D - m)\psi - \frac{1}{4}F_{\mu\nu}F^{\mu\nu}$$

(2) G = SU(2) , $\psi = \begin{pmatrix} p \\ n \end{pmatrix}$

$\mathcal{L} = \bar{\psi}(i\gamma\cdot\partial - m)\psi$ is invariant under the gauge transformation of the first kind

$$\psi' = \Omega\psi \quad , \quad \Omega \in SU(2) \quad , \quad \Omega = \text{const.}$$

$$\Omega = e^{\frac{i}{2}\vec{\tau}\cdot\vec{\Theta}} \tag{1.76}$$

Since the number of generators is three, we need three gauge fields to gauge the isotopic spin.

Covariant derivative:

$$D_\mu\psi_k = (\partial_\mu\delta_{kl} - ig\frac{\tau^i_{kl}}{2}A^i_\mu)\psi_l \tag{1.77}$$

Field strength tensor:

$$F^i_{\mu\nu} = \partial_\mu A^i_\nu - \partial_\nu A^i_\mu + g\,\varepsilon_{ijk}A^j_\mu A^k_\nu \tag{1.78}$$

Gauge invariant Lagrangian:

$$\mathcal{L} = \bar{\psi}(i\gamma\cdot D - m)\psi - \frac{1}{4}F^i_{\mu\nu}F^{i\mu\nu} \tag{1.79}$$

$$= \bar{\psi}(i\gamma\cdot\partial - m)\psi - \frac{1}{4}F^i_{\mu\nu}F^{i\mu\nu} + g\bar{\psi}_k\frac{\tau^i_{kl}}{2}A^i_\mu\gamma^\mu\psi_l$$

(3) G = SU(3) , Chromodynamics, $\psi = \begin{pmatrix} \psi_{red} \\ \psi_{yellow} \\ \psi_{blue} \end{pmatrix}$

The Lagrangian of free quarks with three colors

$$\mathcal{L} = \bar{\psi}(i\gamma\cdot\partial - m)\psi \tag{1.80}$$

is invariant under $\psi \to \psi' = \Omega\psi$, $\Omega \in SU(3)$, Ω = const.

$$\Omega = \exp\left\{ \frac{i}{2} \sum_{i=1}^{8} \theta^i \lambda^i \right\}, \quad T^i = \frac{\lambda^i}{2} \tag{1.81}$$

Covariant derivative:

$$\begin{aligned} D_\mu \psi_k &= \partial_\mu \psi_k - ig \frac{1}{2} \lambda^i_{k\ell} A^i_\mu \psi_\ell \\ &\equiv (\partial_\mu - igA_\mu)_{k\ell} \psi_\ell \end{aligned} \tag{1.82}$$

To gauge the color group $SU(3)_c$, we need eight gauge vector fields (gluon fields, $A_\mu = A^i_\mu \frac{\lambda^i}{2}$). The field strength tensor is

$$F^i_{\mu\nu} = \partial_\mu A^i_\nu - \partial_\nu A^i_\mu + g f_{ijk} A^j_\mu A^k_\nu \tag{1.83}$$

and the gauge invariant Lagrangian

$$\begin{aligned} \mathcal{L}^{QCD} &= \bar{\psi}(i\gamma\cdot D - m)\psi - \frac{1}{4} F^i_{\mu\nu} F^{i\mu\nu} \\ &= \bar{\psi}(i\gamma\cdot\partial - m)\psi - \frac{1}{4} F^i_{\mu\nu} F^{i\mu\nu} + g\bar{\psi}_k \frac{\lambda^i_{k\ell}}{2} A^i_\mu \gamma^\mu \psi_\ell \end{aligned} \tag{1.84}$$

The quark triplet forms the basis for the 3-dimensional (fundamental) representation of SU(3); the generators are the Gell-Mann metrices $T^i = \frac{\lambda^i}{2}$. The field strength tensor $F^i_{\mu\nu}$, however, transforms according to the regular (adjoint) representation of SU(3):

$$\left(T^a_{adj}\right)^{bc} = -i f^{abc} \tag{1.85}$$

$$\begin{aligned} a,b,c = 1, \ldots, n &= (N^2-1) \text{ for } SU(N) \\ &= 8 \text{ for } SU(3) \end{aligned}$$

(For SU(2), it is the pion triplet that transforms according to the regular representation of the Lie algebra $[T^i, T^j] = i\,\varepsilon^{ijk} T^k$, $(T^i_{ad})^{kl} = -i\,\varepsilon^{ikl}$.)

With the definition of the eight 8×8 matrices in (1.85), we can satisfy the Lie algebra (by the Jacobi identity):

$$[T^a_{adj}, T^b_{adj}] = i f^{abc} T^c_{adj} \tag{1.86}$$

and the color vector in $F_{\mu\nu} = \sum_{a=1}^{8} F^a_{\mu\nu} T^a = \vec{F}_{\mu\nu}\cdot\vec{T}$ transforms as

$$\begin{aligned} F'^{\downarrow}_{\mu\nu} &= e^{-i\theta^a(x) T^a_{adj}} F^{\downarrow}_{\mu\nu} \\ &= (1 - i\theta^a(x) T^a_{adj}) F^{\downarrow}_{\mu\nu} + O(\theta^2) \end{aligned}$$

or

$$\delta F^c_{\mu\nu} = -i\,\theta^a (T^a_{adj})^{cd} F^d_{\mu\nu} = f^{cad}\,\theta^a F^d_{\mu\nu} \tag{1.87}$$

The adjoint representation can also be obtained as follows: first, let us go back to the three-dimensional representation and introduce $F^{\mu\nu}$ and Ω as 3×3 matrices for SU(3) , $T^a = \frac{\lambda^a}{2}$:

$$F_{\mu\nu} = F^a_{\mu\nu} T^a , \quad \Omega = e^{-i\theta^a T^a} \tag{1.88}$$

Then it follows that (cf. also eq. (1.67))

$$F'_{\mu\nu} = \Omega F_{\mu\nu} \Omega^{-1} \tag{1.89}$$

The equivalence between (1.89) and (1.87) can be seen by the following calculation:

$$T^a F'^a_{\mu\nu} = e^{-i\theta^a T^a} T^b F^b_{\mu\nu} e^{i\theta^c T^c}$$

$$= F_{\mu\nu}^{b}\,(1 - i\,\theta^{a} T^{a})\,T^{b}\,(1 + i\,\theta^{c} T^{c})$$

$$= F_{\mu\nu}^{b}\,(T^{b} - i\,\theta^{a}[T^{a}, T^{b}]) + O(\theta^{2})$$

$$= F_{\mu\nu}^{b}\,T^{b} + f^{abc}\,\theta^{a}\,T^{c}\,F_{\mu\nu}^{b}$$

or $\quad \delta F_{\mu\nu}^{c} = f^{cab}\,\theta^{a}\,F_{\mu\nu}^{b}$ q.e.d.

Sometimes it is convenient to introduce a generalized cross product for color vectors:

$$A = A^{a} T^{a}, \quad B = B^{a} T^{a}, \quad a = 1, 2, 3, \ldots, n \ (= 8 \text{ for } SU(3))$$

$$\vec{A} \equiv (A_{1}, \ldots, A_{n})$$

$$\vec{B} \equiv (B_{1}, \ldots, B_{n})$$

Then we define $\quad (\vec{A} \times \vec{B})^{a} := f^{abc}\,A^{b} A^{c} \qquad (1.90)$

which implies

$$\begin{aligned} (\vec{A} \times \vec{B})^{a} &= f^{abc}\,A^{b} B^{c} \\ &= -i\;i f^{abc}\,A^{b} B^{c} \\ &= -i\,([T^{b}, T^{c}])^{a}\,A^{b} B^{c} \\ &= -i\,([A, B])^{a} \end{aligned} \qquad (1.91)$$

and, consequently,

$$\vec{A} \times \vec{B} = -\vec{B} \times \vec{A}$$

$$\vec{A} \cdot (\vec{B} \times \vec{C}) = \vec{B} \cdot (\vec{C} \times \vec{A}) = \vec{C} \cdot (\vec{A} \times \vec{B}) \qquad (1.92)$$

$$\vec{A} \times (\vec{B} \times \vec{C}) + \vec{B} \times (\vec{C} \times \vec{A}) + \vec{C} \times (\vec{A} \times \vec{B}) = 0$$

The field strength tensor can thus be written in various forms:

$$\begin{aligned} F_{\mu\nu}^{a} &= \partial_\mu A_\nu^a - \partial_\nu A_\mu^a + g f^{abc} A_\mu^b A_\nu^c \\ &= \partial_\mu A_\nu^a - \partial_\nu A_\mu^a + g(\vec{A}_\mu \times \vec{A}_\nu)^a \\ &= \partial_\mu A_\nu^a - \partial_\nu A_\mu^a - ig[A_\mu, A_\nu]^a \end{aligned} \tag{1.93}$$

or, in vector notation:

$$\vec{F}_{\mu\nu} = \partial_\mu \vec{A}_\nu - \partial_\nu \vec{A}_\mu + g(\vec{A}_\mu \times \vec{A}_\nu) \tag{1.94}$$

Since $F_{\mu\nu} = F_{\mu\nu}^a T^a$, $A_\mu = A_\mu^a T^a$, we also have the matrix-valued field tensor

$$F_{\mu\nu} = \partial_\mu A_\nu - \partial_\nu A_\mu - ig[A_\mu, A_\nu] \tag{1.95}$$

The transformation property

$$\begin{aligned} \delta F_{\mu\nu}^{c} &= f^{cad}\, \Theta^a F_{\mu\nu}^d \\ &= (\vec{\Theta} \times \vec{F}_{\mu\nu})^c \\ \text{or} \quad \delta \vec{F}_{\mu\nu} &= \vec{\Theta} \times \vec{F}_{\mu\nu} \end{aligned} \tag{1.96}$$

identifies $\vec{F}_{\mu\nu}$ again as color vector in eight-dimensional unitary color space. Finally, let us recall that $F_{\mu\nu}$ can be obtained as commutator of two covariant derivatives; $D_\mu = \partial_\mu - ig A_\mu$:

$$[D_\mu, D_\nu]\phi = -ig F_{\mu\nu}\phi \tag{1.97}$$

So far, we have been acting with the covariant derivative D_μ on fields ϕ which transform according to $\delta\phi_k = -i\Theta^a T_{kl}^a \phi_l$.

In QCD, the T^a are the generators in the 3-dimensional fundamental representation, $T^a = \frac{\lambda^a}{2}$. Now we want to show that if $W = W^a T^a$ transforms according to the adjoint representation, then the covariant derivative of W is given by

$$D_\mu W = \partial_\mu W - ig\,[A_\mu, W], \quad \delta\vec{W} = \vec{\Theta} \times \vec{W} \tag{1.98}$$

In particular, we have $(\delta\vec{F} = \vec{\Theta} \times \vec{F})$

$$D_\mu F_{\rho\sigma} = \partial_\mu F_{\rho\sigma} - ig\,[A_\mu, F_{\rho\sigma}] \tag{1.99}$$

Writing $\quad D_\mu W = T^a (D_\mu W)^a$

or

$$\overrightarrow{D_\mu W} = \{(D_\mu W)^a\}_{a=1,\dots,n}$$

we have to prove that

$$(D_\mu W)' = \Omega\,(D_\mu W)\,\Omega^{-1}$$

or

$$\begin{aligned}\delta(\overrightarrow{D_\mu W}) &= \vec{\Theta} \times (\overrightarrow{D_\mu W}) \\ &= \vec{\Theta} \times (\partial_\mu \vec{W} + g\,\vec{A}_\mu \times \vec{W}) \\ &= \vec{\Theta} \times \partial_\mu \vec{W} + g\,\vec{\Theta} \times (\vec{A}_\mu \times \vec{W})\end{aligned}$$

Now

$$\begin{aligned}\delta\,\overrightarrow{D_\mu W} &= \partial_\mu \delta\vec{W} + g\,\delta\vec{A}_\mu \times \vec{W} + g\,\vec{A}_\mu \times \delta\vec{W} \\ &= \partial_\mu(\vec{\Theta} \times \vec{W}) - (\partial_\mu \vec{\Theta}) \times \vec{W} + g(\vec{\Theta} \times \vec{A}_\mu) \times \vec{W} + g\,\vec{A}_\mu \times (\vec{\Theta} \times \vec{W}) \\ &= (\partial_\mu \vec{\Theta}) \times \vec{W} + \vec{\Theta} \times (\partial_\mu \vec{W}) - (\partial_\mu \vec{\Theta}) \times \vec{W} + g(\vec{\Theta} \times \vec{A}_\mu) \times \vec{W} + g\,\vec{A}_\mu \times (\vec{\Theta} \times \vec{W}) \\ &= \vec{\Theta} \times \partial_\mu \vec{W} + g[(\vec{\Theta} \times \vec{A}_\mu) \times \vec{W} + \vec{A}_\mu \times (\vec{\Theta} \times \vec{W})]\end{aligned}$$

Jacobi id. q.e.d.

$$= \vec{\Theta} \times \partial_\mu \vec{W} + g\, \vec{\Theta} \times (\vec{A}_\mu \times \vec{W})$$

Another important relation is

$$[D_\mu, D_\nu] W = -ig\, [F_{\mu\nu}, W] \tag{1.100}$$

Proof:

$$D_\nu \vec{W} = \partial_\nu \vec{W} + g \vec{A}_\nu \times \vec{W} , \quad D_\nu \equiv \partial_\nu + g \vec{A}_\nu \times ,$$

$$D_\mu D_\nu \vec{W} = \partial_\mu (\partial_\nu \vec{W} + g \vec{A}_\nu \times \vec{W}) + g \vec{A}_\mu \times (\partial_\nu \vec{W} + g \vec{A}_\nu \times \vec{W}) ,$$

$$\begin{aligned}
[D_\mu, D_\nu] \vec{W} &= \partial_\mu \partial_\nu \vec{W} + g(\partial_\mu \vec{A}_\nu) \times \vec{W} + g \vec{A}_\nu \times \partial_\mu \vec{W} + g \vec{A}_\mu \times \partial_\nu \vec{W} + g^2 \vec{A}_\mu \times (\vec{A}_\nu \times \vec{W}) \\
&\quad - \partial_\mu \partial_\nu \vec{W} - g(\partial_\nu \vec{A}_\mu) \times \vec{W} - g \vec{A}_\mu \times \partial_\nu \vec{W} - g \vec{A}_\nu \times \partial_\mu \vec{W} - g^2 \vec{A}_\nu \times (\vec{A}_\mu \times \vec{W}) \\
&= g[\partial_\mu \vec{A}_\nu - \partial_\nu \vec{A}_\mu] \times \vec{W} + g^2 [\vec{A}_\mu \times (\vec{A}_\nu \times \vec{W}) - \vec{A}_\nu \times (\vec{A}_\mu \times \vec{W})] \\
&= g[\partial_\mu \vec{A}_\nu - \partial_\nu \vec{A}_\mu + g \vec{A}_\mu \times \vec{A}_\nu] \times \vec{W} \\
&= g\, \vec{F}_{\mu\nu} \times \vec{W}
\end{aligned}$$

Since $(\vec{A} \times \vec{B})^a = -i\, [A, B]^a$, we obtain at last

$$[D_\mu, D_\nu] W = -ig\, [F_{\mu\nu}, W] , \quad \text{q.e.d.}$$

Here are some other useful representations of $D_\mu W$:

$$\begin{aligned}
D_\mu W &= \partial_\mu W - ig\, [A_\mu, W] , \quad A_\mu = A_\mu^a T^a , \; W = W^b T^b \\
&= \partial_\mu W - ig\, [T^b, T^c]\, A_\mu^b W^c \\
&= \partial_\mu W + g\, f^{abc}\, T^a A_\mu^b W^c
\end{aligned} \tag{1.101}$$

or

$$(D_\mu W)^a = \partial_\mu W^a + g f^{abc} A_\mu^b W^c$$
$$= \partial_\mu W^a - ig \underbrace{(if^{abc})}_{=(T_{adj}^b)^{ac}} A_\mu^b W^c \quad ,$$

$$D_\mu^{ac} W^c = [\partial_\mu \delta^{ac} - ig (T_{adj}^b)^{ac} A_\mu^b] W^c$$

which yields

$$D_\mu^{ab} = \partial_\mu \delta^{ab} - ig (T_{adj}^c)^{ab} A_\mu^c \qquad (1.102)$$

a,b,c = 1,2, ... n (= 8 for SU(3))

whereas in the (fundamental) representation of the matter fields ϕ , we found:

$$D_\mu^{kl} = \partial_\mu \delta^{kl} - ig (\tfrac{1}{2}\lambda^a)^{kl} A_\mu^a \qquad (1.103)$$

a = 1,2, ... 8

k,l = 1, ... 3

Given the gauge invariant Lagrangian, we can easily get the Euler-Lagrange equations of motion. First, let us assume that G = SU(N) ; then it is always possible to choose the following normalization of the generators (in the fundamental representation)

$$\mathrm{tr}(T^a T^b) = \tfrac{1}{2}\delta^{ab} \quad ; a,b = 1, .. , N^2-1 \qquad (1.104)$$

so that we can rewrite the gauge field kinetic term as

$$-\tfrac{1}{2}\mathrm{tr}(F_{\mu\nu}F^{\mu\nu}) = -\tfrac{1}{2}F_{\mu\nu}^a F^{b\mu\nu}\,\mathrm{tr}[T^a T^b]$$
$$= -\tfrac{1}{4}F_{\mu\nu}^a F^{a\mu\nu} \qquad (1.105)$$

The Lagrangian is then given in the form

$$\mathcal{L} = \bar{\psi}(i\gamma\cdot D - m)\psi - \frac{1}{2}\,\mathrm{tr}(F_{\mu\nu}F^{\mu\nu}) \tag{1.106}$$

Like in QED, the variation $\frac{\delta\mathcal{L}}{\delta\bar{\psi}}=0$ produces the Dirac equation

$$(i\gamma\cdot D - m)\psi = 0 \;,\quad D_\mu = \partial_\mu - igA^a_\mu T^a \tag{1.107}$$

where ψ now stands for an N-component color spinor.

The equations of motion for the vector field follow from

$$\partial_\mu \frac{\partial\mathcal{L}}{\partial(\partial_\mu A^a_\nu)} = \frac{\partial\mathcal{L}}{\partial A^a_\nu}$$

or

$$\frac{\delta}{\delta A^a_\nu(x)}\int d^4y\;\mathcal{L}(y) = 0$$

Hence,

$$\begin{aligned}\delta S^{Y.M.} &= -\frac{1}{2}\,\delta\int d^4x\;\mathrm{tr}(F_{\mu\nu}F^{\mu\nu})\\ &= -\frac{1}{2}\int d^4x\;\mathrm{tr}(\delta F_{\mu\nu}F^{\mu\nu} + F_{\mu\nu}\,\delta F^{\mu\nu})\\ &= -\int d^4x\;\mathrm{tr}(F_{\mu\nu}\,\delta F^{\mu\nu})\end{aligned} \tag{1.108}$$

Writing explicitly (cf. eq. (1.95))

$$\begin{aligned}F^{\mu\nu} &= \partial^\mu A^\nu - \partial^\nu A^\mu - ig[A^\mu, A^\nu]\\ &= \partial^\mu A^\nu - igA^\mu A^\nu - (\mu\leftrightarrow\nu)\end{aligned}$$

we have $\delta F^{\mu\nu} = \partial^\mu\delta A^\nu - ig\,\delta A^\mu A^\nu - igA^\mu\delta A^\nu - (\mu\leftrightarrow\nu)$

so that (1.108) can be written as $(F_{\mu\nu} = -F_{\nu\mu})$

$$\delta S^{Y.M.} = -2\int d^4x \; tr\left[F_{\mu\nu}(\partial^\mu \delta A^\nu - ig\,\delta A^\mu A^\nu - igA^\mu \delta A^\nu)\right]$$

integ. by parts

$$= 2\int d^4x \; tr\left[(\partial^\mu F_{\mu\nu})\delta A^\nu + ig\,\underbrace{F_{\mu\nu}\delta A^\mu A^\nu}_{=-tr[A^\mu F_{\mu\nu}\delta A^\nu]} + ig F_{\mu\nu} A^\mu \delta A^\nu\right]$$

$$= 2\int d^4x \; tr\left[\left(\partial^\mu F_{\mu\nu} - ig\,[A^\mu, F_{\mu\nu}]\right)\delta A^\nu\right]$$

The fermionic part yields

$$\frac{\partial \mathcal{L}^f}{\partial A^a_\mu} = g\,\bar{\psi}\gamma^\mu T^a \psi$$

so that, altogether:

$$D^\mu F_{\mu\nu} \equiv \partial^\mu F_{\mu\nu} - ig\,[A^\mu, F_{\mu\nu}] = j_{quark\ \nu} \tag{1.109}$$

with

$$j^a_{quark\ \nu} = g\,\bar{\psi}\gamma_\nu T^a \psi \tag{1.110}$$

or

$$\partial^\mu \vec{F}_{\mu\nu} + g(\vec{A}^\mu \times \vec{F}^{\mu\nu} - \bar{\psi}\gamma_\nu \vec{T}\psi) = 0 \tag{1.111}$$

color (i-spin) carried by $\vec{A}_\mu$ quanta — quark color (nucleon i-spin)

We have color current conservation:

$$\vec{j}_\nu = g(\bar{\psi}\gamma_\nu \vec{T}\psi - \vec{A}^\mu \times \vec{F}_{\mu\nu}) \tag{1.112}$$

quark current — gluon current

$$\partial^\nu \partial^\mu \vec{F}_{\mu\nu} = 0 \implies \partial^\nu \vec{j}_\nu = 0 \tag{1.113}$$

$\delta S^{Y.M.} = 0$ yields the equations of motion for the source-free Yang-Mills field:

$$\partial^\mu \mathcal{F}_{\mu\nu} - ig\,[A^\mu, \mathcal{F}_{\mu\nu}] = 0 \tag{1.114}$$

or
$$\partial^\mu \mathcal{F}^a_{\mu\nu} + g\, f^{abc} A^{\mu b} \mathcal{F}^c_{\mu\nu} = 0 \tag{1.115}$$

or
$$\partial^\mu \vec{\mathcal{F}}_{\mu\nu} + g\,\vec{A}^\mu \times \vec{\mathcal{F}}_{\mu\nu} = 0 \tag{1.116}$$

Note that because $\vec{A}_\mu$ does not transform as a vector in color space, the current (1.112) does not transform as a color vector either. However, the associated color charge transforms as a color vector:

$$Q = Q^a T^a = \int_G d^3x\, j_0 \overset{(1.111)}{=} \int_G d^3x\, \partial^\mu \mathcal{F}_{\mu 0}$$

$$= \int_G d^3x\, \partial^j \mathcal{F}_{j0} = \lim_{R\to\infty} \int_{\partial G} d\sigma^j\, \mathcal{F}_{j0} \quad = \text{color charge in } G \tag{1.117}$$

or $Q^a = \int_{\partial G_\infty} d\sigma^j \mathcal{F}^a_{j0}$. These transform covariantly under gauge transformations which become constant asymptotically (i.e., x-independent: $\Omega|_{\partial G_\infty}$ = const.):

$$Q \rightarrow Q' = \int_{\partial G_\infty} d\sigma^j \mathcal{F}'_{j0} = \int_{\partial G_\infty} d\sigma^j\, \Omega\, \mathcal{F}_{j0}\, \Omega^{-1}$$

$$= \Omega \Big(\int_{\partial G_\infty} d\sigma^j\, \mathcal{F}_{0j} \Big) \Omega^{-1} = \Omega\, Q\, \Omega^{-1}$$

From the Jacobi identity,

$$[D_\mu, [D_\rho, D_\sigma]] + [D_\rho, [D_\sigma, D_\mu]] + [D_\sigma, [D_\mu, D_\rho]] = 0 \tag{1.118}$$

and the relation (1.97) $\mathcal{F}_{\mu\nu} = \frac{i}{g}[D_\mu, D_\nu]$, we obtain the

Bianchi identity:

$$D_\mu F_{\rho\sigma} + D_\rho F_{\sigma\mu} + D_\sigma F_{\mu\rho} = 0 \tag{1.119}$$

By introducing the dual of the field strength tensor

$$^*F_{\mu\nu} = \frac{1}{2}\varepsilon_{\mu\nu\rho\sigma}F^{\rho\sigma} \tag{1.120}$$

we get

$$D^\mu \, {}^*F_{\mu\nu} = 0 \tag{1.121}$$

which means

$$\partial^\mu \, {}^*F^a_{\mu\nu} + g f^{abc} A^{\mu b} \, {}^*F^c_{\mu\nu} = 0$$

or (1.122)

$$\partial^\mu \, {}^*F_{\mu\nu} - ig\,[A^\mu, {}^*F_{\mu\nu}] = 0$$

or

$$\partial^\mu \, {}^*\vec{F}_{\mu\nu} + g\,\vec{A}^\mu \times {}^*\vec{F}_{\mu\nu} = 0$$

From

$$\partial^\mu F_{\mu\nu} - ig\,[A^\mu, F_{\mu\nu}] = 0$$

we obtain, using $\partial^\mu \partial^\nu F_{\mu\nu} = 0$,

$$\partial^\nu j_\nu = 0 \tag{1.123}$$

with the conserved color current

$$j_\nu = ig\,[A^\mu, F_{\mu\nu}] = \partial^\mu F_{\mu\nu} \tag{1.124}$$

Sometimes we will need the field strength color vectors explicitly:

$$E^a_n := F^a_{0n} \tag{1.125}$$

$$a = 1,2, \ . \ . \ N^2-1$$

$$B^a_n := -\tfrac{1}{2}\,\varepsilon_{nij}\,F^a_{ij} \qquad n,i,j = 1,2,3$$

where E_n^a is the color electric field and B_n^a is the color magnetic field.

Additional References to
<u>Classical Yang-Mills Theory</u>

E. Abers, B. Lee, Phys. Rep. <u>9</u>, 1 (1973)

P. Becher, M. Böhm, H. Joos, "Eichtheorien der starken und elektroschwachen Wechselwirkung", Teubner, Stuttgart (1981)

C. Itzykson, J.B. Zuber, "Quantum Field Theory", McGraw-Hill, New York (1980)

P. Ramond, "Field Theory", Benjamin Cummings, Reading/Mass. (1981)

P.H. Frampton, Lectures on Gauge Field Theories, UCLA/76/TEP/15

2. Path Integral Quantization of Gauge Theories

In the beginning of this chapter, we want to develop path integral quantization in a very heuristic way, and illustrate it by deriving various Feynman propagators.

Recall that the path integral in non-relativistic quantum mechanics is a method used to obtain transformation amplitudes

$q(t')=q'$ — t', t_N, t_2, t_1, t — $q(t)=q$

$$\langle q',t'|q,t\rangle = N\int \prod_{i=1}^{N} dq(t_i)\, e^{iS(t',t)} \tag{2.1}$$

$$S(t',t) = \int_t^{t'} d\tau\, L \tag{2.2}$$

We have to integrate over all paths satisfying $q(t) = q$, $q(t') = q'$.

The generalization of the formula (2.1) for many coordinates $\{q_j\}$ is given by

$$\langle \{q_j'\},t'|\{q_j\},t\rangle = N\prod_j \prod_{i=1}^{N} \int dq_j(t_i)\, e^{iS(t',t)}$$

with

$$q_j(t) = q_j\ ,\quad q_j(t') = q_j'$$

To make the transition to quantum field theory, let us first consider the case where a single scalar field φ is present. Next, we divide space into small boxes around the coordinates x_j so that we have the correspondence

$$\varphi(x_j, t) \longleftrightarrow q_j$$

meaning that at each spatial point, φ is an independent variable.

Let $|0, \pm\infty\rangle$ be the vacuum states in the remote (infinite) past and future. Then we have for the vacuum-to-vacuum amplitude

$$\langle 0, +\infty | 0, -\infty\rangle \equiv \langle 0_+ | 0_-\rangle = N \prod_j \prod_i \int d\varphi(x_j, t_i)\, e^{i\int_{-\infty}^{\infty} d\tau\, L} \tag{2.3}$$

In covariant notation, the action takes the form

$$L(t) = \int d^3y\, \mathcal{L}(\vec{y}, t) \longrightarrow \int d\tau\, L = \int d^4y\, \mathcal{L}(y)$$

If we denote

$$\prod_j \prod_i \int d\varphi(x_j, t_i) = \prod_x \int d\varphi(x) \quad ,$$

all 4 comp. of x

$$N \prod_x \int d\varphi(x) = \int [d\varphi]$$

we obtain the fundamental path integral formula

$$\langle 0_+ | 0_-\rangle = \int [d\varphi]\, e^{i\int d^4x\, \mathcal{L}} \tag{2.4}$$

In order to get the n-point functions, we couple the field to an external c-number source, giving the generating functional

$$Z[J] := e^{iW[J]} := \langle 0_+ | 0_-\rangle^J = \int [d\varphi]\, e^{i\int d^4x \{\mathcal{L} + J(x)\varphi(x)\}} \tag{2.5}$$

Z[J] is the generating functional for all the Green's functions, including disconnected pieces, while W[J] is the generating functional for the connected parts only.

$$G^N(x_1,\dots,x_N) = \langle 0|T(\varphi(x_1)\dots\varphi(x_N))|0\rangle$$

$$=\left(\frac{1}{i}\right)^N \frac{\delta}{\delta J(x_1)} \cdots\cdots \frac{\delta}{\delta J(x_N)}\; Z[J]\Big|_{J=0} \tag{2.6}$$

All path integrals which we will have to compute will be Gaussian. The basic formula which we will apply in one or another form is then

$$\int_{\mathbb{R}^n} dx_1\dots dx_N\, e^{-x_\ell M_{\ell m} x_m + V_\ell x_\ell} = \pi^{\frac{N}{2}} (\det M)^{-\frac{1}{2}}\, e^{-\frac{1}{4} V_\ell M^{-1}_{\ell m} V_m} \tag{2.7}$$

where M stands for a real, symmetric $n\times n$ matrix with spectrum $\{M'_j\}$, $M'_j>0$, and $(\det M)^{-1/2} = \left(\prod_{j=1}^{n} M'_j\right)^{-1/2}$.

Obviously, we run into problems if M has zero eigenvalues and thus is not invertible. This is exactly what we encounter when a gauge degree of freedom is present. Before we return to this problem, let us look at an example where M is invertible.

Given

$$\mathcal{L} = \frac{1}{2}\partial_\mu\phi\,\partial^\mu\phi - \frac{1}{2}m^2\phi^2 - V(\phi) \tag{2.8}$$

and

$$Z[J] = \int[d\phi]\, e^{i\int d^4x\,\{\mathcal{L} + J(x)\phi(x)\}} \tag{2.9}$$

$$= \int[d\phi]\, e^{-i\int d^4x\, V(\phi)}\, e^{i\int d^4x\{-\frac{1}{2}\phi(\Box+m^2)\phi + J\phi\}} \tag{2.10}$$

$$= e^{-i\int d^4x\, V\left(\frac{1}{i}\frac{\delta}{\delta J(x)}\right)}\, Z_0[J] \tag{2.11}$$

where the generating functional for the free case (V = 0) is given by

$$Z_0[J] = \int[d\phi]\, e^{-\frac{i}{2}\int d^4x\,\{\phi(\Box+m^2)\phi + J\phi\}} \tag{2.12}$$

$$= \text{const.} \cdot e^{-\frac{i}{2}\int d^4x\, d^4y\, J(x) \langle x | \frac{1}{\Box + m^2} | y \rangle J(y)} \tag{2.13}$$

In the last step we used formula (2.7). Clearly, we encounter the scalar particle propagator which is determined by the quadratic term in ϕ :

$$\langle x | \frac{1}{\Box + m^2} | y \rangle = - \int \frac{d^4p}{(2\pi)^4} \frac{e^{-ip\cdot(x-y)}}{p^2 - m^2 + i\varepsilon} \tag{2.14}$$

In ordinary QED, we have the free Lagrangian

$$\mathcal{L} = -\frac{1}{4} F_{\mu\nu} F^{\mu\nu} = -\frac{1}{4}(\partial_\mu A_\nu - \partial_\nu A_\mu)(\partial^\mu A^\nu - \partial^\nu A^\mu)$$

$$= -\frac{1}{4}\left[\partial_\mu A_\nu \partial^\mu A^\nu - \partial_\mu A_\nu \partial^\nu A^\mu - \partial_\nu A_\mu \partial^\mu A^\nu + \partial_\nu A_\mu \partial^\nu A^\mu\right]$$

$$\overset{\text{integ. by parts}}{=} -\frac{1}{2}\left[-(\partial_\mu \partial^\mu A_\nu) A^\nu + (\partial^\mu \partial^\nu A_\mu) A^\nu\right]$$

$$= \frac{1}{2} A_\mu \left[g^{\mu\nu} \Box - \partial^\mu \partial^\nu\right] A_\nu \tag{2.15}$$

and therefore, the corresponding action is given by

$$S = \frac{1}{2} \int d^4x\, A_\mu(x) \left[g^{\mu\nu} \Box - \partial^\mu \partial^\nu\right] A_\nu(x)$$

$$= -\frac{1}{2} \int \frac{d^4k}{(2\pi)^4} \tilde{A}_\mu(k) \left[g^{\mu\nu} k^2 - k^\mu k^\nu\right] \tilde{A}_\nu(-k) \tag{2.16}$$

The operator $(g^{\mu\nu}k^2 - k^\mu k^\nu) := M^{\mu\nu}(k)$ has no inverse, because it has eigenfunctions k^ν, with eigenvalue zero :

$$M^{\mu\nu}(k)\, k_\nu = (k^\mu k^2 - k^\mu k^2) = 0 \tag{2.17}$$

This fact corresponds to the freedom of making a gauge transformation in the integrand of the action, $A_\mu(x) \longrightarrow A_\mu(x) + \partial_\mu \Lambda(x) \equiv A_\mu^\Lambda(x)$,

$$S[A_\mu] = S[A_\mu^\Lambda] \tag{2.18}$$

If we include arbitrary field configurations $A_\mu(x)$ in the path integral

$$\int [dA_\mu^a]\, e^{i S[A_\mu^a]} \tag{2.19}$$

we are integrating over many redundant gauge equivalent fields which make the integral diverge. It is then our goal to select only one representative from every set of gauge equivalent potentials (orbits) which contribute to the path integral (2.19). For this reason, one chooses a gauge condition $f[A_\mu] = 0$ (e.g., $f[A_\mu(x)] = \partial^\mu A_\mu(x) - C$, $f[A_\mu(x)] = \partial_i A^i(x)$) which is satisfied by only one A_μ in every orbit, i.e., the gauge condition is such that the equation

$$f[A_\mu^\Lambda] = 0 \tag{2.20}$$

has a unique solution $\Lambda(x)$ for given $A_\mu(x)$. All the above arguments can be carried through for non-abelian fields so that we can formulate our uniqueness condition as

$$f^a[A_\mu^{[\omega]}] = f^a[A_\mu] \implies \omega^a = 0 \tag{2.21}$$

where

$$A_\mu^{[\Omega]} := A_\mu^{[\omega]} := \Omega A_\mu \Omega^{-1} - \frac{i}{g}(\partial_\mu \Omega)\Omega^{-1} \tag{2.22}$$

and

$$\Omega = e^{-i\omega^a T^a} \equiv e^{-i\vec{\omega}\cdot\vec{T}}$$

Corresponding to $\int df\, \delta(f) = 1$, we have

$$\int [df] \prod_x \delta(f[A_\mu^\Omega(x)]) = 1 \tag{2.23}$$

Changing to Ω as integration variable yields

$$\int [d\Omega]\, \Delta_f[A_\mu^\Omega]\, \delta(f[A_\mu^\Omega]) = 1 \tag{2.24}$$

Here it is important to note that the (Faddeev-Popov) determinant $\Delta_f[A_\mu^\Omega]$ is actually gauge invariant, i.e., independent of the gauge function Ω so that we are allowed to write (2.24) in the form

$$\Delta_f^{-1}[A_\mu] = \int [d\Omega]\, \delta(f[A_\mu^\Omega]) \tag{2.25}$$

That Δ_f^{-1} is indeed gauge invariant follows from

$$\begin{aligned} \Delta_f^{-1}[A_\mu^{\Omega'}] &= \int [d\Omega]\, \delta(f[A_\mu^{[\Omega'\Omega]}]) \\ &= \int [d(\Omega'\Omega)]\, \delta(f[A_\mu^{[\Omega'\Omega]}]) \\ &= \int [d\Omega'']\, \delta(f[A_\mu^{[\Omega'']}]) \\ &= \Delta_f^{-1}[A_\mu] \end{aligned}$$

$d\Omega$ is invariant measure

q.e.d. (2.26)

Furthermore, it can be shown that

$$[dA_\mu^\Omega] = [dA_\mu] \quad \text{i.e.,} \quad \det\left(\frac{\delta A_\mu^\Omega}{\delta A_\mu}\right) = 1 \tag{2.27}$$

and

$$S[A_\mu^\Omega] = S[A_\mu]$$

so that our vacuum persistence amplitude is given by substituting 1 of eq. (2.24) into (2.19),

$$\langle 0_+ | 0_- \rangle = Z = \int [dA_\mu]\, e^{iS[A]}\, 1$$

$$= \int [dA_\mu]\, \Delta_f[A_\mu] \int [d\Omega]\, \delta(f[A_\mu^\Omega])\, e^{iS[A_\mu]} \tag{2.28}$$

Performing a gauge transformation $A_\mu^\Omega \longrightarrow A_\mu$, and using the fact that $[dA_\mu]$, Δ_f, and S are gauge invariant, we obtain

$$\langle 0_+ | 0_- \rangle = \int [d\Omega] \underbrace{\int [dA_\mu]\, \Delta_f[A_\mu]\, \delta(f[A_\mu])\, e^{iS[A_\mu]}}_{\text{indep. of } \Omega .} \tag{2.29}$$

So $\int [d\Omega]$ just gives an (infinite) constant which we are going to omit in the future.

Hence, the correct path integral for pure gauge theories is given by

$$\int [dA_\mu]\, \Delta_f[A_\mu]\, \delta(f[A_\mu])\, e^{iS[A_\mu]} \tag{2.30}$$

This expression is independent of the special choice of f, since $\int [dA]\, e^{iS}$ as well as $\int d\Omega$ = const. do not depend on f.

Now it is convenient to perform a transformation of variables in (2.25)

$$\Delta_f^{-1}[A_\mu] = \int [d\Omega]\, \delta(f[A_\mu^\Omega])$$

$$= \int [d\omega^a]\, \delta(f[A_\mu^\omega])$$

namely,

$$\left(\omega^a(x)\right)_{a=1,\dots,N^2-1} \longrightarrow \left(f^a(x)\right)_{a=1,\dots,N^2-1},$$

$$f^a(x) = f^a[A^\omega(x)]$$

so that

$$[d\omega] = \det\left(\frac{\delta\omega}{\delta f}\right)[df]$$

We then obtain

$$\Delta_f^{-1}[A] = \int [df] \det\left(\frac{\delta\omega}{\delta f}\right)\delta[f] = \det\left(\frac{\delta\omega}{\delta f}\right)\Big|_{f=0}$$

or

$$\Delta_f[A] = \det\left(\frac{\delta f^a(x)}{\delta\omega^b(y)}\right)\Big|_{f=0} \tag{2.31}$$

Applying the chain rule yields

$$\frac{\delta f^a(x)}{\delta\omega^b(y)} = \frac{\partial f^a(x)}{\partial A_\mu^c(x)} \frac{\delta A_\mu^c(x)}{\delta\omega^b(y)} \tag{2.32}$$

For the covariant gauge $f^a[A(x)] = \partial_\mu A^{a\mu}(x)$:

$$\frac{\partial f^a(x)}{\partial A_\mu^c(x)} = \delta^{ac}\partial_x^\mu \tag{2.33}$$

In chapter 1, eq. (1. 63), we already showed the A_μ^c responds under gauge transformation as

$$\begin{aligned}
\delta A_\mu^c &= -\frac{1}{g}\partial_\mu\omega^c + f^{cab}\omega^a A_\mu^b \\
&= -\frac{1}{g}[\partial_\mu\omega^c - g f^{cab}\omega^a A_\mu^b] \\
&= -\frac{1}{g}[\partial_\mu\delta^{ac} + g f^{cba} A_\mu^b]\omega^a \\
&= -\frac{1}{g}[\partial_\mu\vec{\omega} + g\vec{A}_\mu \times \vec{\omega}]^c \\
&= -\frac{1}{g}(D_\mu\vec{\omega})^c \\
&= -\frac{1}{g} D_\mu^{cb}\omega^b
\end{aligned} \tag{2.34}$$

In eq. (2.32) we need

$$\frac{\delta A^c_\mu(x)}{\delta \omega^b(y)} = -\frac{1}{g} D^{cb}_\mu(x)\,\delta(x-y) \tag{2.35}$$

So we obtain at last, for covariant gauges,

$$\det\left(\frac{\delta f}{\delta \omega}\right) = \det\left[-\frac{1}{g}\partial^\mu_x D^{ab}_\mu(x)\,\delta(x-y)\right] \tag{2.36}$$

Since our expression (2.30) is independent of f , we are free to choose a new gauge function

$$f'^b[A(x)] = f^b[A(x)] - c^b(x) \tag{2.37}$$

Furthermore, we can integrate over the A^a_μ-independent functions $c^b(x)$ with an arbitrary weight. This affects only the normalization constant in front of the path integral. Consequently, we arrive at

$$\int[dA^a_\mu]\,\delta(f[A])\det\left(\frac{\delta f}{\delta\omega}\right)e^{iS[A_\mu]}$$

$$\longrightarrow \int[dA^a_\mu]\,\delta(f[A]-c)\det\left(\frac{\delta f}{\delta\omega}\right)e^{iS[A_\mu]}$$

$$\longrightarrow \int[dc^a]\,\underbrace{e^{-\frac{i}{2\alpha}\int d^4x\, c^a(x)c^a(x)}}_{\text{weight factor}}\int[dA^a_\mu]\,\delta(f[A]-c)\det\left(\frac{\delta f}{\delta\omega}\right)e^{iS}$$

$$= \int[dA^a_\mu]\det\left(\frac{\delta f}{\delta\omega}\right)e^{iS[A]-\frac{i}{2\alpha}\int d^4x\, f^a[A]f^a[A]} \tag{2.38}$$

So we get as our final formula for the generating functional (with external sources J)

$$Z[J^a_\mu] = \int[dA^a_\mu]\det\left(\frac{\delta f}{\delta\omega}\right)e^{iS[A]-\frac{i}{2\alpha}\int d^4x\, f^a f^a + i\int d^4x\, J^a_\mu A^{a\mu}} \tag{2.39}$$

and upon using (for anti-commuting c-numbers η, η^*)

$$\int [d\eta \, d\eta^*] \, e^{i\int \eta^* M \eta} = e^{+\mathrm{Tr}\,\ln M} = \det M \tag{2.40}$$

we obtain (for $M := \frac{\delta f}{\delta \omega}$)

$$Z[J] = \int [dA \, d\eta \, d\eta^*] \exp\Big\{ iS[A] - \frac{i}{2\alpha}\int d^4x \, f^a f^a + i\int d^4x \, J^a_\mu(x) A^{\mu a}(x) + i\int d^4x \, d^4y \, \eta^{*a}(x) \frac{\delta f^a(x)}{\delta \omega^b(y)} \eta^b(y) \Big\} \tag{2.41}$$

The $\eta(x)$, $\eta^*(x)$ are fictitious complex massless scalar fields obeying Fermi-Dirac statistics (Faddeev-Popov ghosts).

Note that in the limiting case $\alpha \to 0$, only those functions A^a_μ contribute to the path integral which satisfy $f[A] = 0$:

$$\delta[f] = \prod_x \delta(f(x)) = \lim_{\alpha \to 0} \frac{1}{\sqrt{\pi\alpha}} e^{-\frac{1}{\alpha}\int d^4x \, f^2(x)} \tag{2.42}$$

Now we have reached our goal: the quadratic terms in the normal Lagrangian $\mathcal{L}$ together with $\frac{1}{2\alpha}(f^a)^2$ give us the desired non-singular operator whose inversion produces the photon (gluon, ...) propagator in the corresponding gauge.

Let us verify this statement in QED. ($f = \partial_\mu A^\mu$)
Observe that $\Delta_f[A] = \det\left(-\frac{1}{g}\Box\right) \longrightarrow -\frac{1}{g}\prod k^2$ is divergent, but independent of A and can therefore be absorbed into the normalization constant.

$$\begin{aligned} Z^{QED}[J] &= \int [dA] \exp\Big\{ i\int d^4x \Big[-\frac{1}{4}(\partial_\mu A_\nu - \partial_\nu A_\mu)(\partial^\mu A^\nu - \partial^\nu A^\mu) + \frac{1}{2\alpha}\partial_\mu A^\mu \partial_\nu A^\nu + J_\mu A^\mu \Big] \Big\} \\ &\overset{(2.16)}{=} \int [dA] \exp\Big\{ -\frac{i}{2}\int d^4x \Big[A_\mu \Big(-g^{\mu\nu}\Box + \partial^\mu \partial^\nu \big(1 - \frac{1}{\alpha}\big)\Big) A_\nu - J_\mu A^\mu \Big] \Big\} \\ &= \exp\Big\{ -\frac{i}{2}\int d^4x \, d^4y \, J_\mu(x) \, D_+^{\mu\nu}(x-y|\alpha) \, J_\nu(y) \Big\} \end{aligned} \tag{2.43}$$

where $$D_+^{\mu\nu}(x-y|\alpha) = \langle x|\left[-g^{\mu\nu}\Box + \partial^\mu\partial^\nu(1-\tfrac{1}{\alpha})\right]^{-1}|y\rangle$$

or $$\left[-g^{\mu\nu}\Box + \partial^\mu\partial^\nu(1-\tfrac{1}{\alpha})\right] \underbrace{D_{+\nu\rho}(x-y|\alpha)}_{=\int\frac{d^4k}{(2\pi)^4} D_{+\nu\rho}(k)e^{ik(x-y)}} = \delta^\mu_\rho\,\delta(x-y) \tag{2.44}$$

so that in momentum space

$$\left[-k^2 g^{\mu\nu} + k^\mu k^\nu(1-\tfrac{1}{\alpha})\right] D_{+\nu\rho}(k) = \delta^\mu_\rho \tag{2.45}$$

Thus, the photon propagator is given by

$$D_{+\mu\nu}(k) = \frac{-g_{\mu\nu} + \frac{k_\mu k_\nu}{k^2}(1-\alpha)}{k^2+i\varepsilon} \tag{2.46}$$

since

$$\left[-k^2 g^{\mu\nu} + k^\mu k^\nu(1-\tfrac{1}{\alpha})\right]\frac{1}{k^2+i\varepsilon}\left\{-g_{\nu\rho} + \frac{k_\nu k_\rho}{k^2}(1-\alpha)\right\}$$

$$= \frac{1}{k^2+i\varepsilon}\Big[k^2 g^\mu_\rho - (1-\alpha)k^\mu k_\rho - (1-\tfrac{1}{\alpha})k^\mu k_\rho$$

$$+ k^\mu k_\rho(1-\alpha)(1-\tfrac{1}{\alpha})\Big]$$

$$= \frac{1}{k^2+i\varepsilon}k^2 g^\mu_\rho \xrightarrow[\varepsilon\to 0]{} g^\mu_\rho \qquad \text{q.e.d.}$$

Altogether: $$D_{+\mu\nu}(x-y|\alpha) = -\int\frac{d^4k}{(2\pi)^4}e^{ik(x-y)}\frac{g_{\mu\nu} - \frac{k_\mu k_\nu}{k^2}(1-\alpha)}{k^2+i\varepsilon} \tag{2.47}$$

whereby $\alpha \to 0$ is called the Landau gauge

$$D^L_{+\mu\nu}(x-y) = -\int\frac{d^4k}{(2\pi)^4}e^{ik(x-y)}\frac{g_{\mu\nu} - \frac{k_\mu k_\nu}{k^2}}{k^2+i\varepsilon} \tag{2.48}$$

and $\alpha = 1$ denotes the photon propagator in the Feynman gauge

$$D^{F}_{+\mu\nu}(x-y) = -\int \frac{d^4k}{(2\pi)^4} e^{ik(x-y)} \frac{g_{\mu\nu}}{k^2+i\varepsilon} \tag{2.49}$$

So, with immense effort, we have derived the customary QED propagator for the photon. The formalism comes into its own when we deal with more complicated Lagrangians, like QCD.

In the latter case, we have (2.39) without external sources:

$$\langle 0_+ | 0_- \rangle = \int [dA^a_\mu] \, \Delta_f [A^a_\mu] \, e^{i\int d^4x \{\mathcal{L}^{Y.M.}(x) + \mathcal{L}'(x)\}}$$

where

$$\mathcal{L}^{Y.M.} = -\frac{1}{2} \, tr(F_{\mu\nu} F^{\mu\nu}) = -\frac{1}{4} F^a_{\mu\nu} F^{\mu\nu a}$$

$$\mathcal{L}' = -\frac{1}{2\alpha} (\partial_\mu A^{a\mu})^2$$

and (2.36): $$\Delta_f [A_\mu] = \det \Big[-\frac{1}{g} \partial^\mu_x \underbrace{D^{ab}_\mu(x)}_{=\partial^x_\mu \delta^{ab} - ig(\vec{T}\cdot\vec{A}_\mu)^{ab}} \delta(x-y) \Big] \tag{2.50}$$

$$= \det \Big[\big(-\frac{1}{g} \Box_x \delta^{ab} + i \partial^\mu_x (\vec{T}\cdot\vec{A}_\mu(x))^{ab}\big) \delta(x-y) \Big]$$

$$= \det \Big\{ -\frac{1}{g} \Box_x \big(1 - ig \Box_x^{-1} \partial^\mu_x (\vec{T}\cdot\vec{A}_\mu(x))^{ab}\big) \delta(x-y) \Big\}$$

$$= \underbrace{\det \big(-\frac{1}{g} \Box_x\big)}_{\text{const. : drop}} \det(\mathcal{O}) \tag{2.51}$$

and $$\det \mathcal{O} = e^{\mathrm{Tr} \ln \mathcal{O}}, \quad \mathcal{O} = 1 - ig \Box^{-1} \partial^\mu (\vec{T}\cdot\vec{A}_\mu). \tag{2.52}$$

Hence, our Yang-Mills vacuum persistence amplitude can be written as

$$\langle 0_+|0_-\rangle = \int [dA^a_\mu]\, e^{i\int d^4x\,\{\mathcal{L}^{YM} + \mathcal{L}'\}}\, e^{\mathrm{Tr}\,\ln[1 - ig\Box^{-1}\partial_\mu \vec{T}\cdot\vec{A}^\mu]} \tag{2.53}$$

The last term stems from the contribution of the Faddeev-Popov determinant. The diagramatical interpretation of that term is easiest when we introduce the scalar propagator

$$D^0_+(x-y) = \langle x|-\Box^{-1}|y\rangle \tag{2.54}$$

Then, using

$$\ln(1+x) = -\sum_{n=1}^{\infty} \frac{(-x)^n}{n} \tag{2.55}$$

we obtain the loop contributions

$$\begin{aligned} \exp\{\mathrm{Tr}\,\ln[1 - ig\Box^{-1}\partial_\mu \vec{T}\cdot\vec{A}^\mu]\} &= \exp\{\mathrm{Tr}\,\ln[1 + igD^0_+\partial^\mu(\vec{T}\cdot\vec{A}_\mu)]\} \\ &= \exp\Big\{-\sum_{n=1}^{\infty}\frac{(-ig)^n}{n}\int d^4x_1 \ldots d^4x_n\, \mathrm{tr}\, D^0_+(x_n - x_1)\,\partial^{\mu_1}_{x_1}\vec{A}_{\mu_1}(x_1)\cdot\vec{T}\ldots\ldots \\ &\qquad \ldots\ldots D^0_+(x_{n-1} - x_n)\,\partial^{\mu_n}_{x_n}\vec{A}_{\mu_n}(x_n)\cdot\vec{T}\Big\} \end{aligned} \tag{2.56}$$

Another way to understand the origin of this term is to return to (2.50) which we rewrite as

$$\Delta_f[A^a_\mu] = e^{\mathrm{Tr}\,\ln M} = \det M = \int [d\eta\, d\eta^*]\, e^{i\int \eta^* M \eta} \tag{2.57}$$

with

$$M(x,y)_{ab} = \frac{1}{g}\partial^\mu_x\left(-\partial^x_\mu \delta^{ab} + ig(\vec{T}\cdot\vec{A}_\mu)^{ab}\right)\delta(x-y)$$

It is then obvious from the exponential in (2.57) (absorbing $\frac{1}{g}$ in the normalization constant) that the ghost contributions can be obtained from the coupling

$$i g\, \eta_a^*(x)\, \partial_x^\mu \left((\vec{T}\cdot\vec{A}_\mu(x))_{ab}\, \eta_b(x) \right) = g\, \eta_a^*(x)\, \partial_x^\mu \left(f^{abc} A_\mu^c\, \eta_b \right) \qquad (2.58)$$

Hence, the Faddeev-Popov determinant gives the Feynman ghost loops. These extra terms are necessary to insure a unitary field theory. We could now construct all Feynman rules from the generating functional (with external sources), cf. eq. (2.41); we will, however, postpone doing so until a later chapter.

Additional References to
Path Integral Quantization of Gauge Theories

P. Ramond, "Field Theory", Benjamin Cummings, Reading/Mass. (1981)

E. Abers, B. Lee, Phys. Rep. 9, 1 (1973)

P. Becher, M. Böhm, H. Joos, "Eichtheorien der starken und elektroschwachen Wechselwirkung", Teubner, Stuttgart (1981)

K. Huang, "Quarks, Leptons and Gauge Fields", World Scientific, Singapore(1982)

F.J. Yndurâin, "Quantum Chromodynamics", Springer-Verlag, New York (1983)

3. Background Field Methods

The background field method dates back to Schwinger's famous paper of 1951, entitled "On Gauge Invariance and Vacuum Polarization." In this article, among others, a technique is developed for treating radiative corrections to the photon propagation function which is gauge invariant at all stages.

However, to familiarize the reader with the currently favored path integral approach to background field methods--which is also advantageous beyond one-loop processes-- we will start with the Feynman-Schwinger-Symanzik generating functional for non-gauge theories:

$$\langle 0_+ | 0_- \rangle^K = e^{iW[K]} = Z[K] = \int [d\phi] e^{iS[\phi] + i\int K\phi} \tag{3.1}$$

As usual, $\langle 0_+ | 0_- \rangle^K$ denotes the vacuum persistence amplitude from the distant past to the distant future in presence of the external source K(x). The quantity $S[\phi]$ is the action of the "quantum field" $\phi(x)$, $S[\phi] = \int d^4x \, \mathcal{L}\{\phi(x), \partial_\mu\phi(x)\}$, and the functional integral is extended over all field configurations $\phi(x)$. Finally, to generate the various Green's functions of the theory and ultimately S-matrix elements, the external source K(x) in (3.1) has been coupled to the quantum field $\phi(x)$: $\int K\phi := \int d^4x \, K(x)\phi(x)$.

$e^{iW[K]}$ is known to be the generating functional of the connected and disconnected Green's functions in presence of K(x), which can be obtained by functionally differentiating Z[K] :

$$\langle 0_+ | T(\phi(x_1) \ldots \phi(x_n)) | 0_- \rangle^K = \left(\frac{1}{i}\frac{\delta}{\delta K(x_1)}\right) \cdots \left(\frac{1}{i}\frac{\delta}{\delta K(x_n)}\right) e^{iW[K]} \tag{3.2}$$

Since the disconnected pieces of the Green's functions do not contribute to the S-matrix, it is preferable to employ $W[K]$ itself which can be shown to generate the connected Green's functions:

$$\langle 0_+ | T(\phi(x_1) \ldots \phi(x_n)) | 0_- \rangle_c^K = \left(\frac{1}{i}\frac{\delta}{\delta K(x_1)}\right) \cdots \left(\frac{1}{i}\frac{\delta}{\delta K(x_n)}\right) W[K] \tag{3.3.}$$

In the sequel, we will frequently be interested in the mean field produced in the vacuum by the external source:

$$\bar{\phi}(x) := \langle \phi(x) \rangle^K := \frac{\langle 0_+ | \phi(x) | 0_- \rangle^K}{\langle 0_+ | 0_- \rangle^K} \tag{3.4}$$

This quantity can be obtained from Schwinger's action principle, i.e., by investigating eq. (3.1) under the response of $K(x) \longrightarrow K(x) + \delta K(x)$. Then the left-hand side of (3.1) can be rewritten as:

$$\delta \langle 0_+ | 0_- \rangle^K = \int d^4x \, \delta K(x) \left(\frac{\delta}{\delta K(x)} \langle 0_+ | 0_- \rangle^K \right) \tag{3.5}$$

while the right-hand side changes according to

$$\delta \langle 0_+ | 0_- \rangle^K = \int [d\phi] \, e^{iS[\phi] + i\int (K+\delta K)\phi} - \int [d\phi] \, e^{iS[\phi] + i\int K\phi}$$

which, when expanded to first order in δK, produces

$$i \int d^4x \, \delta K(x) \left(\int [d\phi] \, \phi(x) \, e^{iS[\phi] + i\int K\phi} \right) \tag{3.6}$$

Evidently, this equation together with eq. (3.5) gives

$$\frac{1}{i} \frac{\delta}{\delta K(x)} \langle 0_+ | 0_- \rangle^K = \int [d\phi] \, \phi(x) \, e^{iS[\phi] + i\int K\phi} \tag{3.7}$$

or, if we divide by $\langle 0_+|0_-\rangle^K$, this yields a chain of useful identities:

$$\frac{\frac{1}{i}\frac{\delta}{\delta K(x)}\langle 0_+|0_-\rangle^K}{\langle 0_+|0_-\rangle^K} = \frac{\delta}{\delta K(x)}\,\frac{1}{i}\ln\langle 0_+|0_-\rangle^K = \frac{\delta W[K]}{\delta K(x)} = \bar{\Phi}(x)$$

$$= \langle\phi(x)\rangle^K = \frac{\int[d\phi]\,\phi(x)\,e^{i\int d^4x\,(\mathcal{L}(\phi)+K\phi)}}{\int[d\phi]\,e^{i\int d^4x\,(\mathcal{L}(\phi)+K\phi)}} \tag{3.8}$$

The mean field

$$\bar{\Phi}(x) = \frac{\delta}{\delta K(x)}\,W[K] \tag{3.9}$$

is obviously a functional of K(x). In the following, we will always allow relations such as (3.9) to be invertible, i.e.,

$$\bar{\Phi} = \bar{\Phi}(K) \quad\Longrightarrow\quad K = K(\bar{\Phi})$$

Let us now introduce the effective action $\Gamma[\bar{\Phi}]$, also known as generating functional for one-particle irreducible Green's functions, by means of a Legendre transform of $W[K]$ with respect to the "coordinate" K:

$$\Gamma[\bar{\Phi}] = W[K(\bar{\Phi})] - \int d^4x\,K(\bar{\Phi}(x))\,\bar{\Phi}(x) \tag{3.10}$$

Differentiation of $\Gamma[\bar{\Phi}]$ with respect to $\bar{\Phi}$ (x) reproduces the source function K(x), i.e.,

$$\frac{\delta\Gamma[\bar{\Phi}]}{\delta\bar{\Phi}(x)} = \int d^4y\,\underbrace{\frac{\delta W}{\delta K(y)}}_{(3.9)}\,\frac{\delta K(y)}{\delta\bar{\Phi}(x)} - \int d^4y\,\frac{\delta K(y)}{\delta\bar{\Phi}(x)}\,\bar{\Phi}(y) - K(x)$$

or

$$\frac{\delta \Gamma[\bar{\Phi}]}{\delta \bar{\Phi}(x)} = -K(x) \tag{3.11}$$

which identifies $\Gamma[\bar{\Phi}]$ as effective action. Equation (3.11) defines an implicit source-field relation which has its classical analogue in

$$\frac{\delta}{\delta \phi^{cl}(x)} S[\phi^{cl}] = -K(x) \tag{3.12}$$

yielding the equation of motion for the classical field ϕ^{cl} in the presence of the external current K(x).

Next, we will consider a quantity similar to (3.1), in which the action is a functional of the quantum field ϕ (x) plus a so-called background field B(x), which is left arbitrary for the time being:

$$\tilde{Z}[K,B] = e^{i\tilde{W}[K,B]} = \int [d\phi]\, e^{iS[\phi+B] + i\int K\phi} \tag{3.13}$$

Note that only the quantum field ϕ (x) is coupled to the external source K(x).

Instead of (3.9), we now obtain for the mean field:

$$\tilde{\phi}(x) = \frac{\delta \tilde{W}}{\delta K(x)} = \frac{\delta}{\delta K(x)} \frac{1}{i} \ln \tilde{Z}[K,B] \tag{3.14}$$

from which follows K = K($\tilde{\phi}$, B), and the conventional effective action (3.10) becomes replaced by

$$\tilde{\Gamma}[\tilde{\phi},B] = \tilde{W}[K(\tilde{\phi},B),B] - \int d^4x\, K(\tilde{\phi},B)(x)\, \tilde{\phi}(x) \tag{3.15}$$

which is called the background field generating functional. The essence of these various definitions becomes evident when we shift the integration variable $\phi(x)$ in (3.13), i.e., replace

$$\phi(x) \longrightarrow \phi(x) - B(x) \tag{3.16}$$

The result is

$$\tilde{Z}[K,B] = e^{i\tilde{W}[K,B]} = \int [d\phi] e^{iS[\phi] + i\int K\phi} e^{-i\int KB}$$

or

$$\tilde{Z}[K,B] = Z[K]\, e^{-i\int KB} \tag{3.17}$$

Likewise, upon taking the logarithm of this expression, we have

$$\tilde{W}[K,B] = W[K] - \int KB \tag{3.18}$$

Recalling the relations (3.9) and (3.14), we have in addition

$$\tilde{\phi}(x) = \bar{\phi}(x) - B(x) \tag{3.19}$$

Finally, starting with eq. (3.15), the background field effective action is simply given by

$$\begin{aligned} \tilde{\Gamma}[\tilde{\phi},B] &= \tilde{W}[K,B] - \int K\tilde{\phi} \\ &\overset{(3.18)}{\underset{(3.19)}{=}} \left(W[K] - \int KB\right) - \int K(\bar{\phi} - B) \\ &= W[K] - \int K\bar{\phi} \\ &\underset{(3.10)}{=} \Gamma[\bar{\phi}] \end{aligned}$$

or

$$\tilde{\Gamma}[\tilde{\phi},B] = \Gamma[\tilde{\phi}+B] = \Gamma[\bar{\phi}]\Big|_{\bar{\phi}=\tilde{\phi}+B} \tag{3.20}$$

which says that the background field effective action $\tilde{\Gamma}[\tilde{\phi},B]$ is exactly the usual effective action Γ, as defined in (3.10), with, however, the argument $\bar{\phi}$ replaced by $\tilde{\phi}+B$.

In particular, we obtain for $\tilde{\phi}=0$:

$$\tilde{\Gamma}[0,B]=\Gamma[B] \qquad ; \tag{3.21}$$

hence the effective action $\Gamma[B]$ is given by computing $\tilde{\Gamma}[0,B]$, where the latter has no dependence on $\tilde{\phi}$. Consequently, while $\left(\frac{\delta}{\delta\tilde{\phi}}\right)^n \tilde{\Gamma}[\tilde{\phi},B]$ would generate one-particle irreducible Green's functions in presence of the background field $B(x)$, $\tilde{\Gamma}[0,B]$ is the sum of all one-particle irreducible vacuum graphs in presence of $B(x)$. This, then, is the advantage of the background field method: it permits computation of the effective action by summing only vacuum graphs, i.e., graphs with no external lines.

There are two different, equally important methods to evaluate $\tilde{\Gamma}[0,B]$. The first comes to bear if we are interested in weakly changing, i.e., almost constant fields $B(x)$. Here, we treat $B(x)$ exactly. The second method of approximation uses the smallness of the coupling constant involved to allow the background field to be treated perturbatively. Here, the background field is left arbitrary.

Now let us look at QCD in the background field gauge. In the conventional approach, one starts with the generating functional

$$Z[J]=e^{iW[J]}=\int[dQ]\det\left[\frac{\delta G^a}{\delta\omega^b}\right](Q)\,e^{i\int d^4x\left\{\mathcal{L}(Q)-\frac{G^aG^a}{2\alpha}+J^a_\mu Q^a_\mu\right\}} \tag{3.22}$$

with

$$\mathcal{L}(Q)=-\frac{1}{4}F^a_{\mu\nu}F^a_{\mu\nu}$$

and

$$F^a_{\mu\nu}=\partial_\mu Q^a_\nu-\partial_\nu Q^a_\mu+gf^{abc}Q^b_\mu Q^c_\nu$$

or

$$F_{\mu\nu} = \partial_\mu Q_\nu - \partial_\nu Q_\mu + g A_\mu \times A_\nu$$

Let us recall that

$$A = A^a T^a \quad , \quad a = 1, \dots , (N^2-1) \quad \text{for } SU(N)$$

and $(A \times B)^a := f^{abc} A^b B^c = (-i) i f^{abc} A^b B^c$

$$= (-i)\,([T^b, T^c])^a A^b B^c = -i\,([A,B])^a$$

G^a is the gauge fixing term, e.g., $G^a = \partial_\mu Q_\mu^a$

The response of the gauge field under a gauge transformation is

$$\delta Q_\mu^a(x) = \frac{1}{g} \partial_\mu \omega^a(x) - f^{abc} \omega^b Q_\mu^c(x)$$

$$= \frac{1}{g} [\partial_\mu \delta^{ab} - g f^{abc} Q_\mu^c] \omega^b(x)$$

$$= \frac{1}{g} [\partial_\mu \delta^{ab} + g f^{acb} Q_\mu^c] \omega^b(x)$$

$$= \frac{1}{g} D_\mu^{ab}(Q)\, \omega^b(x) \tag{3.23}$$

where

$$D_\mu^{ab}(Q) = \partial_\mu \delta^{ab} + g f^{acb} Q_\mu^c$$

$$= \partial_\mu \delta^{ab} - g f^{cab} Q_\mu^c$$

$$= \partial_\mu \delta^{ab} - ig (T_{adj}^c)^{ab} Q_\mu^c$$

$$\equiv \partial_\mu \delta^{ab} - ig (Q_\mu)^{ab} \tag{3.24}$$

So we obtain for the derivative of the gauge-fixing term

$$\frac{\delta G^a(x)}{\delta \omega^b(y)}[Q] = \frac{\partial G^a(Q(x))}{\partial Q_\mu^c(x)} \frac{\delta Q_\mu^c(x)}{\delta \omega^b(y)}$$

$$\underset{(3.23)}{=} \frac{1}{g} \frac{\partial G^a(Q(x))}{\partial Q^c_\mu(x)} D^{cb}_\mu(Q(x))\, \delta(x-y) \tag{3.25}$$

We now could proceed and compute the various quantities, namely $G^{(n)}$ from Z, $G^{(n)}_c$ from W, and finally, the 1PI graphs from

$$\Gamma[\bar{Q}] = W[J] - \int d^4x\, J^a_\mu \bar{Q}^a_\mu \,, \quad J = J(\bar{Q}) \tag{3.26}$$

$$\bar{Q}^a_\mu = \frac{\delta W}{\delta J^a_\mu} \,. \tag{3.27}$$

However, we want to set up the background field approach. This is obtained from the above conventional theory by shifting the integration variable in the functional integral--just as we have done in the non-gauge theory. Hence, replace Q by

$$Q \longrightarrow Q + A$$

where only Q gets coupled to the external source J. So, in analogy to (3.22), the modified generating functional is given by

$$\tilde{Z}[J,A] = e^{i\tilde{W}[J,A]} := \int [dQ] \det\left[\frac{\delta G}{\delta \omega}\right](A+Q)\, e^{i\int d^4x \left\{\mathcal{L}(A+Q) - \frac{G^a G^a}{2\alpha} + J^a_\mu Q^a_\mu\right\}} \tag{3.28}$$

where

$$\frac{\delta G^a(x)}{\delta \omega^b(y)}[A+Q] = \frac{1}{g} \frac{\partial G^a(Q(x))}{\partial Q^c_\mu(x)} D^{cb}_\mu(A(x)+Q(x))\, \delta(x-y) \tag{3.29}$$

and, in analogy to (3.26, 27):

$$\tilde{\Gamma}[\tilde{Q},A] = \tilde{W}[J,A] - \int d^4x\, J^a_\mu \tilde{Q}^a_\mu \,, \quad J = J(\tilde{Q},A) \tag{3.30}$$

$$\tilde{Q}^a_\mu := \frac{\delta \tilde{W}}{\delta J^a_\mu} = \tilde{Q}^a_\mu(J,A) \tag{3.31}$$

The various field variables used here are

Q^a_μ = quantum field, integration variable in (3.28)

A^a_μ = background field

$\bar{Q}^a_\mu = \frac{\delta W}{\delta J^a_\mu}$ = argument of the conventional action $\Gamma[\bar{Q}]$

$\tilde{Q}^a_\mu = \frac{\delta \tilde{W}}{\delta J^a_\mu}$ = argument of the background field effective action $\tilde{\Gamma}[\tilde{Q},A]$

The great advantage of the background field method is that there exists a choice of gauge fixing term G^a for which the background field effective action $\tilde{\Gamma}[0,A]$ is a gauge invariant functional of A. This background gauge condition is

$$G^a[Q(x)] := \partial_\mu Q^a_\mu + g f^{abc} A^b_\mu Q^c_\mu \equiv (\partial_\mu Q_\mu + g A_\mu \times Q_\mu)^a \tag{3.32}$$

$$= (\partial_\mu \delta^{ac} + g f^{abc} A^b_\mu) Q^c_\mu$$

$$= (\partial_\mu \delta^{ac} - g f^{bac} A^b_\mu) Q^c_\mu$$

$$= (\partial_\mu \delta^{ac} - ig (T^b_{adj})^{ac} A^b_\mu) Q^c_\mu$$

$$= (\partial_\mu \delta^{ac} - ig (A_\mu)^{ac}) Q^c_\mu$$

$$= D^{ac}_\mu (A) Q^c_\mu \tag{3.33}$$

which yields

$$\frac{\partial G^a(x)}{\partial Q^c_\mu(x)} [Q(x)] = D^{ac}_\mu (A(x))$$

In the functional integral (3.28), we need, however, the shifted value $G^a(A, A+Q) = D^{ac}_\mu(A)\,(A+Q)^c_\mu$:

$$\frac{\delta G^a(x)}{\delta \omega^b(y)}[A+Q] = D^{ac}_\mu(A(x)) \frac{1}{g} D^{cb}_\mu(A(x)+Q(x))\,\delta(x-y)$$

$$= \frac{1}{g}[\partial_\mu - igA_\mu]^{ac}[\partial_\mu - ig(Q_\mu+A_\mu)]^{cb}\,\delta(x-y) \tag{3.34}$$

We will now show that with the choice of gauge stated in eq. (3.32), the background field generating functional $\tilde{Z}[J,A]$ as well as $\tilde{W}[J,A]$ are invariant under the transformations

$$\delta A^a_\mu = -f^{abc}\omega^b A^c_\mu + \frac{1}{g}\partial_\mu\omega^a \qquad (3.35)$$
$$(\delta A_\mu = -\omega\times A_\mu + \frac{1}{g}\partial_\mu\omega)$$
$$\delta J^a_\mu = -f^{abc}\omega^b J^c_\mu \qquad (3.36)$$
$$(\delta J_\mu = -\omega\times J_\mu)$$

To prove this, let us make a change of integration variables in (3.28):

$$Q^a_\mu \longrightarrow Q^a_\mu - f^{abc}\omega^b Q^c_\mu \qquad (3.37)$$

or $$\delta Q^a_\mu = -f^{abc}\omega^b Q^c_\mu$$
$$(\delta Q_\mu = -\omega\times Q_\mu)$$

(1) Adding (3.35) and (3.37), we obtain

$$\delta(A+Q)^a_\mu = -f^{abc}\omega^b (A+Q)^c_\mu + \frac{1}{g}\partial_\mu\omega^a \qquad (3.38)$$

which is just a gauge transformation on the field variable $(A+Q)^a_\mu$, and so the action $S[A+Q] = \int d^4x\, \mathcal{L}(A+Q)$ is clearly invariant.

(2) $G \overset{(3.32)}{=} \partial\cdot Q + g\,A\times Q$

$$\begin{aligned}
\delta G &= \partial\cdot\delta Q + g\,\delta A\times Q + g\,A\times\delta Q \\
&= -\cancel{\partial\omega\times Q} - \omega\times\partial Q - g(\omega\times A)\times Q + \cancel{\partial\omega\times Q} - g\,A\times(\omega\times Q) \\
&= -\omega\times\partial Q - g[(\omega\times A)\times Q + A\times(\omega\times Q)] \\
&\overset{\text{Jac. id.}}{=} -\omega\times\partial Q - g\,\omega\times(A\times Q)
\end{aligned} \qquad (3.39)$$

where we employed the Jacobi identity

$$A\times(B\times C) + B\times(C\times A) + C\times(A\times B) = 0$$

Now we need to show $\delta G^2 = 0$:

$$\delta G^2 = 2G\cdot\delta G$$

$$\overset{(3.39)}{=} 2\,(\partial Q + g\,A\times Q)\cdot(-\omega\times\partial Q - g\,\omega\times(A\times Q))$$

$$= 2\,\{-\partial Q\cdot(\omega\times\partial Q) - g\,\partial Q\cdot[\omega\times(A\times Q)]$$

$$- g\,(A\times Q)\cdot(\omega\times\partial Q) - g^2(A\times Q)\cdot[\omega\times(A\times Q)]\,\}$$

$$= 2\,\{-g\,(A\times Q)\cdot(\partial Q\times\omega) - g\,(A\times Q)\cdot(\omega\times\partial Q)\}$$

$$= 0$$

Equations (3.36) and (3.37) represent an adjoint group rotation, so the term $J^a_\mu\, Q^a_\mu \sim \mathrm{Tr}(J_\mu Q_\mu) = \mathrm{Tr}JQ$ is clearly invariant:

(3) $$\delta\,(JQ) = \delta J\cdot Q + J\cdot\delta Q$$

$$= -(\omega\times J)\cdot Q + J\cdot(-\omega\times Q)$$

$$= (J\times\omega)\cdot Q - J\cdot(\omega\times Q)$$

$$= J\cdot(\omega\times Q) - J\cdot(\omega\times Q)$$

$$= 0$$

Likewise, the determinant in (3.28) is invariant

(4) $$\det\left[\frac{\delta G}{\delta\omega}\right] \xrightarrow{(3.34)} \det D\ D = \det U\,D'\,D'\,U^{\dagger} = \det D'\,D'$$

Now it immediately follows that $\tilde{\Gamma}[\tilde{Q},A]$ is also invariant, i.e.,

$$\tilde{\Gamma}[\tilde{Q},A] = \tilde{W}[J,A] - \int d^4x\, J^a_\mu\, \tilde{Q}^a_\mu\ , \quad J = J(\tilde{Q}),\ \tilde{Q}^a_\mu = \frac{\delta\tilde{W}}{\delta J^a_\mu}$$

remains unchanged under

$$\delta A^a_\mu = -f^{abc}\,\omega^b A^c_\mu + \frac{1}{g}\,\partial_\mu\omega^a$$

$$\left(\delta A_\mu = -\omega\times A_\mu + \frac{1}{g}\,\partial_\mu\omega\right)$$

$$\delta\tilde{Q}^a_\mu = -f^{abc}\,\omega^b\, Q^c_\mu$$

$$\left(\delta\tilde{Q}_\mu = -\omega\times\tilde{Q}_\mu\right)$$

since $\tilde{Q}$ is just the conjugate variable to J,

$$\tilde{Q} = \frac{\delta \tilde{W}}{\delta J}$$, and $\tilde{W}$ is invariant under $\delta A = -\omega \times A + \frac{1}{g}\partial\omega$

and $\delta J = -\omega \times J$ so that with

$J^a_\mu \; \tilde{Q}^a_\mu \sim \mathrm{Tr}\, J\, \tilde{Q}$ we obtain

$$\begin{aligned}
\delta(J\cdot\tilde{Q}) &= \delta J\cdot\tilde{Q} + J\cdot\delta\tilde{Q} \\
&= -(\omega\times J)\cdot\tilde{Q} - J\cdot(\omega\times Q) \\
&= 0
\end{aligned}$$

We will now turn to the relation between $\tilde{\Gamma}[0, A]$ and $\Gamma[\bar{Q}]$.

For this reason, we return to eq. (3.28):

$$\tilde{Z}[J,A] = \int [dQ]\, \underbrace{\det\left[\frac{\delta G}{\delta\omega}\right](A+Q)}_{=\frac{1}{g}D(A)\,D(A+Q)}\, e^{i\int d^4x\{\mathcal{L}(A+Q) - \frac{1}{2\alpha}[D^{ac}_\mu(A)Q^c_\mu]^2 + J^a_\mu Q^a_\mu\}}$$

$$\overset{Q\to Q-A}{=} \int [dQ]\,\det\left[\frac{\delta G}{\delta\omega}\right](Q)\, e^{i\int d^4x\{\mathcal{L}(Q) - \frac{1}{2\alpha}[D^{ac}_\mu(A)(Q-A)^c_\mu]^2 + J^a_\mu Q^a_\mu - J^a_\mu A^a_\mu\}}$$

$$\equiv \Big(\int [dQ]\, \underbrace{\det\left[\frac{\delta G}{\delta\omega}\right](Q)}_{=\frac{1}{g}D(A)\,D(Q)}\, e^{i\int d^4x\{\mathcal{L}(Q) - \frac{1}{2\alpha}G^aG^a + J^a_\mu Q^a_\mu\}} \Big)\, e^{-i\int d^4x\, J^a_\mu A^a_\mu}$$

$$= Z[J]\, e^{-i\int d^4x\, J^a_\mu A^a_\mu} \tag{3.40}$$

where $Z[J]$ is the coventional generating functional of eq. (3.22), evaluated, however, with the gauge fixing term

$$\begin{aligned}
G^a &= D^{ac}_\mu(A)(Q-A)^c_\mu \\
&\overset{(3.32)}{=} \partial_\mu(Q^a_\mu - A^a_\mu) + g f^{abc} A^b_\mu (Q^c_\mu - \cancel{A^c_\mu})
\end{aligned}$$

$$= \partial_\mu Q_\mu^a - \partial_\mu A_\mu^a + g f^{abc} A_\mu^b Q_\mu^c \tag{3.41}$$

This also enters the det in Z[J]!

In shorthand notation, we end up with

$$\underset{\substack{\uparrow \text{ gauge fixing} \\ G = \partial Q + g A\times Q}}{\tilde{Z}[J,A]} = \underset{\substack{\uparrow \\ G = \partial Q - \partial A + g A\times Q}}{Z[J]} \, e^{-i\int d^4x\, J_\mu^a A_\mu^a} \tag{3.42}$$

From

$$W = -i \ln Z \, , \quad \tilde{W} = -i \ln \tilde{Z}$$

it follows that W and $\tilde{W}$ are related by

$$\tilde{W}[J,A] = W[J] - \int d^4x\, J_\mu^a A_\mu^a \tag{3.43}$$

and since $\bar{Q} = \frac{\delta W}{\delta J}$ and $\tilde{Q} = \frac{\delta \tilde{W}}{\delta J}$, we find that $\tilde{Q}_\mu^a = \bar{Q}_\mu^a - A_\mu^a$

Finally, performing a Legendre transform on (3.43),

$$\tilde{\Gamma}[\tilde{Q},A] = \tilde{W}[J,A] - \int d^4x\, J_\mu^a \tilde{Q}_\mu^a \, ,$$

$$\tilde{Q}_\mu^a = \frac{\delta \tilde{W}}{\delta J_\mu^a} = \tilde{Q}_\mu^a(A,J) \Rightarrow J = J(\tilde{Q},A)$$

$$= W[J] - \int d^4x\, J_\mu^a (A + \tilde{Q})_\mu^a$$

$$\overset{(3.26)}{=} \Gamma[\bar{Q} = \tilde{Q} + A] \tag{3.44}$$

we have a relation between the background field effective action and the conventional effective action:

$$\tilde{\Gamma}[\tilde{Q},A] = \Gamma[\bar{Q}]\Big|_{\bar{Q} = \tilde{Q} + A} \tag{3.45}$$

The gauge invariant effective action is just

$$\tilde{\Gamma}[0,A] = \Gamma[\bar{Q}]\Big|_{\bar{Q}=A} \tag{3.46}$$

$$\uparrow \qquad\qquad\qquad \uparrow$$

$$G = \partial Q + gA\times Q \qquad G = \partial Q - \partial A + gA\times Q$$

(background field gauge)

The background field effective action $\tilde{\Gamma}[0,A]$ generates 1PI Green's functions. These are calculated with the aid of Feynman rules derived from the shifted action $S[Q + A]$, the gauge fixing term as indicated in (3.46) and the determinant which is written in terms of an anti-commuting ghost field:

$$\frac{\delta G(x)}{\delta\omega(y)}[A+Q] \overset{(3.34)}{=} \frac{1}{g}[\partial - igA][\partial - ig(Q+A)]\delta(x-y) \quad ,$$

$$\int[d\theta\, d\theta^*]\, e^{i\theta^* M\theta} = \det M$$

which means, in our particular case,

$$\theta^* \frac{\delta G}{\delta\omega}[A+Q]\,\theta = \int dx\,dy\, \theta^*(x)\big[\partial_x^2 - ig\partial_x(Q+A) - igA\partial_x - g^2A(Q+A)\big]\delta(x-y)\,\theta(y)$$

$$= \int dx\,dy\, \theta^*(x)\big[\partial_x^2 + ig\overleftarrow{\partial}_x(Q+A) - igA\overrightarrow{\partial}_x - g^2A(Q+A)\big]\delta(x-y)\,\theta(y)$$

$$\uparrow$$

$$\int dy\, \partial_x\delta(x-y)\,\theta(y)$$

$$= -\int dy\,\partial_y\delta(x-y)\,\theta(y) = \partial_x\theta(x)$$

$$= \int dx\, \theta^*(x)\big[\Box + ig\overleftarrow{\partial}_\mu(Q_\mu + A_\mu) - igA_\mu\overrightarrow{\partial}_\mu - g^2A_\mu(Q_\mu + A_\mu)\big]\theta(x)$$

$$= \int dx\, \theta^*_a(x)\big[\Box\delta^{ab} + ig\overleftarrow{\partial}_\mu(Q^c_\mu + A^c_\mu)\underbrace{(T^c)^{ab}}_{=-if^{cab} = if^{acb}}$$

$$- ig A_\mu^c \overrightarrow{\partial}_\mu \underbrace{(T^c)^{ab}}_{= i f^{acb}}$$

$$- g^2 A_\mu^c (A_\mu^d + Q_\mu^d) \underbrace{(T^c T^d)^{ab}}\ ] \ \Theta_b(x)$$

$$= (T^c)^{ax} (T^d)^{xb} = (-i)^2 f^{cax} f^{dxb}$$

$$= - f^{acx} f^{xdb}$$

Hence, we find for the ghost Lagrangian

$$\mathcal{L}_{\text{ghost}} = - \Theta_a^* \left[\Box \delta^{ab} - g \overleftarrow{\partial}_\mu f^{acb} (A_\mu^c + Q_\mu^c) + g f^{acb} A_\mu^c \overrightarrow{\partial}_\mu \right.$$

$$\left. + g^2 f^{acx} f^{xdb} A_\mu^c (A_\mu^d + Q_\mu^d) \right] \Theta_b \tag{3.47}$$

Note that vertices involving Q-fields are used inside diagrams, while vertices involving A-fields are used for external lines. All propagators appearing in these 1PI diagrams are Q-field propagators. This is the basic idea, since the A-field propagator is undefined because the A-field gauge invariance has not been broken. Since only 1PI graphs are being considered, vertices involving only one Q-field will never contribute and so can be ignored.

When calculating the effective action $\tilde{\Gamma}[0,A]$, divergences occur which have to be renormalized:

$$\begin{aligned} (A_\mu)_0 &= Z_A^{1/2} A_\mu \\ g_0 &= Z_g \, g \\ \alpha_0 &= Z_\alpha \, \alpha \end{aligned} \tag{3.48}$$

These are the field, coupling constant and gauge fixing renormalization constants and they alone are sufficient to renormalize the theory. The Q-field and the ghost field do not have to be renormalized, since they only appear inside loops.

Above we showed that gauge invariance is manifest in the background field approach and so there exists a relation between the renormalization constants Z_g and Z_A. This can best be seen from the field renormalization

$$(F_{\mu\nu}^a)_0 = Z_A^{1/2} \left[\partial_\mu A_\nu^a - \partial_\nu A_\mu^a + g Z_g Z_A^{\frac{1}{2}} f^{abc} A_\mu^b A_\nu^c \right]$$

and this will take on the gauge covariant form if

$$(F^a_{\mu\nu})_o = \text{const.}\ F^a_{\mu\nu}$$

Hence, gauge invariance of $\tilde{\Gamma}[0,A]$ implies

$$Z_g = Z_A^{-1/2} \tag{3.49}$$

so that $g_o A_o = gA$ (e.g., $e_o H_o = eH$).

It is fairly easy to incorporate fermions into our theory.

First, recall the conventional approach with the generating functional

$$Z[J,\eta,\bar\eta] = e^{iW[J,\eta,\bar\eta]}$$
$$= \int [dQ\, d\psi\, d\bar\psi] \det\left[\frac{\delta G}{\delta\omega}\right](Q)\, e^{i\int d^4x \{\mathcal{L}^{Y.M.}(Q) + \bar\psi(i\partial\!\!\!/ - gQ^{\mu i}\gamma_\mu T^i - m)\psi - \frac{1}{2\alpha}G^a G^a + JQ + \bar\eta\psi + \bar\psi\eta\}} \tag{3.50}$$

and the effective action

$$\Gamma[\bar Q, \overline{(\psi)}, \overline{(\bar\psi)}] = W[J,\eta,\bar\eta] - \int (J\bar Q + \bar\eta\,\overline{(\psi)} + \overline{(\bar\psi)}\,\eta) \tag{3.51}$$

whereby

$$\bar Q = \frac{\delta W}{\delta J} \tag{3.52}$$

and

$$\overline{(\bar\psi)} = -\frac{\delta W}{\delta\eta}\,, \quad \overline{(\psi)} = \frac{\delta W}{\delta\bar\eta}$$

On the way to a gauge invariant effective action, we shift Q by $Q \to Q + A$, so that

$$\tilde Z[J,\eta,\bar\eta;A] = e^{i\tilde W[J,\eta,\bar\eta;A]}$$
$$= \int [dQ\, d\psi\, d\bar\psi] \underbrace{\det\left[\frac{\delta G}{\delta\omega}\right](Q+A)}_{\frac{1}{g}D(A)D(Q+A)}\, e^{i\int d^4x \{\mathcal{L}^{Y.M.}(Q+A) + \bar\psi(i\partial\!\!\!/ - g(A+Q)^{\mu i}\gamma_\mu T^i - m)\psi - \frac{1}{2\alpha}\{D(A)Q\}^2 + JQ + \bar\eta\psi + \bar\psi\eta\}} \tag{3.53}$$

As before, we can show that this background field generating functional is invariant under

$$\begin{aligned} \delta A &= -\omega \times A + \frac{1}{g}\partial\omega \\ \delta J &= -\omega \times J \\ \eta' &= e^{-i\omega}\eta \qquad (\omega \equiv \omega^a T^a) \\ \bar\eta' &= \bar\eta\, e^{i\omega} \end{aligned} \tag{3.54}$$

To prove the invariance, it is only necessary to change variables according to (cf. (3.37))

$$\delta Q = -\omega \times Q\,, \quad \psi' = e^{-i\omega}\psi\,, \quad \bar{\psi}' = \bar{\psi}\, e^{i\omega} \tag{3.55}$$

A Legendre transform then brings us to

$$\tilde{\Gamma}[\tilde{Q}, \tilde{\psi}, \tilde{\bar{\psi}}; A] = \tilde{W}[J, \eta, \bar{\eta}; A] - \int d^4x \{ J\tilde{Q} + \bar{\eta}\tilde{\psi} + \tilde{\bar{\psi}}\eta \} \tag{3.56}$$

with
$$\tilde{Q} = \frac{\delta \tilde{W}}{\delta J} \tag{3.57}$$

and
$$\tilde{\bar{\psi}} = -\frac{\delta \tilde{W}}{\delta \eta}\,, \quad \tilde{\psi} = \frac{\delta \tilde{W}}{\delta \bar{\eta}}$$

As before, $\tilde{\Gamma}$ can be shown to be invariant under

$$\begin{aligned} \delta A &= -\omega \times A + \frac{1}{g}\partial\omega \\ \delta\tilde{Q} &= -\omega \times \tilde{Q} \\ \tilde{\psi}' &= e^{-i\omega}\tilde{\psi} \\ \tilde{\bar{\psi}}' &= \tilde{\bar{\psi}}\, e^{i\omega} \end{aligned} \tag{3.58}$$

Now let us return to eq. (3.53), which we rewrite as

$$\tilde{Z}[J, \eta, \bar{\eta}; A] = e^{i\tilde{W}[J, \eta, \bar{\eta}; A]} \tag{3.59}$$

$$\overset{Q \to Q - A}{=} \int [dQ\, d\psi\, d\bar{\psi}]\, \underbrace{\det\left[\frac{\delta G}{\delta\omega}\right](Q)}_{= \frac{1}{g} D(A) D(Q)}\, e^{i\int d^4x \{ \mathcal{L}^{Y.M.}(Q) + \bar{\psi}(i\partial\!\!\!/ - gQ^{\mu i}\gamma_\mu T^i - m)\psi - \frac{1}{2\alpha}[D(A)(Q-A)]^2 + JQ + \bar{\psi}\eta + \bar{\eta}\psi \}}$$

$$\cdot\, e^{-i\int JA}$$

and a comparison with eq. (3.50) permits us to write

$$\tilde{Z}[J,\eta,\bar{\eta};A] = Z[J,\eta,\bar{\eta}]\,e^{-i\int JA} \tag{3.60}$$

$$\tilde{W}[J,\eta,\bar{\eta};A] = W[J,\eta,\bar{\eta}] - \int JA$$

$$\uparrow \qquad\qquad\qquad \uparrow$$

$$G_1 = \partial Q + gA\times Q \qquad G_2 = \partial Q - \partial A + gA\times Q$$

where the left-hand side is taken in the background field gauge.
We also find

$$\begin{aligned}\tilde{Q} &= \bar{Q} - A\\ \tilde{\psi} &= \overline{(\psi)}\\ \tilde{\bar{\psi}} &= \overline{(\bar{\psi})}\end{aligned} \tag{3.61}$$

At last we come to the relation between Γ and $\tilde{\Gamma}$:

$$\begin{aligned}\tilde{\Gamma}[\tilde{Q},\tilde{\bar{\psi}},\tilde{\psi};A] &= \tilde{W}[J,\eta,\bar{\eta};A] - \int(J\tilde{Q} + \bar{\eta}\tilde{\psi} + \tilde{\bar{\psi}}\eta)\\ &= W[J,\eta,\bar{\eta};A] - \int\{J\cdot(\tilde{Q}+A) + \bar{\eta}\tilde{\psi} + \tilde{\bar{\psi}}\eta\}\\ &= \Gamma\left[\bar{Q} = \tilde{Q}+A\,,\ \overline{(\psi)} = \tilde{\psi}\,,\ \overline{(\bar{\psi})} = \tilde{\bar{\psi}}\right]\end{aligned} \tag{3.62}$$

and with $\tilde{Q} = \tilde{\bar{\psi}} = \tilde{\psi} = 0$, we obtain the gauge invariant effective action in the background field gauge

$$\tilde{\Gamma}_{G_1}[0,0,0;A] = \Gamma_{G_2}[A,0,0] \tag{3.63}$$

Let us illustrate the use of all these rather formal expressions within the context of QED. Here we can simplify our various equations, since

$$f^{ijk} = 0$$

and $\det[D(A)D(A+Q)] = \det\partial^2 = \text{const}$

so that we may write

$$\tilde{Z}[J,\eta,\bar{\eta};A] = e^{i\tilde{W}[J,\eta,\bar{\eta};A]}$$

$$= \int_{Q \to Q+A} [dQ\, d\Psi\, d\bar{\Psi}]\, e^{\, i \int d^4x \{ -\frac{1}{4} F^2(Q+A) + \bar{\Psi}(i\not\partial - e(\not Q + \not A) - m)\Psi - \frac{1}{2\alpha}(\partial_\mu Q^\mu)^2 + J_\mu Q^\mu + \bar{\Psi}\eta + \bar{\eta}\Psi \}} \tag{3.64}$$

Here, the background gauges reduce to

$$G_1(Q) = \partial_\mu Q^\mu \qquad G_2(Q) = \partial_\mu (Q^\mu - A^\mu) \tag{3.65}$$

The generating functional $\tilde{Z}$ is invariant under

$$\delta A = \frac{1}{e}\partial\Lambda \qquad \delta J = 0 \qquad \eta' = e^{-i\Lambda}\eta \qquad \bar{\eta}' = \bar{\eta}\, e^{i\Lambda} \tag{3.66}$$

and, after a change of variables:

$$\delta Q = 0, \quad \Psi' = e^{-i\Lambda}\Psi, \quad \bar{\Psi}' = \bar{\Psi} e^{i\Lambda}$$

we find that $\tilde{\Gamma}[\tilde{Q}, \tilde{\Psi}, \tilde{\bar{\Psi}}; A]$ remains unchanged under the transformation

$$\delta A = \frac{1}{e}\partial\Lambda \qquad \delta\tilde{Q} = 0 \qquad \tilde{\Psi}' = e^{-i\Lambda}\tilde{\Psi} \qquad \tilde{\bar{\Psi}}' = \tilde{\bar{\Psi}} e^{i\Lambda} \tag{3.67}$$

Next, let us employ the gauge invariance of $\tilde{\Gamma}$:

$$\delta\tilde{\Gamma} = \int dx\, dy \left[\frac{\delta\tilde{\Gamma}}{\delta A_\mu(x)} \frac{\delta A_\mu(x)}{\delta\Lambda(y)} - \frac{\delta\tilde{\Gamma}}{\delta\tilde{\Psi}_\alpha(x)} \frac{\delta\tilde{\Psi}_\alpha(x)}{\delta\Lambda(y)} + \frac{\delta\tilde{\bar{\Psi}}_\alpha(x)}{\delta\Lambda(y)} \frac{\delta\tilde{\Gamma}}{\delta\tilde{\bar{\Psi}}_\alpha(x)} \right] \delta\Lambda(y) = 0 \tag{3.68}$$

The signs in (3.68) are correct, since in general, when

$$F[\bar{\chi}, \chi] = \int dx\, dy\, \bar{\chi}_i(x)\, M_{ij}(x,y)\, \chi_j(y)$$

$\bar{\chi}, \chi =$ Graßmann fields
$M =$ anti-symmetrical

$$\frac{\delta F}{\delta \bar{\chi}_k(z)} = \int dy\ M_{kj}(z,y)\,\chi_j(y)$$

$$\frac{\delta F}{\delta \chi_k(z)} = -\int dx\ \bar{\chi}_i(x)\,M_{ik}(z)\ ,\ \left\{\frac{\delta}{\delta \chi}, \bar{\chi}\right\} = 0$$

which means for

$$F[\bar{\chi}(\lambda),\chi(\lambda)] = \int dx\,dy\ \bar{\chi}_i(x|\lambda)\,M_{ij}(x,y)\,\chi_j(y|\lambda)\ ;\ \lambda(x)\in\mathbb{R}\ :$$

$$\frac{\delta F}{\delta\lambda(w)} = \int dx\,dy\left\{\frac{\delta\bar{\chi}_i(x|\lambda)}{\delta\lambda(w)}M_{ij}(x,y)\,\chi_j(y|\lambda) + \bar{\chi}_i(x|\lambda)\,M_{ij}(x,y)\frac{\delta\chi_j(y|\lambda)}{\delta\lambda(w)}\right.$$

$$= \int dx\ \frac{\delta\bar{\chi}_i(x|\lambda)}{\delta\lambda(w)}\,\frac{\delta F}{\delta\bar{\chi}_i(x)} - \int dy\,\frac{\delta F}{\delta\chi_j(y)}\,\frac{\delta\chi_j(y|\lambda)}{\delta\lambda(w)}$$

$$= \int dx\left\{\frac{\delta\bar{\chi}_i(x|\lambda)}{\delta\lambda(w)}\,\frac{\delta F}{\delta\bar{\chi}_i(x)} - \frac{\delta F}{\delta\chi_j(x)}\,\frac{\delta\chi_j(x|\lambda)}{\delta\lambda(w)}\right\} \qquad (3.69)$$

Now we are going to use the following equations in (3.68):

$$A_\mu(x) = A^0_\mu(x) + \frac{1}{e}\,\partial_\mu\lambda(x)$$

(1) $$\frac{\delta A_\mu(x)}{\delta\lambda(y)} = \frac{1}{e}\,\partial^x_\mu\,\delta(x-y)$$

$$\tilde{\psi}(x) = e^{-i\lambda(x)}\,\tilde{\psi}^0(x)$$

$$= \tilde{\psi}^0(x) - i\lambda(x)\,\tilde{\psi}^0(x) + O(\lambda^2)$$

$$\frac{\delta\tilde{\psi}(x)}{\delta\lambda(y)} = -i\,\delta(x-y)\,\tilde{\psi}^0(x) + O(\lambda)$$

(2) $$= -i\,\delta(x-y)\,\tilde{\psi}(x) + O(\lambda)$$

$$\tilde{\bar{\psi}}(x) = \tilde{\bar{\psi}}^0(x)\,e^{i\lambda(x)}$$

$$= \tilde{\bar{\psi}}^0(x) + i\lambda(x)\,\tilde{\bar{\psi}}^0(x) + O(\lambda^2)$$

(3) $$\frac{\delta \tilde{\tilde{\varphi}}(x)}{\delta \Lambda(y)} = i\,\delta(x-y)\,\tilde{\tilde{\varphi}}^{0}(x) + O(\lambda) = i\,\delta(x-y)\,\tilde{\tilde{\varphi}}(x) + O(\lambda)$$

After these preparations, we go back to eq. (3.68):

$$0 = \int dx\,dy \Big[\underbrace{\frac{1}{e} \frac{\delta \tilde{\Gamma}}{\delta A_{\mu}(x)} \partial^{x}_{\mu} \delta(x-y)} + i \frac{\delta \tilde{\Gamma}}{\delta \tilde{\varphi}_{\alpha}(x)} \tilde{\varphi}_{\alpha}(x)\, \delta(x-y) + i\, \tilde{\tilde{\varphi}}_{\alpha}(x) \frac{\delta \tilde{\Gamma}}{\delta \tilde{\tilde{\varphi}}_{\alpha}(x)} \delta(x-y) \Big] \Lambda(y)$$

$$\to \Big(-\partial^{x}_{\mu} \frac{\delta \tilde{\Gamma}}{\delta A_{\mu}(x)} \Big) \delta(x-y)$$

Performing the y-integration, we obtain

$$0 = \int dx \Big[-\frac{1}{e} \partial^{x}_{\mu} \frac{\delta \tilde{\Gamma}}{\delta A_{\mu}(x)} + i \frac{\delta \tilde{\Gamma}}{\delta \tilde{\varphi}_{\alpha}(x)} \tilde{\varphi}_{\alpha}(x) + i\, \tilde{\tilde{\varphi}}_{\alpha}(x) \frac{\delta \tilde{\Gamma}}{\delta \tilde{\tilde{\varphi}}(x)} \Big] \Lambda(x), \quad \forall \Lambda$$

so that $(x \to y)$

$$-\frac{1}{e} \partial^{y}_{\mu} \frac{\delta \tilde{\Gamma}}{\delta A_{\mu}(y)} + i \frac{\delta \tilde{\Gamma}}{\delta \tilde{\varphi}_{\alpha}(y)} \tilde{\varphi}_{\alpha}(y) + i\, \tilde{\tilde{\varphi}}_{\alpha}(y) \frac{\delta \tilde{\Gamma}}{\delta \tilde{\tilde{\varphi}}_{\alpha}(y)} = 0 \tag{3.70}$$

Let us take the functional derivative $\frac{\delta}{\delta \tilde{\tilde{\varphi}}_{\beta}(x)}$ of this equation. The result is

$$-\frac{1}{e} \partial^{y}_{\mu} \frac{\delta}{\delta \tilde{\tilde{\varphi}}_{\beta}(x)} \frac{\delta \tilde{\Gamma}}{\delta A_{\mu}(y)} + i \Big(\frac{\delta}{\delta \tilde{\tilde{\varphi}}_{\beta}(x)} \frac{\delta \tilde{\Gamma}}{\delta \tilde{\varphi}_{\alpha}(y)} \Big) \tilde{\varphi}_{\alpha}(y)$$

$$+ i\, \delta_{\alpha\beta}\, \delta(x-y) \frac{\delta \tilde{\Gamma}}{\delta \tilde{\tilde{\varphi}}_{\alpha}(y)} - \underset{(*)}{i\, \tilde{\tilde{\varphi}}_{\alpha}(y) \frac{\delta}{\delta \tilde{\varphi}_{\beta}(x)} \frac{\delta \tilde{\Gamma}}{\delta \tilde{\tilde{\varphi}}_{\alpha}(y)}} = 0 \tag{3.71}$$

If we differentiate this equation once more, this time with respect to $\delta / \delta \tilde{\varphi}_{\gamma}(z)$, we get

$$-\frac{1}{e} \frac{\delta}{\delta \tilde{\varphi}_{\gamma}(z)} \frac{\delta}{\delta \tilde{\tilde{\varphi}}_{\beta}(x)} \frac{\delta \tilde{\Gamma}}{\delta A_{\mu}(y)} + 0^{(*)} + i \frac{\delta}{\delta \tilde{\tilde{\varphi}}_{\beta}(x)} \frac{\delta \tilde{\Gamma}}{\delta \tilde{\varphi}_{\alpha}(y)} \delta_{\alpha\gamma} \delta(z-y)$$

$$+ i\, \delta_{\alpha\beta}\, \delta(x-y) \frac{\delta}{\delta \tilde{\varphi}_{\gamma}(z)} \frac{\delta \tilde{\Gamma}}{\delta \tilde{\tilde{\varphi}}_{\alpha}(y)} = 0 \tag{3.72}$$

The two terms in equation (3.71, 72) indicated by * do not contribute, as can easily be seen from the Ansatz for $\tilde{\Gamma}$:

$$\tilde{\Gamma} = \int dy\, dz\; \tilde{\bar{\psi}}_\beta(y)\, \tilde{S}^{-1}_{\beta\gamma}(y-z)\, \tilde{\psi}_\gamma(z) + \tag{3.73}$$

$$+ \int dx\, dy\, dz\; \tilde{\bar{\psi}}_\alpha(z)\, A_\mu(y)\, \tilde{\Gamma}^\mu_{\alpha\beta}(x,y,z)\, \tilde{\psi}_\beta(z) + \cdots$$

When we use this form for $\tilde{\Gamma}$ in (3.72), we obtain

$$-\frac{1}{e}\partial^y_\mu \tilde{\Gamma}^\mu_{\beta\gamma}(x,y,z) - i\delta(y-z)\,\tilde{S}^{-1}_{\beta\gamma}(x-y) + i\delta(x-y)\,\tilde{S}^{-1}_{\beta\gamma}(y-z) = 0$$

Finally, we end up with the Ward-Takahashi relation

$$\partial^y_\mu \tilde{\Gamma}^\mu(x,y,z) = -ie\left[\delta(y-z) - \delta(x-y)\right]\tilde{S}^{-1}(x-z) \tag{3.74}$$

(Note that in this version, the photon line is attached to the y coordinate.)

As another illustration of our background field approach, we now want to consider the effective QCD-Lagrangian with covariant constant colormagnetic background field. So, let us begin with the generating functional in the background gauge:

$$Z[J;A] = \int [dQ\, d\psi\, d\bar{\psi}]\, \det\left[\tfrac{1}{g} D(A) D(Q)\right] \cdot$$

$$\cdot \exp\left\{ i \int d^4x \left[\mathcal{L}^{Y.M.}(Q) - \frac{1}{2\alpha}\{D(A)(Q-A)\}^2 + \bar{\psi}(i \not{D}(Q) - m)\psi + J^a_\mu Q^a_\mu \right]\right\}$$

$$= \int [dQ\, d\psi\, d\bar{\psi}\, d\varphi\, d\varphi^*]\, \exp\Big\{ i \underbrace{\int d^4x \Big[\mathcal{L}^{Y.M.}(Q) - \frac{1}{2\alpha}\{D(A)(Q-A)\}^2}_{=:\, S_\alpha}$$

$$\underbrace{+\bar{\psi}(i\not{D}(Q) - m)\psi}_{=:\, S_f} \;\underbrace{+ \frac{1}{g}\varphi^* D(A) D(Q)\varphi}_{=:\, S_g} + J^a_\mu Q^a_\mu \Big]\Big\}$$

The associated one-loop effective action is given by

$$\Gamma_1[\bar{Q};A] = -i \ln \int [dQ\, d\psi\, d\bar{\psi}\, d\varphi\, d\varphi^*] \exp\Big\{ i \int \Big(Q \frac{\delta^2 S_\alpha}{\delta Q^2} Q + \bar{\psi} \frac{\delta^2 S_f}{\delta \bar{\psi} \delta \psi} \psi + \varphi^* \frac{\delta^2 S_g}{\delta \varphi^* \delta \varphi} \varphi \Big) \Big\}$$

all derivatives taken at $Q = \bar{Q}$
$\psi = \bar{\psi} = 0,\ \varphi = \varphi^* = 0$

$$= -i \ln \Big\{ \int [dQ]\, e^{i \int Q \frac{\delta^2 S_\alpha}{\delta Q^2}[\bar{Q}] Q} \cdot \int [d\psi\, d\bar{\psi}]\, e^{i \int \bar{\psi} (i \not{D}(Q) - m) \psi} \cdot \int [d\varphi\, d\varphi^*]\, e^{\frac{i}{g} \int \varphi^* D(A) D(\bar{Q}) \varphi} \Big\}$$

Performing the standard integrals, we end up with

$$\Gamma_1[\bar{Q};A] = \frac{i}{2} \ln \det \Big[\frac{\delta^2 S_\alpha[\bar{Q},A]}{\delta \bar{Q}^2} \Big] - i \ln \det \big[i \not{D}(\bar{Q}) - m \big] - i \ln \det \big[D(A) D(\bar{Q}) \big] \tag{3.75}$$

where the classical action is given by

$$S_\alpha[\bar{Q},A] = \int d^4x \Big\{ \mathcal{L}(\bar{Q}) - \frac{1}{2\alpha} [D(A)(\bar{Q}-A)]^2 \Big\} \tag{3.76}$$

Our previous formula (3.45) tells us that

$$\tilde{\Gamma}[\tilde{Q};A] = \Gamma[\bar{Q} = \tilde{Q} + A; A]$$

which yields the gauge invariant effective action for $\tilde{Q} = 0$:

$$\tilde{\Gamma}[0;A] = \Gamma[\bar{Q} = A; A] \tag{3.77}$$

or, explicitly:

$$\tilde{\Gamma}_1[A] = \Gamma_1[A;A]$$
$$= \frac{i}{2} \ln \det \left[\frac{\delta^2 S_\alpha}{\delta \bar{Q}^2} [\bar{Q}=A;A] \right]$$
$$- i \ln \det [i \not{D}(A) - m]$$
$$- i \ln \det [D(A)^2] \tag{3.78}$$

This expression is invariant under

$$\delta A = -\omega \times A + \frac{1}{g} \partial \omega$$

The result of the computation of the various determinants has been given in the literature. For G = SU(2), the leading-log model as described by the effective action (3.78) reduces to ($F_{12} = -F_{21} = B$)

$$\mathcal{L}_{eff}(B) = -\frac{1}{2} B^2 \left[1 + b_0 g^2 \ln \frac{gB}{\mu^2} \right] \tag{3.79}$$

with

$$b_0 = 2(4\pi)^{-2} \frac{11}{3} .$$

Additional References to
Background Field Methods

L.F. Abbott, Nucl. Phys. B185, 189 (1981);

G. 't Hooft, in Acta Universitatis Wratislavensis no. 38, 12th Winter School in Theoretical Physics in Karpacz; Functional and probabilistic methods in quantum field theory vol. I (1975)

Applications:

W. Dittrich, M. Reuter, Phys. Lett. 128B, 321 (1983)

W. Dittrich, M. Reuter, Phys. Lett. 144B, 99 (1984)

4. Instantons and Θ-Vacua

Let us begin this chapter with a brief introduction to homotopy theory. Consider two manifolds X and Y together with a set of continuous mappings $f, g, ..$

$$f : X \to Y, \quad x \mapsto f(x) = y; \quad x \in X, \; y \in Y$$

Then two mappings are defined to be homotopic if they can be continuously distorted into one another. That is, f is homotopic to g if there exists an intermediate family of continuous mappings $H(x,t)$, $0 \leq t \leq 1$,

$$H : X \times I \longrightarrow Y \quad , \quad I = [0,1] \tag{4.1}$$

such that

$$\begin{aligned} H(x,0) &= f(x), \\ H(x,1) &= g(x). \end{aligned} \tag{4.2}$$

H is then called a homotopy between f and g.

Here is an illustration of the formulae (4.1,2):

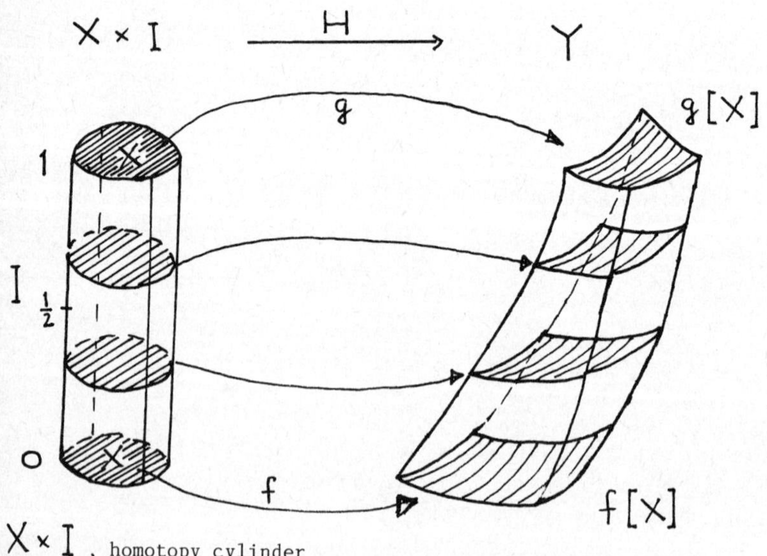

$X \times I$, homotopy cylinder

Next we define a product of paths:

$$f * g : \quad [0,1] \longrightarrow Y$$

or

$$x \longmapsto (f * g)(x) := \begin{cases} f(2x) & \text{for } x \in [0, \frac{1}{2}] \\ g(2x-1) & \text{for } x \in [\frac{1}{2}, 1] \end{cases}$$

The inverse of f is $f^{-1} : \quad [0,1] \longrightarrow Y$,

$$x \longmapsto f^{-1}(x) := f(1-x), \ \forall \ x \in [0,1]$$

The homotopy relation usually indicated by $\sim$ is an equivalence relation on the set of continuous mappings $X \longrightarrow Y$:

(a) reflexive : $f \sim f$

(b) symmetric : $f_1 \sim f_2 \Rightarrow f_2 \sim f_1$

(c) transitive: $f_1 \sim f_2 \wedge f_2 \sim f_3 \Rightarrow f_1 \sim f_3$

Let us verify (b) and (c) with the aid of equations (4.1,2).

(b') $F : \quad f_1 \sim f_2 \quad \Rightarrow \quad G : \quad f_2 \sim f_1$

$$G(x,t) := F(x, 1-t), \quad t \in [0,1]$$

$$\begin{array}{lcl} F : \ f_1 \sim f_2 & & G : \ f_2 \sim f_1 \\ F(x,0) = f_1(x) & \overset{!}{\Longrightarrow} & G(x,0) = f_2(x) \\ F(x,1) = f_1(x) & & G(x,1) = f_1(x) \end{array}$$

since
$$G(x,0) = F(x,1) = f_2(x)$$
$$G(x,1) = F(x,0) = f_1(x)$$

(c') $F : \ f_1 \sim f_2 \wedge G : f_2 \sim f_3 \Rightarrow H : f_1 \sim f_3$

$$H(x,t) := \begin{cases} F(x,2t), & 0 \le t \le \frac{1}{2} \\ G(x,2t-1), & \frac{1}{2} \le t \le 1 \end{cases}$$

$$F : f_1 \sim f_2 \quad \wedge \quad G : f_2 \sim f_3 \quad \Longrightarrow \quad H : f_1 \sim f_3$$

$$\begin{array}{lll} F(x,0) = f_1(x) & G(x,0) = f_2(x) & H(x,0) = f_1(x) \\ F(x,1) = f_2(x) & G(x,1) = f_3(x) & H(x,1) = f_3(x) \end{array}$$

$$\text{since} \quad \begin{array}{l} H(x,0) = F(x,0) = f_1(x) \\ H(x,\tfrac{1}{2}) = F(x,1) = f_2(x) \\ H(x,\tfrac{1}{2}) = G(x,0) = f_2(x) \\ H(x,1) = G(x,1) = f_3(x) \end{array}$$

So, homotopically equivalent maps form an equivalence class. Hence the equivalence relation $\sim$ decomposes the set of mappings $\{ f : X \rightarrow Y \}$ into equivalence classes called homotopy classes H_{F_i}.

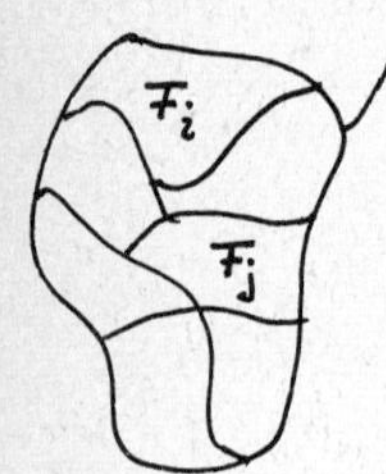

$\{ f : X \rightarrow Y \}$

$$H_{F_i} = \{ f : X \rightarrow Y \mid f \sim F_i \}$$

where F_i denotes a representative of the equivalence class H_{F_i}.

Next we want to show that homotopy classes can assume a group structure under multiplication. This can best be seen by taking a specific example for X, namely, the closed line interval $I = [0,1]$ with endpoints identified. This manifold is topologically equivalent to a circle. We want to restrict ourselves to continuous maps f which satisfy $f(0) = f(1) = y_o$, a fixed point in Y; then the product of two classes H_{F_i} and H_{F_j} is defined as the set of all maps homotopic to F_iF_j, where $F_i(F_j)$ is a representative of $H_{F_i}(H_{F_j})$: $H_{F_i}H_{F_j} = \{ f \sim F_iF_j \}$.

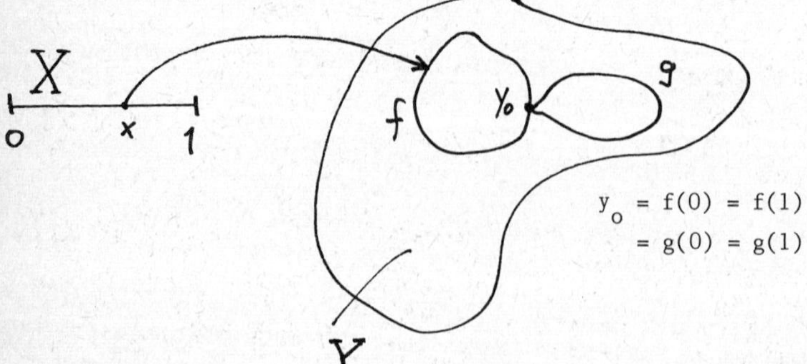

Notice that the product is independent of our choice of the representatives because

$$(f_1 \sim f_2) \wedge (g_1 \sim g_2) \Rightarrow f_1 g_1 \sim f_2 g_2$$

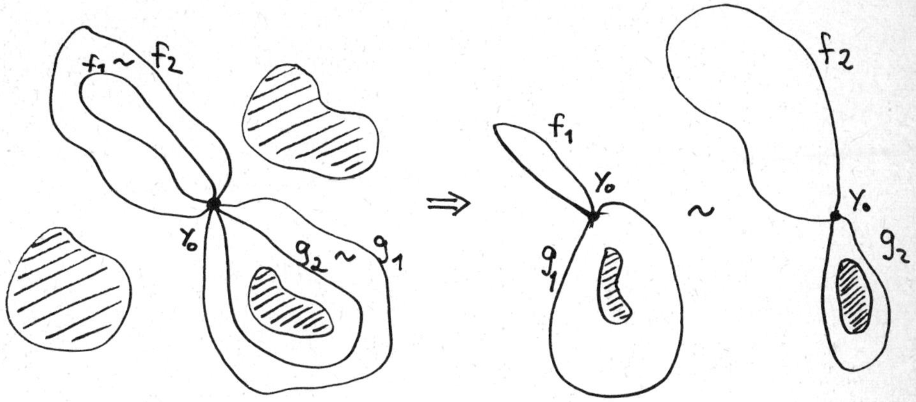

Also note that the identity element $e = ff^{-1}$ is the class of mappings homotopic to the constant mapping C :

$$x \mapsto C(x) = y_o, \; \forall \; x \in [0,1].$$

Now it is easy to convince oneself that the homotopy classes with the composition law defined above form a group. Our particular example is called the first homotopy group of Y :

$$\Pi_1(Y)$$

from $X = S^1$

with group properties

(1) $H_{F_i}(H_{F_j} H_{F_k}) = (H_{F_i} H_{F_j}) H_{F_k}$

(2) N.E. : C (all mappings which are homotopic to the constant mapping)

(3) INV.E. : $H^{-1}_{F_i} = H_{F_i^{-1}}$

Besides, $\pi_1(S^1)$ is a commutative group.

Group manifolds are topological spaces, and it is possible to calculate the first homotopy group for any group G. Let us demonstrate this explicitly for G = U(1), so that we have

$$X = S^1 , \quad Y = U(1)$$

i.e., $\quad f : S^1 \rightarrow U^1 \quad$ or $\quad S^1 \rightarrow S^1$

What, then, is the right-hand side of $\pi_1(U_1) = ?$

The pictorial situation looks like this:

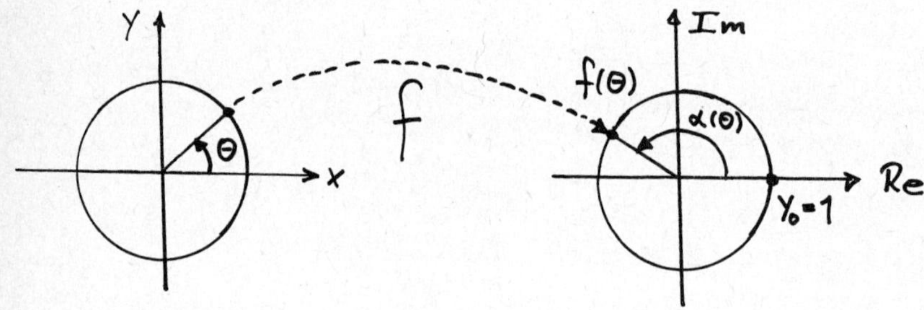

$X = S^1 = \{(\cos\Theta, \sin\Theta) | \Theta \in [0, 2\pi]\}$ $\qquad Y = U(1) = \{e^{i\alpha} | \alpha \in \mathbb{R}\}$

We shall now classify these mappings into homotopy classes.

$$f : [0, 2\pi] \longrightarrow U(1), \qquad \Theta \longmapsto f(\Theta) = e^{i\alpha(\Theta)} \tag{4.3}$$

with $\quad f(o) = f(2\pi) = y_o = 1 \in U(1)$

so that $\quad f(o) = e^{i\alpha(0)} = e^{i\alpha(2\pi)} = f(2\pi)$

which yields $\alpha(2\pi) = \alpha(o) + 2\pi n, \; n \in \mathbb{Z}$ (4.4)

Hence, $\Theta \longmapsto f(\Theta) = e^{i\alpha(\Theta)}$ becomes decomposed into homotopy classes which are classified by an integer $n \in \mathbb{Z}$. We give some examples:

1) $f_1(\Theta) = e^{i \sin\Theta}$, $\alpha_1(\Theta) = \sin\Theta$

$f_2(\Theta) = 1$, $\alpha_2(\Theta) = 0$

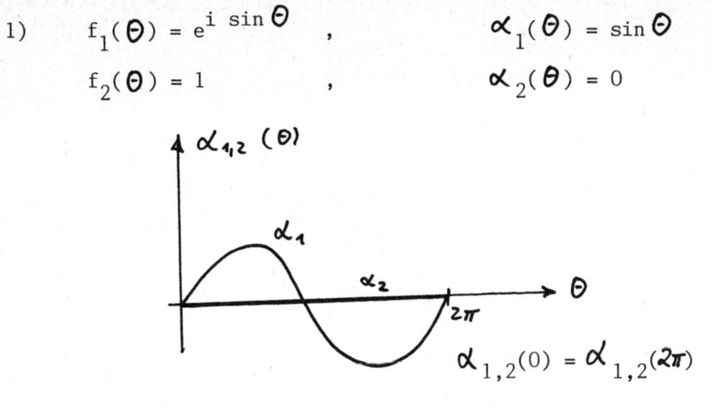

$\alpha_{1,2}(0) = \alpha_{1,2}(2\pi)$,

$n_{1,2} = 0$

f_1 and f_2 are homotopic since they can be continuously deformed into each other via a sequence of mappings, namely a homotopy given by

$$F(\Theta, t) := e^{i\,t\,\sin\Theta}$$

so that $F(\Theta, 0) = 1 = f_2(\Theta)$

and $F(\Theta, 1) = e^{i \sin\Theta} = f_1(\Theta)$

So, any mapping with $\alpha(0) = (2\pi) = 0$ (which includes the identity mapping) belongs to the same homotopy class:

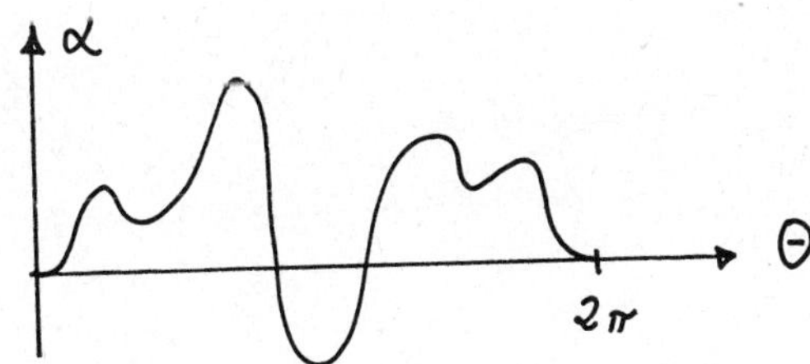

2) $f_1(\Theta) = e^{i\Theta}$, $\alpha_1(\Theta) = \Theta$

$f_2(\Theta) = 1$, $\alpha_2(\Theta) = 0$

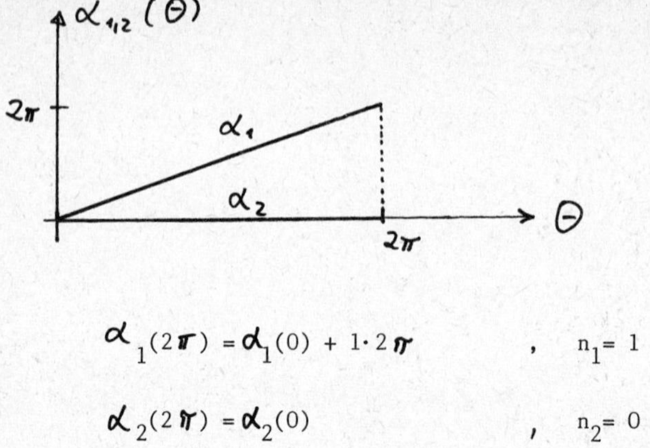

$$\alpha_1(2\pi) = \alpha_1(0) + 1\cdot 2\pi \qquad , \quad n_1 = 1$$

$$\alpha_2(2\pi) = \alpha_2(0) \qquad , \quad n_2 = 0$$

This example yields two mappings, f_1 and f_2, which are not homotopic, i.e., there is no way to continuously deform f_1 into f_2 although f_1 of the present example is not so much different from f_1 of the first example. $f_1(\Theta) = e^{i\Theta}$ identifies a new homotopy class which does not include the identity.

3) $\quad f_1(\Theta) = e^{in\Theta} \qquad , \alpha_1(\Theta) = n\,\Theta$

$\quad\;\; f_2(\Theta) = e^{im\Theta} \qquad , \alpha_2(\Theta) = m\,\Theta \qquad , m,n \in \mathbb{Z}$

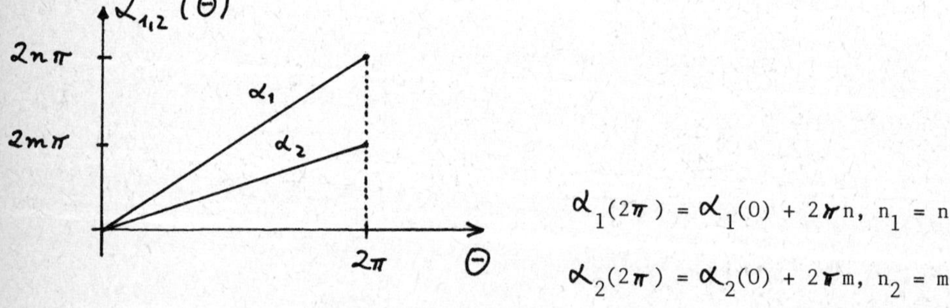

$$\alpha_1(2\pi) = \alpha_1(0) + 2\pi n, \quad n_1 = n$$

$$\alpha_2(2\pi) = \alpha_2(0) + 2\pi m, \quad n_2 = m$$

f_1 and f_2 are not homotopic for $m \neq n$, and they are homotopic if, and only if, $m = n$.

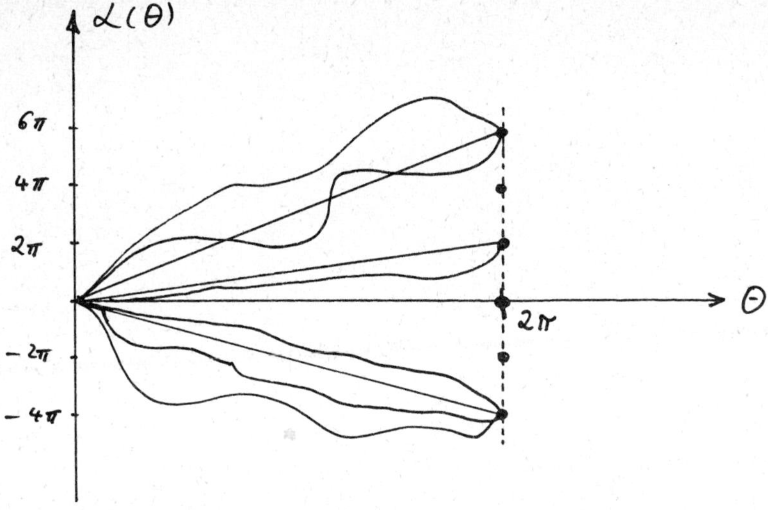

It is clear that any such mapping $f : \theta \to f(\theta)$, because it has to satisfy eq. (4.4), is homotopic to one, and only one, of the mappings

$$f_n(\theta) = e^{i\,n\theta}, \qquad n \in \mathbb{Z} \tag{4.5}$$

so that

$$\Pi_1(\,U(1)\,) = \mathbb{Z} \tag{4.6}$$

The group property can be illustrated as follows:

Inverse element of $\{ f \,|\, f \sim f_n \}$ is given by $\{ f \,|\, f \sim f_{-n} \}$:

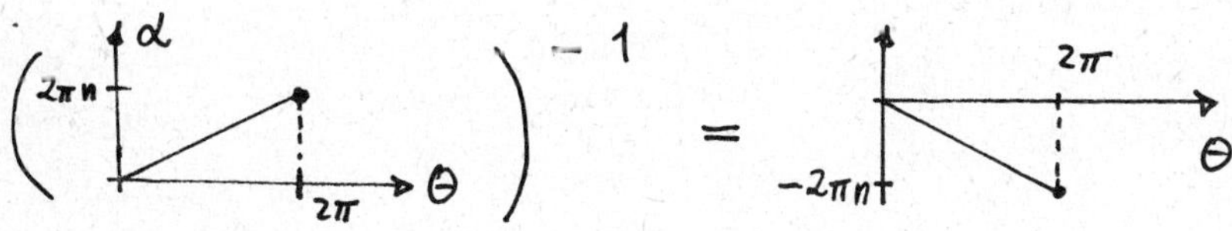

Neutral element: $\{ f \,|\, f \sim f_o \}$, $f_o(\theta) = 1 = y_o$

Composition (multiplication):

$$f_1 \sim f_n \;, \quad f_2 \sim f_m \;, \qquad f_1 f_2 \sim f_n f_m = f_{n+m}$$

In pictures

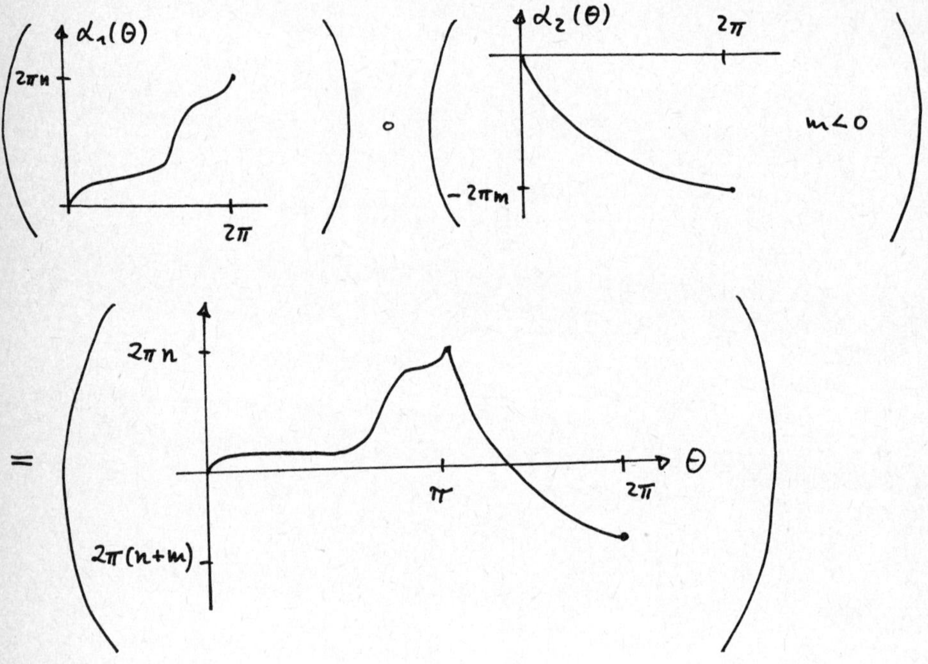

Obviously, when Θ varies from 0 to 2π, so that $n\Theta = 0, .., 2\pi n$, $f_n(\Theta) = e^{i\,n\Theta}$ goes around the U(1) circle n-times. Hence, one round trip in $X = S^1$ corresponds to n round trips in $Y = U(1)$, or n points in $X = S^1$ have the same image point in U(1), e.g.,

$f_4(\Theta) = e^{i\,4\Theta}$:

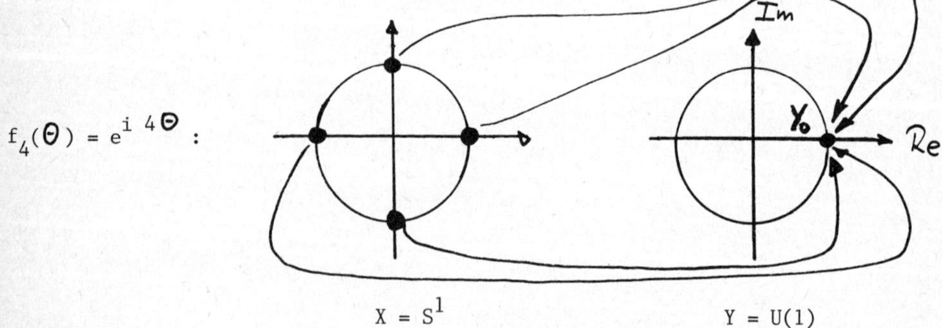

$X = S^1$ $\qquad$ $Y = U(1)$

For obvious reasons, n is called the winding number, ν, and any mapping homotopic to $f_n(\theta)$ also winds around the group space U(1) n times.

After having explained what the first homotopy group is all about, we give the definition of the nth homotopy group π_n (Y): the nth homotopy group π_n (Y) is the set of all equivalence classes of maps of the unit sphere S^n into a topological space Y. Two mappings are said to be homotopic if they can be continuously deformed into each other, i.e., the mappings

$$f : S^n \rightarrow Y$$

are decomposed into equivalence classes which are the group elements of

$$\pi_n(Y)$$

Unfortunately, all of this is hard to visualize for the higher homotopy groups. Nevertheless, here are some results which can be understood, to some extent, by analogy with π_1:

$$\pi_1(U(1)) = \pi_1(S^1) = \mathbb{Z}$$

$$\pi_1(SU(2)) = \pi_1(S^3) = 0$$

all mappings are homotopic; SU(2) is simply connected.

$$\pi_1(U(2)) = \mathbb{Z}$$

$$\pi_n(S^m) = 0, \quad n < m$$

$$\pi_n(S^n) = \mathbb{Z}$$

$$\pi_n(S^1) = 0, \quad n > 1$$

$$\pi_3(SU(2)) = \pi_3(S^3) = \mathbb{Z}$$

$$\pi_3(SO(4)) = \mathbb{Z} \times \mathbb{Z}$$

Following this introduction into homotopy theory, we are now prepared for a discussion of finite euclidean action configurations, instantons. In the sequel, we will concentrate on euclidean SU(2) gauge theory. This can be achieved by the following replacements:

$$g = (+,-,-,-) \longrightarrow g = (+,+,+,+)$$

$$x^o = -i x^4 \,, \quad \partial_o = i\partial_4 \,, \quad D_o = i\, D_4 \,, \quad A_o = iA_4$$

$$\vec{E}_k^{\text{Mink}} = i\, \vec{E}_k^{\text{Eucl.}} \,, \quad k = 1,2,3;$$

$$\vec{B}_k^{\text{Mink}} = \vec{B}_k^{\text{Eucl.}}$$

This means that the action will be rewritten as $(x^4 \equiv \tau)$

$$iS = i\int d^3x\, dt\, \mathcal{L}(t,\vec{x})$$

$$\longrightarrow \int d^3x\, d\tau\, \mathcal{L}(-i\tau,\vec{x}) = -\int d^4x_E\, \mathcal{L}_E(x_E) =: -S_E$$

with

$$\mathcal{L}_E(x_E) = -\mathcal{L}(-i\tau,\vec{x})$$

and

$$\int [dA]\, e^{iS} \longrightarrow \int [dA]\, e^{-S_E} \tag{4.7}$$

The measure contains the gauge fixing as well as the Faddeev-Popov compensating determinant.

Note: In the following, we will omit the index E again.

In Euclidean space-time, the Lagrangian reads

$$\mathcal{L} = \frac{1}{4} F^a_{\mu\nu} F^a_{\mu\nu} = \frac{1}{2}\left[E^a_n E^a_n + B^a_n B^a_n \right] \tag{4.8}$$

with $E^a_n = F^a_{on}$, $B^a_n = -\frac{1}{2}\mathcal{E}_{nij}\, F^a_{ij}$

The energy-momentum tensor is

$$\Theta_{\alpha\beta} = F^a_{\mu\alpha} F^a_{\mu\beta} - g_{\alpha\beta}\mathcal{L}$$
$$= \frac{1}{4}(F^a_{\mu\alpha} + \tilde{F}^a_{\mu\alpha})(F_{\mu\beta} - \tilde{F}^a_{\mu\beta}) \tag{4.9}$$

with $\tilde{F}_{\mu\nu} := \frac{1}{2}\,\mathcal{E}_{\mu\nu\alpha\beta}\, F_{\alpha\beta}$

Obviously, $\Theta_{\alpha\beta} = 0$ for self-dual or anti-self-dual fields

$$F^a_{\alpha\beta} = \pm\, \tilde{F}^a_{\alpha\beta} \tag{4.10}$$

Another important gauge and Lorentz invariant quantity is given by the pseudoscalar density

$$D := \frac{1}{4}\, F^a_{\mu\nu}\, \tilde{F}^a_{\mu\nu} = -E^a_n\, B^a_n \tag{4.11}$$

Why did we not consider this term as a kinetic term for the gauge field? Simply because eq. (4.11) can be expressed as a pure divergence:

$$D = \frac{1}{2g^2}\,\partial_\mu J_\mu \tag{4.12}$$

with

$$\frac{1}{2g^2} J_\mu = \mathcal{E}_{\mu\alpha\beta\gamma}\, \mathrm{tr}\left[A_\alpha \partial_\beta A_\gamma + \frac{3g}{2i} A_\alpha A_\beta A_\gamma\right] \tag{4.13}$$

Proof:

$$D = \frac{1}{4}\,\mathcal{E}_{\mu\nu\rho\sigma}\, \mathrm{tr}\,(F_{\mu\nu} F_{\rho\sigma})\ , \quad F_{\mu\nu} = \partial_\mu A_\nu - ig A_\mu A_\nu - (\mu \leftrightarrow \nu)$$

$$= \mathcal{E}_{\mu\nu\rho\sigma}\, \mathrm{tr}\left[(\partial_\mu A_\nu - ig A_\mu A_\nu)(\partial_\rho A_\sigma - ig A_\rho A_\sigma)\right]$$

$$= \varepsilon_{\mu\nu\rho\sigma}\, tr\left[\partial_\mu A_\nu \partial_\rho A_\sigma - ig\, \partial_\mu A_\nu A_\rho A_\sigma - ig A_\mu A_\nu \partial_\rho A_\sigma - g^2 A_\mu A_\nu A_\rho A_\sigma\right]$$

The cyclic property of the trace can be used to eliminate the AAAA-term. So, we are left with

$$D = \varepsilon_{\mu\nu\rho\sigma}\, tr\left[\partial_\mu A_\nu \partial_\rho A_\sigma - 2ig A_\mu A_\nu \partial_\rho A_\sigma\right]$$

Now

$$\varepsilon_{\mu\nu\rho\sigma}\, tr[A_\mu A_\nu \partial_\rho A_\sigma] = \frac{1}{3} \varepsilon_{\mu\nu\rho\sigma}\, \partial_\rho\, tr[A_\mu A_\nu A_\sigma]$$

and

$$\varepsilon_{\mu\nu\rho\sigma}\, tr[\partial_\mu A_\nu \partial_\rho A_\sigma] = \varepsilon_{\mu\nu\rho\sigma}\, \partial_\rho\, tr[A_\sigma \partial_\mu A_\nu]$$

$$= \varepsilon_{\mu\nu\rho\sigma}\, \partial_\rho\, tr[\partial_\mu A_\nu A_\sigma]$$

$$(\varepsilon_{\mu\nu\rho\sigma}\, tr[A_\sigma \partial_\rho \partial_\mu A_\nu] = 0 \; !)$$

so that

$$D = \partial_\rho\, \varepsilon_{\mu\nu\rho\sigma}\, tr\left[\partial_\mu A_\nu A_\sigma - \frac{2ig}{3} A_\mu A_\nu A_\sigma\right]$$

$$= \partial_\rho\, \varepsilon_{\rho\sigma\mu\nu}\, tr\left[A_\sigma \partial_\mu A_\nu + \frac{2g}{3i} A_\sigma A_\mu A_\nu\right]$$

q.e.d.

Integrating over D, we obtain the topological charge (Pontryagin index) of the euclidean field configuration $A^a_\mu(x)$:

$$q[A] := \frac{g^2}{8\pi^2} \int d^4x\, D(x) = \frac{g^2}{32\pi^2} \int d^4x\, F^a_{\mu\nu} \tilde{F}^a_{\mu\nu}$$

$$= \frac{g^2}{16\pi^2} \int d^4x\, tr(F_{\mu\nu} \tilde{F}^{\mu\nu})$$

$$= \frac{1}{16\pi^2} \oint_{S^3_\infty} d\omega_\mu J_\mu + q_{sing.} \qquad (4.14)$$

where we have to admit the possibility that Gauß' theorem need not be applicable for singular field configurations.

The importance of the instanton solutions becomes evident when quantizing QCD via functional integral methods. In performing the functional integral, one has to integrate over all classical field configurations $A^a_\mu(x)$ in imaginary time which interpolate between the vacua at $x_0 = \pm\infty$. The dominant contributions to this integration stem from those field configurations for which $S_E[A^a_\mu]$ is stationary. Here lies the special relevance of the (anti-) self-dual solutions of the classical equation

$$D_\mu F_{\mu\nu} = 0, \qquad (4.15)$$

i.e., of those fields which satisfy

$$F^a_{\mu\nu} = \pm \tilde{F}^a_{\mu\nu} \qquad (4.16)$$

$$(\text{i.e., } \vec{E} = \pm \vec{B})$$

If this condition is met, then the field equations (4.15) are automatically satisfied, since $\tilde{F}_{\mu\nu}$ always satisfies Bianchi's identity:

$$D_\mu \tilde{F}_{\mu\nu} = 0 \qquad (4.17)$$

so that $D_\mu \tilde{F}_{\mu\nu} = D_\mu F_{\mu\nu} = 0.$

Before actually seeking solutions to the field equations (4.15), let us show that the action for arbitrary euclidean solutions is bounded from below by $|q|$.

In euclidean space, we have

$$\mathrm{tr}\left[(F_{\mu\nu} \mp \tilde{F}_{\mu\nu})(F_{\mu\nu} \mp \tilde{F}_{\mu\nu})\right] \geq 0 \tag{4.18}$$

since it is the sum of squares. It follows that

$$\mathrm{tr}\left[F_{\mu\nu}F_{\mu\nu} + \tilde{F}_{\mu\nu}\tilde{F}_{\mu\nu}\right] \geq \pm 2\,\mathrm{tr}\left[F_{\mu\nu}\tilde{F}_{\mu\nu}\right]$$

or, using $\mathrm{tr}\left[\tilde{F}_{\mu\nu}\tilde{F}_{\mu\nu}\right] = \mathrm{tr}\left[F_{\mu\nu}F_{\mu\nu}\right]$ (4.19)

(note in eucl. space: $\varepsilon_{\varrho\sigma\mu\nu}\,\varepsilon_{\mu\nu\alpha\beta} = 2[\delta_{\varrho\alpha}\delta_{\sigma\beta} - \delta_{\varrho\beta}\delta_{\sigma\alpha}]$)

$$\mathrm{tr}\left[F_{\mu\nu}F_{\mu\nu}\right] \geq \pm\,\mathrm{tr}\left[F_{\mu\nu}\tilde{F}_{\mu\nu}\right] \tag{4.20}$$

This implies for the action

$$S = \frac{1}{2}\int d^4x\, \mathrm{tr}\left[F_{\mu\nu}F_{\mu\nu}\right] \geq \pm\frac{1}{2}\int d^4x\, \mathrm{tr}\left[F_{\mu\nu}\tilde{F}_{\mu\nu}\right] \tag{4.21}$$

so that, following from $\mathrm{tr}\left[F_{\mu\nu}F_{\mu\nu}\right] \geq 0$, we are certain that $S \geq 0$; however, (4.21) tells us that $S \geq +\frac{1}{2}\int d^4x \ldots$ and $S \geq -\frac{1}{2}\int d^4x \ldots$ which means

$$S \geq \left|\frac{1}{2}\int d^4x\, \mathrm{tr}\left[F_{\mu\nu}\tilde{F}_{\mu\nu}\right]\right| = \frac{8\pi^2}{g^2}|q| \tag{4.22}$$

or

$$S[A^a_\mu] \geq \frac{8\pi^2}{g^2}|q[A^a_\mu]| \tag{4.23}$$

which establishes a lower bound for the value of the Yang-Mills euclidean action. Clearly, equality is established, i.e., the lower bound is reached,

$$S = \frac{8\pi^2}{g^2}|q| \tag{4.24}$$

when $F_{\mu\nu} = \pm\tilde{F}_{\mu\nu}$ corresponding to (anti-)self-dual solutions. Let us also recall, cf. eq. (4.9), that these solutions have zero euclidean energy momentum tensor.

We now turn to the topological interpretation of q .

Let us recall that for non-singular functions (q sing = 0, cf. eq. (4.14)), we can rewrite the euclidean action as

$$S = \frac{1}{2}\int d^4x \; tr[F_{\mu\nu} F_{\mu\nu}] \geq \frac{8\pi^2}{g^2} |q|$$

$$= \frac{1}{2g^2} \oint_{S^3_\infty} d\omega_\mu J_\mu \tag{4.25}$$

Thus, the minimum value of the action depends on the behavior of the gauge fields at infinity. Evidently, for S to be finite, $F_{\mu\nu}$ must decrease sufficiently fast at euclidean infinity

$$F_{\mu\nu}(x) \xrightarrow[|x|\to\infty]{} 0 \tag{4.26}$$

which means that the potential tends to a configuration ("pure gauge")

$$A_\mu \longrightarrow -\frac{i}{g}(\partial_\mu \Omega)\Omega^{-1} \quad \text{for } x^2 \to \infty \tag{4.27}$$

which is obtained from $A_\mu = 0$ by a gauge transformation. In this way, we obtain $F_{\mu\nu} = 0$ asymptotically.

The boundary condition (4.27) associates an SU(2) group element $\Omega(x)$ with each point on the 3-sphere S^3_∞ in euclidean E^4. Hence, any solution with the behavior (4.27) defines a mapping

$$S^3_\infty \to SU(2), \quad x \mapsto \Omega(x) \tag{4.28}$$

or, since the SU(2) group manifold is topologically the same as the 3-sphere in E^4, the solution $A_\mu(x)$ determines a mapping

$$S^3_\infty \longrightarrow S^3 \tag{4.29}$$

All mappings of the 3-sphere onto itself are decomposed into homotopy classes which will be classified by an index $n \in \mathbb{Z}$:

$$\pi_3(SU(2)) = \pi_3(S^3) = \mathbb{Z} . \tag{4.30}$$

The integer n is simply the number of times S^3 gets covered by the mapping of S^3_∞. The trivial gauge field $A_\mu = 0$ maps the euclidean sphere S^3_∞ into a single point of the internal group space S^3 and is therefore defined by the identity with winding number equal to zero, etc.

Now we want to demonstrate that for gauge fields $A^a_\mu(x)$ with the asymptotic behavior (4.27), we obtain indeed

$$q[A^a_\mu] = n \in \mathbb{Z} \tag{4.31}$$

Recall that for a pure gauge $A_\mu = -\frac{i}{g}(\partial_\mu \Omega)\Omega^{-1}$, the current J_μ is

$$J_\mu = 2g^2 \varepsilon_{\mu\alpha\beta\gamma} \operatorname{tr}[A_\alpha \partial_\beta A_\gamma + \frac{2g}{3i} A_\alpha A_\beta A_\gamma]$$

$$= -2\varepsilon_{\mu\alpha\beta\gamma} \operatorname{tr}\{(\partial_\alpha\Omega)\Omega^{-1}\underbrace{\partial_\beta[(\partial_\gamma\Omega)\Omega^{-1}]}-\tfrac{2}{3}(\partial_\alpha\Omega)\Omega^{-1}(\partial_\beta\Omega)\Omega^{-1}(\partial_\gamma\Omega)\Omega^{-1}\}$$

$$=(\partial_\beta\partial_\gamma\Omega)\Omega^{-1}+(\partial_\gamma\Omega)\partial_\beta\Omega^{-1} = -(\partial_\gamma\Omega)\Omega^{-1}(\partial_\beta\Omega)\Omega^{-1}$$

$$\left(\Omega\Omega^{-1}=1 \Rightarrow (\partial_\beta\Omega)\Omega^{-1}+\Omega\partial_\beta\Omega^{-1}=0 \Rightarrow \partial_\beta\Omega^{-1}=-\Omega^{-1}(\partial_\beta\Omega)\Omega^{-1}\right)$$

or
$$J_\mu = -2\varepsilon_{\mu\alpha\beta\gamma} \operatorname{tr}[-(\partial_\alpha\Omega)\Omega^{-1}(\partial_\gamma\Omega)\Omega^{-1}(\partial_\beta\Omega)\Omega^{-1} - \tfrac{2}{3}(\partial_\alpha\Omega)\Omega^{-1}(\partial_\beta\Omega)\Omega^{-1}(\partial_\gamma\Omega)\Omega^{-1}]$$

$$= -2\, \varepsilon_{\mu\alpha\beta\gamma}\, tr\left[(\partial_\alpha \Omega)\Omega^{-1}(\partial_\beta \Omega)\Omega^{-1}(\partial_\gamma \Omega)\Omega^{-1} - \tfrac{2}{3}\ \%\ \right]$$

$$= -\frac{2}{3}\, \varepsilon_{\mu\alpha\beta\gamma}\, tr\left[(\partial_\alpha \Omega)\Omega^{-1}(\partial_\beta \Omega)\Omega^{-1}(\partial_\gamma \Omega)\Omega^{-1}\right] \qquad (4.32)$$

Thus, for a solution which satisfies the boundary condition (4.27) at infinity, the surface integral becomes (for regular functions $\partial_\alpha \partial_\beta A_\mu = \partial_\beta \partial_\alpha A_\mu$, $q_{sing} = 0$)

$$q \underset{(4.14)}{=} \frac{1}{16\pi^2} \oint_{S^3_\infty} d\omega_\mu J_\mu$$

$$= -\frac{1}{24\pi^2} \oint_{S^3_\infty} d\omega_\mu\, \varepsilon_{\mu\alpha\beta\gamma}\, tr\left[\partial_\alpha \Omega \Omega^{-1}\, \partial_\beta \Omega\Omega^{-1}\, \partial_\gamma \Omega\Omega^{-1}\right] \qquad (4.33)$$

which depends entirely on the group element $\Omega(x)$. It is remarkable that the (minimum) value of the euclidean action depends on the properties of $\Omega(x)$ only and not on the details of the field configurations at finite x.

Note that S^3_∞ is parametrized by three angles, $\varphi_1, \varphi_2, \varphi_3$, while the group elements Ω are characterized by three angles, Θ_1, Θ_2, Θ_3 : $\Omega(x) = \Omega(\,\Theta_\alpha(x)\,)$.

Then

$$\partial_\mu \Omega \equiv \frac{\partial \Omega}{\partial x_\mu} = \frac{\partial \Theta^a}{\partial x_\mu}\frac{\partial \Omega}{\partial \Theta^a} \equiv \partial_\mu \Theta^a\, \partial_a \Omega \qquad (4.34)$$

Equation (4.33) then becomes

$$q = -\frac{1}{24\pi^2} \oint d\omega_\mu\, \varepsilon_{\mu\alpha\beta\gamma}\, \partial_\alpha \Theta_a\, \partial_\beta \Theta_b\, \partial_\gamma \Theta_c \,\cdot$$

$$\cdot\ tr\left[\partial_a \Omega\Omega^{-1}\, \partial_b \Omega\Omega^{-1}\, \partial_c \Omega\Omega^{-1}\right]$$

and since $\varepsilon_{\mu\alpha\beta\gamma}$ is cyclic in ($\alpha\beta\gamma$) and $tr\,[abc]$ is cyclic in (abc), we obtain

$$q = -\frac{3!}{24\pi^2} \oint d\omega_\mu\, \varepsilon_{\mu\alpha\beta\gamma}\, \partial_\alpha \Theta_1\, \partial_\beta \Theta_2\, \partial_\gamma \Theta_3$$

$$\cdot\; tr\left[\partial_1\Omega\,\Omega^{-1}\,\partial_2\Omega\,\Omega^{-1}\,\partial_3\Omega\,\Omega^{-1}\right]$$

so that

$$q = -\frac{1}{4\pi^2} \oint_{S^3_\infty} d\omega_\mu\, \varepsilon_{\mu\alpha\beta\gamma}\, \partial_\alpha\Theta_1 \partial_\beta\Theta_2 \partial_\gamma\Theta_3\; tr\left[\partial_1\Omega\,\Omega^{-1}\,\partial_2\Omega\,\Omega^{-1}\,\partial_3\Omega\,\Omega^{-1}\right] \tag{4.35}$$

Now we need the following formula (without proof):

$$d\Theta_1\, d\Theta_2\, d\Theta_3 = d\omega_\mu\, \varepsilon_{\mu\alpha\beta\gamma}\, \partial_\alpha\Theta_1\, \partial_\beta\Theta_2\, \partial_\gamma\Theta_3 \tag{4.36}$$

which is essentially

$$d\Theta_1\, d\Theta_2\, d\Theta_3 = \text{(Jacobian)} \cdot d\varphi_1\, d\varphi_2\, d\varphi_3$$

So we can write q in the form

$$q = -\frac{1}{4\pi^2}\, n \int_{S^3_{SU(2)}} d\Theta_1 d\Theta_2 d\Theta_3\; tr\left[\partial_1\Omega\,\Omega^{-1}\,\partial_2\Omega\,\Omega^{-1}\,\partial_3\Omega\,\Omega^{-1}\right] \tag{4.37}$$

Note that instead of integrating over S^3_∞, we are now integrating over the group manifold of the gauge group $S^3_{SU(2)}$; $n \in \mathbb{Z}$ tells us on how many points of S^3 one point of $S^3_{SU(2)}$ becomes mapped. If we choose for Θ_a the Euler angles

$$\Omega(\Theta_a) = e^{\frac{i}{2}\Theta_1\sigma_3}\, e^{\frac{i}{2}\Theta_2\sigma_2}\, e^{\frac{i}{2}\Theta_3\sigma_3}$$

then one obtains

$$tr\left[\partial_1\Omega\,\Omega^{-1}\,\partial_2\Omega\,\Omega^{-1}\,\partial_3\Omega\,\Omega^{-1}\right] = -\frac{1}{4}\sin\Theta_2 \tag{4.38}$$

so that

$$q = -\frac{1}{4\pi^2}\, n \cdot 2 \cdot \left(-\frac{1}{4}\right) \int d\theta_1\, d\theta_2\, d\theta_3 \,\sin\theta_2$$

The extra factor of 2 takes into account the improper rotation ($-\Omega$) of SU(2), since the Euler angle only describes proper rotations Ω. Here, then, is the final expression which terminates our proof:

$$q = \frac{n}{8\pi^2} \int_0^{2\pi} d\theta_1 \int_0^{2\pi} d\theta_3 \int_0^{\pi} d\theta_2 \,\sin\theta_2 = n, \text{ q.e.d.} \tag{4.39}$$

Again, n measures the number of times the group manifold SU(2) is covered when x spans the whole surface S^3_∞ (the 3-sphere or any of its homotopic deformations).

By way of illustration, we now consider the simplest nontrivial case, n = 1, the instanton. Instantons are localized, non-singular solutions of the classical euclidean field equations (4.15), with one unit of topological charge. Instantons are self-dual fields and have a vanishing euclidean energy-momentum tensor, $\Theta_{\alpha\beta} = 0$, cf. eq. (4.9); in particular, $T_{00} = \mathcal{H} = 0$. The associated action is ($q = 1$)

$$S\,[\text{instanton}] = \frac{8\pi^2}{g^2} \tag{4.40}$$

The first known instanton solution of euclidean SU(2) theory employed a spherically symmetric ansatz,

$$A_\mu = \frac{r^2}{r^2+\lambda^2}\left[-\frac{i}{g}(\partial_\mu\Omega)\Omega^{-1}\right] \xrightarrow[r^2\to\infty]{} -\frac{i}{g}(\partial_\mu\Omega)\Omega^{-1} \tag{4.41}$$

$$\Omega = \frac{\vec{x}\cdot\vec{\sigma} \mp i x_0}{\sqrt{r^2}} \quad , \quad \Omega^{-1} = \frac{\vec{x}\cdot\vec{\sigma} \pm i x_0}{\sqrt{r^2}} \tag{4.42}$$

$$r^2 \equiv x_0^2 + \vec{x}^2$$

or, in components,

$$A_0^a = \mp \frac{1}{g} \frac{2 x_a}{r^2+\lambda^2} \tag{4.43}$$

$$A_i^a = -\frac{1}{g} \varepsilon_{ian} \frac{2 x_n}{r^2+\lambda^2} \pm \frac{1}{g} \delta_{ai} \frac{2 x_0}{r^2+\lambda^2}$$

This solution is self-dual

$$E_n^a = \pm B_n^a = \frac{1}{g} \delta_{an} \frac{4\lambda^2}{(r^2+\lambda^2)^2} \tag{4.44}$$

The topological charge density is concentrated around $x_\mu = 0$ and has a size in space-time given by λ^2. For $\lambda^2 \to \infty$, the instanton extends over all E^4 (large instanton). For $\lambda^2 \approx 0$, the effects of the (small) instanton are concentrated near $x^2 = 0$. Note that λ is an arbitrary scale, since pure SU(2) gauge theory has no built-in scale ; this is the reason for the arbitrariness in the size of the instanton.

Given the topological charge density, eq. (4.44), we can easily compute the topological charge

$$D(x) = - E_n^a B_n^a = \mp E_n^a E_n^a = \mp \delta_{na}\delta_{na} \frac{16}{(x^2+\lambda^2)^4} \cdot \frac{\lambda^4}{g^2}$$

$$q = \frac{g^2}{8\pi^2} \int d^4x \; D(x)$$

$$= \mp \frac{3\cdot 16}{8\pi^2} \lambda^4 \int d^4x \frac{1}{(r^2+\lambda^2)^4}$$

$$= \mp 6\lambda^4 \int_0^\infty d(r^2)\, r^2 \frac{1}{(r^2+\lambda^2)^4}$$

$$= \mp 6\lambda^4 \int_0^\infty dy \frac{y}{(y+\lambda^2)^4}$$

$$= \mp 6 \int_0^\infty dz \frac{z}{(1+z)^4}$$

$$= \mp 6 \left[\frac{1}{3} \frac{1}{(1+z)^3} - \frac{1}{2} \frac{1}{(1+z)^2} \right]_0^\infty$$

$$= \mp 6 \left[-\frac{1}{3} + \frac{1}{2} \right]$$

q.e.d. (4.45)

$$= \mp 1$$

Let us now proceed and try to explain how the topological charge is connected with tunneling events between classical distinct gauge field vacua. Note that we are back in Minkowski space.

To explore this aspect, let us choose the temporal gauge

$$A_o^a = 0 \tag{4.46}$$

Then the Minkowski-space vacuum solutions ($F_{\mu\nu}^{vac} = 0$) are time-independent pure gauge fields

$$A_i(\vec{x}) = -\frac{i}{g}\left(\partial_i \Omega(\vec{x})\right) \Omega^{-1}(\vec{x}) \tag{4.47}$$

(Time-dependent gauge transformations would generate nonzero A_o^a .) Furthermore, we can argue that for our static vacua $\Omega(\vec{x})$, it holds that

$$\Omega(\vec{x}) \longrightarrow 1 \qquad \text{for} \quad |\vec{x}| \longrightarrow \infty \tag{4.48}$$

i.e., in all directions. This compactifies $\mathbf{R}^3$ to a three-sphere S^3 .

Thus, $\Omega(\vec{x})$ defines a mapping of three-space with all points at infinity identified into the SU(2) group and, since SU(2) is also topologically equivalent to S^3, $\Omega(\vec{x})$ defines a mapping

$$\Omega : R^3 \cup \{\infty\} \cong S^3 \longrightarrow SU(2) \cong S^3 \quad , \vec{x} \mapsto \Omega(\vec{x}) \tag{4.49}$$

These mappings fall into homotopy classes and so do the pure gauge potentials (4.47)

$$\Pi_3(S^3) = Z$$

Thus, the classical vacuum becomes decomposed in an infinite number of topologically equivalent vacua which are classified by an index $n \in Z$. Gauge transformations $\Omega(\vec{x})$ of different vacua cannot be deformed continuously into eachother.

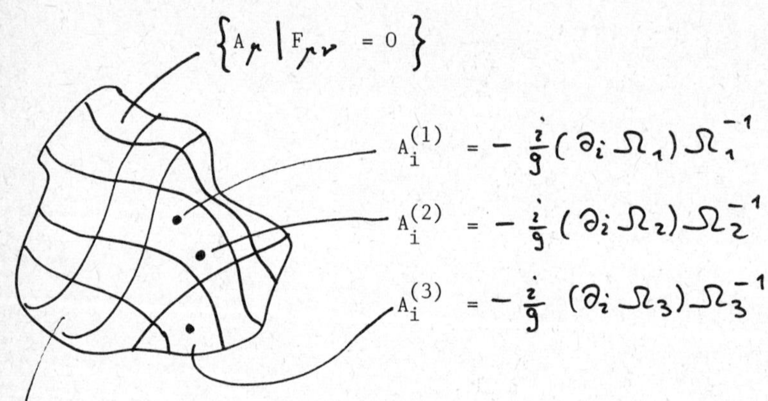

In our picture, Ω_1 and Ω_2 are homotopic to eachother; Ω_3, however, is not homotopic to Ω_1, Ω_2.

The perturbative gauge theory vacuum has $n = 0$, for this is the only one which contains the configuration $A_i^a(\vec{x}) = 0$ for all $\vec{x} \in \mathcal{R}^3$. All other potentials of the $n = 0$ vacuum are continuously deformable to the identity $\Omega_0(\vec{x}) = 1$.

A gauge transformation which cannot be connected continuously with $\Omega_0(\vec{x}) = 1$ is the following gauge transformation that belongs to the class with n = 1:

$$\Omega_1(\vec{x}) = \sigma_\mu n_\mu \,, \qquad \sigma_\mu = (1, i\vec{\sigma})$$
$$n_\mu = \left(\frac{r^2 - \lambda^2}{r^2 + \lambda^2} \,, -2\lambda \frac{\vec{x}}{r^2 + \lambda^2} \right) , \quad r^2 = \vec{x}^{\,2} \tag{4.50}$$

As $\vec{x}$ ranges over all of 3-space, the unit vector n_μ traces out a sphere S^3 in four-dimensional space. So, indeed Ω_1 maps a 3-space onto $S^3 \cong SU(2)$, covering it once. This will become even clearer if we rewrite (4.50) as (a stereographic projection)

$$\Omega_1(\vec{x}) = \frac{r^2 - \lambda^2}{r^2 + \lambda^2} - i\, 2\lambda \frac{\vec{x}\cdot\vec{\sigma}}{r^2 + \lambda^2} \tag{4.51}$$

A gauge transformation in the n-th homotopy class is given by

$$\Omega_n(\vec{x}) = \left[\Omega_1(\vec{x}) \right]^n \tag{4.52}$$

Let us now return to our originial question of how to connect different Minkowski vacua via tunneling events (instantons in euclidean space). To this end, we introduce a second nonconserved topological charge

$$Q_T(t) = \frac{1}{16\pi^2} \int d^3x \; J_0(\vec{x}, t) \tag{4.53}$$

where

$$D \equiv \frac{1}{2g^2} \partial_\mu J_\mu = \frac{1}{4} F^a_{\mu\nu} \tilde{F}^a_{\mu\nu}$$

$$J_\mu = 2g^2 \varepsilon_{\mu\alpha\beta\gamma} \, \mathrm{tr} \left[A_\alpha \partial_\beta A_\gamma + \frac{2g}{3i} A_\alpha A_\beta A_\gamma \right]$$

Note that only for $D = 0$ does it follow that $Q_T(t) = \underset{t}{\text{const.}}$

$Q_T(t)$ is a number which characterizes a field configuration, $A_\mu(t,\vec{x})$, in normal three-space at a fixed time, t. For an arbitrary field, $Q_T\left[A^a_\mu, t\right]$ will neither be an integer nor will it have any topological meaning. But for pure gauges

$$A_\mu(x) = -\frac{i}{g}\left(\partial_\mu \Omega(x)\right)\Omega^{-1}(x)$$

we will now prove that

$$Q_T(t) = n \in \mathbb{Z}, \tag{4.54}$$

i.e., $Q_T(t)$ denotes the homotopy class index of the gauge transformation Ω.

Recall that, for a pure gauge (4.32),

$$J_\mu = -\frac{2}{3}\,\varepsilon_{\mu\alpha\beta\gamma}\,\mathrm{tr}\left[\partial_\alpha \Omega\Omega^{-1}\partial_\beta\Omega\Omega^{-1}\partial_\gamma\Omega\Omega^{-1}\right]$$

so that

$$J_0 = -\frac{2}{3}\,\varepsilon_{0\alpha\beta\gamma}\left[\cdots\right]$$

$$= -\frac{2}{3}\,\varepsilon_{ijk}\,\mathrm{tr}\left[\partial_i\Omega\Omega^{-1}\partial_j\Omega\Omega^{-1}\partial_k\Omega\Omega^{-1}\right]$$

Let x_0 be fixed, and assume that

$$\Omega = \Omega(x_0, \vec{x}) \longrightarrow 1 \quad \text{for} \quad \vec{x}^2 \to \infty$$

Then

$$Q_T(t) = -\frac{1}{24\pi^2}\int d^3x\;\varepsilon_{ijk}\,\mathrm{tr}\left[\partial_i\Omega\Omega^{-1}\partial_j\Omega\Omega^{-1}\partial_k\Omega\Omega^{-1}\right]$$

Introducing Euler angles and going through the same procedure as before, (eq. (4.33, etc.)), we have $\left(\partial_a\Omega \equiv \frac{\partial\Omega}{\partial\theta^a}\right)$

$$Q_T(t) = -\frac{1}{24\pi^2}\, 2 \int d^3x\ \varepsilon_{ijk}\ \partial_i \theta_a\ \partial_j \theta_b\ \partial_k \theta_c$$

$$\cdot\, tr[\partial_a \Omega\, \Omega^{-1}\, \partial_b \Omega\, \Omega^{-1}\, \partial_c \Omega\, \Omega^{-1}]$$

$$= -\frac{2\cdot 3!}{24\pi^2} \int d^3x\ \varepsilon_{ijk}\ \partial_i \theta_1\ \partial_j \theta_2\ \partial_k \theta_3\ tr[\partial_1 \Omega\, \Omega^{-1}\partial_2 \Omega\, \Omega^{-1}\partial_3 \Omega\, \Omega^{-1}]$$

$$= -\frac{n}{2\pi^2} \int d\theta_1\, d\theta_2\, d\theta_3\ \left(-\frac{1}{4} \sin \theta_2\right)$$

$$= n, \qquad \text{q.e.d.}$$

Note that $Q_T(t)$ can be changed by a local gauge transformation, whereas q is absolutely gauge invariant:

$$q' = \frac{g^2}{16\pi^2} \int d^4x\ tr[F'_{\mu\nu}\, \tilde{F}'_{\mu\nu}] = \frac{g^2}{16\pi^2} \int d^4x\ tr[\Omega F_{\mu\nu} \Omega^{-1} \Omega \tilde{F}_{\mu\nu} \Omega^{-1}]$$

$$= q \tag{4.55}$$

Here, then, an expression appears which sets up the relationship between euclidean solutions, Minkowski-space tunneling and the two types of topological charge, q and Q_T:

$$q[A^a_\mu] = \frac{g^2}{8\pi^2} \int d^4x\ D(A^a_\mu(x))$$

$$Q_T([A^a_\mu], t) = \frac{1}{16\pi^2} \int d^3x\ J_0(A^a_\mu(x))$$

Assuming that $\vec{J}$ vanishes faster than $O(\frac{1}{r^2})$ as $r \to \infty$, we have

$$q[A^a_\mu] = Q_T([A^a_\mu], t=+\infty) - Q_T([A^a_\mu], t=-\infty) \tag{4.56}$$

Proof:

$$\partial_\mu J^\mu = \partial_0 J^0 + \partial_k J^k = 2g^2 D$$

$$\Rightarrow \quad \partial_0 \int d^3x\, J_0(x) + \underbrace{\int d^3x\, \partial_k J^k}_{=0} = 2g^2 \int d^3x\, D$$

$$\Rightarrow \quad \int_{-\infty}^{\infty} dt\, \partial_0 \underbrace{\int d^3x\, J_0(x)}_{=16\pi^2 Q_T(t)} = 2g^2 \int_{-\infty}^{\infty} dt \int d^3x\, D \equiv \frac{8\pi^2}{g^2} q$$

which means

$$Q_T(+\infty) - Q_T(-\infty) = q \quad , \text{ q.e.d.}$$

The interpretation of this formula is that a euclidean solution $A_\mu(x_0,\vec{x})$ interpolates between the real-time Minkowski potentials $A_\mu(t=\pm\infty,\vec{x})$ in the distant past and future. The imaginary time x_0 is the interpolating parameter. In general, this is not a tunneling process, since $A_\mu(-\infty)$ and $A_\mu(+\infty)$ are not separated by a barrier. But, when the fields $A_\mu(\pm\infty)$ are pure gauges, i.e., vacua so that the $Q_T(t=\pm\infty)$ are their respective homotopy class labels, then the $A_\mu(\pm\infty)$ are inequivalent for $Q_T(-\infty) \neq Q_T(+\infty)$, i.e., they are separated by a barrier. Hence, solutions with $q \neq 0$ describe tunneling events between topologically inequivalent vacua. The instanton (in the $A_0^a = 0$ gauge) can be shown to have exactly this behavior

$$\begin{aligned} A_i^{\text{Inst}} &\longrightarrow -\frac{i}{g}\left(\partial_i \Omega_1(\vec{x})\right)\Omega_1^{-1}(\vec{x}) \qquad \text{for } x_0 = it \to -\infty \\ A_i^{\text{Inst}} &\longrightarrow 0 \qquad \text{for } x_0 = it \to +\infty \end{aligned} \tag{4.57}$$

i.e., the exact solution of the imaginary time equations of motion--the instanton--interpolates between the $n = 1$ and $n = 0$ vacuum field configuration. This is another way of saying that we are going under an energy barrier; however, that is precisely what an instanton does.

Now, what does the physical vacuum look like? After all, we have an infinite number of n-vacua. In Hilbert space formulation, we represent the vacuum associated with the homotopy class index n by the state vector $|n\rangle$, with $\langle n|m\rangle = \delta_{nm}$. Gauge transformations can change n, e.g., $\Omega_1(x)$ which changes the topological charge by one unit, is represented by the operator G :

$$G|n\rangle = |n+1\rangle \,, \quad |n\rangle = G^n |n=0\rangle \equiv G^n |0\rangle \tag{4.58}$$

The topological charge operator $Q_T = \frac{1}{16\pi^2}\int d^3x\, J_0$ has the property

$$Q_T|n\rangle = n|n\rangle \Rightarrow \left[Q_T, G\right] = 0 \tag{4.59}$$

$|n\rangle$ is not invariant under G, cf. eq. (4.58); however, the physical vacuum should be invariant under G and so we superimpose the n-vacua to form the so-called Θ vacuum

$$|\Theta\rangle = \sum_n e^{in\Theta} |n\rangle \,, \quad \Theta \in [0, 2\pi] \tag{4.60}$$

which is, indeed, invariant under G (up to a phase):

$$G|\Theta\rangle = \sum_n e^{in\Theta} G|n\rangle = \sum_n e^{in\Theta} |n+1\rangle$$

$$= e^{-i\Theta} \sum_n e^{i(n+1)\Theta} |n+1\rangle = e^{-i\Theta} |\Theta\rangle \tag{4.61}$$

Different Θ-vacua are orthogonal

$$\langle\Theta|\Theta'\rangle = \sum_{n,m} e^{i(n\Theta - m\Theta')} \underbrace{\langle n|m\rangle}_{=\delta_{nm}}$$

$$= \sum_n e^{in(\Theta-\Theta')} = 2\pi\,\delta(\Theta-\Theta') \tag{4.62}$$

So, there exists a continuum of Θ-vacua and each one is a suitable ground state for the physical SU(2) gauge theory.

No gauge invariant operator A, $[A,G] = 0$, can generate transitions between different Θ-vacua:

$$\langle\Theta|A|\Theta'\rangle = \sum_{n,n'} e^{in\Theta' - in'\Theta} \langle n'|A|n\rangle$$

$$= \sum_{n,n'} e^{i(n+m)\Theta' - i(n'+m)\Theta} \langle n'+m | A | n+m \rangle$$

$$(m \in \mathbb{Z} \text{ fixed})$$

$$= e^{im(\Theta'-\Theta)} \sum_{n,n'} e^{in\Theta' - in'\Theta} \langle n' | A | n \rangle$$

$$\left([A,G]=0 \Rightarrow \langle n'+m | A | n+m \rangle = \langle n' | A | n \rangle \right)$$

$$= e^{im(\Theta'-\Theta)} \langle \Theta | A | \Theta' \rangle$$

so that we obtain

$$\langle \Theta | A | \Theta' \rangle \left(1 - e^{-im(\Theta-\Theta')} \right) = 0 \quad \text{for all } m \in \mathbb{Z}$$

$$\Rightarrow \quad \langle \Theta | A | \Theta' \rangle = \langle \Theta | A | \Theta \rangle \, \delta(\Theta - \Theta') \quad \text{q.e.d.} \tag{4.63}$$

If $|\Theta\rangle$, $\Theta \neq 0$, is the physical vacuum of the theory, then the Lagrangian of the pure SU(2) gauge theory contains an additional term proportional to Θ :

$$\mathcal{L}_{eff} = \frac{1}{4} F^a_{\mu\nu} F^a_{\mu\nu} + i \frac{g^2}{8\pi^2} \Theta F^a_{\mu\nu} \tilde{F}^a_{\mu\nu} \tag{4.64}$$

To prove eq. (4.64), we start with the transition amplitude between n-vacua:

$$\lim_{x_0 \to \infty} \langle m | e^{-x_0 H} | n \rangle = \int [dA]_{n-m} \, e^{-\int d^4_x \mathcal{L}}$$

where we integrate over euclidean gauge field configurations which interpolate between the vacua $|n\rangle$ and $|m\rangle$, i.e., $Q_T(-\infty) = n$, $Q_T(+\infty) = m$ with $q = m-n$. We absorbed gauge fixing and ghost terms in the functional integration measure $[dA]$.

Then we consider the transition amplitude between Θ-vacua:

$$\frac{1}{2\pi}\lim_{x_0\to\infty}\langle\Theta'|e^{-x_0 H}|\Theta\rangle = \frac{1}{2\pi}\sum_{n,m} e^{i(n\Theta-m\Theta')}\lim_{x_0\to\infty}\langle m|e^{-x_0 H}|n\rangle$$

$$\overset{m'=m-n}{=} \frac{1}{2\pi}\sum_{n,m'} e^{i(n\Theta-m'\Theta'-n\Theta')}\lim_{x_0\to\infty}\underbrace{\langle m'+n|e^{-x_0 H}|n\rangle}_{=\langle m'|e^{-x_0 H}|0\rangle \text{ since } [H,G]=0}$$

$$= \frac{1}{2\pi}\sum_{m'}\sum_{n} e^{in(\Theta-\Theta')}e^{-im'\Theta}\lim_{x_0\to\infty}\langle m'|e^{-x_0 H}|0\rangle$$

$$= \delta(\Theta-\Theta')\sum_{m} e^{-im\Theta}\lim_{x_0\to\infty}\langle m|e^{-x_0 H}|0\rangle$$

$$= \delta(\Theta-\Theta')\sum_{m} e^{-im\Theta}\int[dA]_m\, e^{-\int d^4x\,\mathcal{L}}$$

At this point, we use the formula for the toplological charge

$$m = q = \frac{g^2}{32\pi^2}\int d^4x\, F^a_{\mu\nu}\tilde{F}^a_{\mu\nu}$$

so that we can continue to write for our transition amplitude between Θ -vacua

$$\ldots = \delta(\Theta-\Theta')\sum_{m}\int[dA]_m\, e^{-\int d^4x\,\{\mathcal{L}+\frac{g^2}{8\pi^2}\Theta\frac{i}{4}F^a_{\mu\nu}\tilde{F}^a_{\mu\nu}\}}$$

$$= \delta(\Theta-\Theta')\int_{\text{all fields}}[dA]\, e^{-\int d^4x\,\mathcal{L}_{eff}}$$

q.e.d.

If there were no instantons, there would be no vacuum tunneling. All the n-vacua would be degenerate in energy ($E_n = 0$). Instantons remove this degeneracy and imply that the Θ -vacua have energy $E(\Theta)$ which can be computed with the aid of $\mathcal{L}_{eff}$:

$$E(\Theta) = -\lim_{T\to\infty}\frac{1}{T}\int_{A(-\frac{T}{2})=A(+\frac{T}{2})}[dA]\, e^{-\int_{-T/2}^{T/2}d\tau\int d^3x\,\mathcal{L}_{eff}} \qquad (4.65)$$

This formula follows from ordinary quantum mechanics:

$$\langle x_2, +\tfrac{T}{2} | x_1, -\tfrac{T}{2} \rangle = \langle x_2 | e^{-HT} | x_1 \rangle = \int\limits_{\substack{x(-T/2)=x_1 \\ x(+T/2)=x_2}} [dx]\, e^{-S[x]}$$

Now,

$$\langle x_1 | e^{-HT} | x_1 \rangle = \sum_{n,n'} \langle x_1 | n \rangle \langle n | e^{-HT} | n' \rangle \langle n' | x_1 \rangle$$

$$= \sum_n |\langle x_1 | n \rangle|^2 e^{-E_n T}$$

where now the $|n\rangle$ are normalized energy eigenstates ($H|n\rangle = E_n|n\rangle$) and E_0 is so that the ground state energy is extracted from the limiting process:

$$\lim_{T\to\infty} \int_{-\infty}^{\infty} dx_1 \langle x_1 | e^{-HT} | x_1 \rangle = e^{-E_0 T} \int dx_1 |\langle x_1 | 0 \rangle|^2$$

$$= e^{-E_0 T}$$

and therefore, the logarithm of this equation is given by

$$E_0 = -\lim_{T\to\infty} \frac{1}{T} \ln \int_{-\infty}^{\infty} dx_1 \langle x_1 | e^{-HT} | x_1 \rangle$$

$$= -\lim_{T\to\infty} \frac{1}{T} \ln \int_{-\infty}^{\infty} dx_1 \int\limits_{x(-T/2)=x(T/2)=x_1} [dx]\, e^{-S[x]}$$

$$= -\lim_{T\to\infty} \frac{1}{T} \ln \int\limits_{\substack{x(-T/2)=x(+T/2) \\ \text{arbitrary}}} [dx]\, e^{-S[x]}$$

q.e.d.

The quantum mechanical analog can best be described by a particle in a periodic potential where tunneling processes occur between the classical vacua:

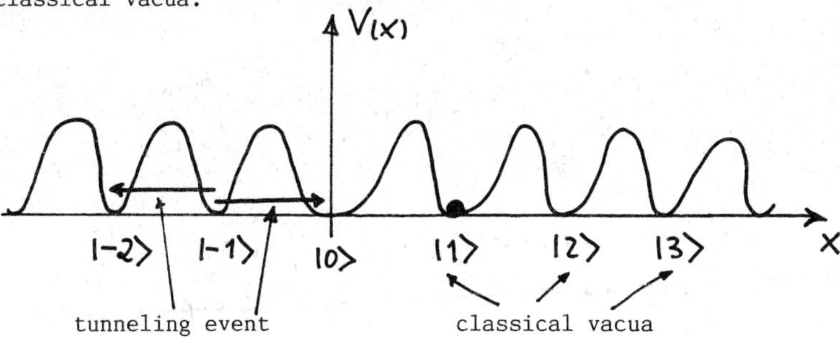

A situation like this:

$= \left| \text{system in a single n-vacuum} \right\rangle$

is not translational (gauge) invariant.

Tunneling processes are needed to restore translational (gauge) invariance.

Note that our additional gauge invariant Θ-term in $\mathcal{L}_{eff}$ eq. (4.64), violates P and T invariance.

$$F^a_{\mu\nu}\tilde{F}^a_{\mu\nu} = -4E^a_n B^a_n$$

which responds under P and T according to

$$P: \quad E^a_n \to -E^a_n, \qquad B^a_n \to +B^a_n$$

$$T: \quad E^a_n \to +E^a_n, \qquad B^a_n \to -B^a_n$$

as

$$\begin{aligned} P(F^a_{\mu\nu}\tilde{F}^a_{\mu\nu})P^{-1} &= -(F^a_{\mu\nu}\tilde{F}^a_{\mu\nu}) \\ T(F^a_{\mu\nu}\tilde{F}^a_{\mu\nu})T^{-1} &= -(F^a_{\mu\nu}\tilde{F}^a_{\mu\nu}) \end{aligned} \tag{4.66}$$

If $\Theta \neq 0$, it would induce a CP violating term in the effective π -N coupling and produce an electric dipole moment for the neutron. The result of such a calculation gives

$$d_n \approx 4 \times 10^{-16} |\Theta| \quad e \cdot \mathrm{cm}$$

Comparison with the experimental bound

$$d_n^{\mathrm{expt}} \leq 1.6 \times 10^{-24} \quad e \cdot \mathrm{cm}$$

yields $|\Theta| \leq 10^{-8}$, (4.67)

a small number indeed. Various theoretical arguments have been given in favor of a non-vanishing value of Θ .

Finally, let us introduce massless fermions into our SU(2) gauge theory. The effect is that massless quarks suppress tunneling processes. Here we present a proof of this fact that is based on the Atiyah-Singer index theorem which will be discussed in great detail later on.

So let us assume $n' \neq 0$ and look at the transition amplitude (in Minkowski space)

$$\langle n=n' | n=0 \rangle = \langle n' | e^{-itH} | 0 \rangle$$

$$= \int_{\mathcal{F}_{q[A]=n'}} [dA\, d\psi\, d\bar{\psi}]\, e^{i\int d^4x \{\bar{\psi}(i\not{D}-m)\psi - \frac{1}{4}F^2\}}$$

where $\mathcal{F}_{q[A]=n'}$ indicates that the path integral is to be taken only over those tunneling configurations for which $q[A] = n'$. Again we absorbed gauge fixing terms and ghosts in the functional measure. The fermionic part can easily be evaluated so that

$$\langle n=n' | n=0 \rangle = \int_{\mathcal{F}_{q[A]=n'}} [dA]\, \det(i\not{D}-m)\, e^{-\frac{i}{4}\int F^2}$$

For m = 0, and, after performing a Wick rotation, this yields

(in euclidean space)

$$\langle n=n' | n=0 \rangle = \int_{\mathcal{F}_{q[A]=n'}} [dA] \det(i \not{D}) \, e^{-\frac{1}{4}\int F^2} \tag{4.68}$$

Now the Atiyah-Singer theorem states that (cf. Chapter 12)

$$n_+[A] - n_-[A] = q[A] = n' \neq 0 \tag{4.69}$$

where $n_\pm$ are, respectively, the number of zero-modes of chirality ± 1 in the presence of the background field A^a_μ.
Equation (4.69) implies that n_+ and/or $n_- \neq 0$, which means that A^a_μ with $q[A] \neq 0$ has at least one zero mode ϕ satisfying

$$\not{D} \phi = 0 \qquad \text{(i.e., } \lambda = 0\text{)}$$

and, consequently,

$$\det \not{D} = \prod_i \lambda_i = 0 \qquad \text{where spectrum } \not{D} = \{\lambda_i\}$$

which concludes our proof of $\langle n = n' | n = 0 \rangle = 0$.

Additional References to Instantons and Θ-Vacua

A. Actor, Rev. Mod. Phys. 51, 461 (1979)

M.J. Teper, Rutherford Preprint RL-80-004

P. Ramond, "Field Theory", Benjamin Cummings, Reading/Mass. (1981)

P. Becher, M. Böhm, H. Joos, "Eichtheorien der starken und elektroschwachen Wechselwirkung", Teubner, Stuttgart (1981)

B. Felsager, "Geometry, Particles and Fields", Odense University Press (1981)

C. Itzykson, J.B. Zuber, "Quantum Field Theory", McGraw-Hill, New York (1980)

F.J. Yndurain, "Quantum Chromodynamics", Springer-Verlag, New York (1983)

K. Huang, "Quarks, Leptons and Gauge Fields", World Scientific, Singapore (1982)

C. Nash, S. Sen, "Topology and Geometry for Physicists", Academic Press, London (1983)

5. Schwinger Model $(QED)_2$

Prior to discussing $(QED)_2$, let us recall the Lagrangian of ordinary quantum electrodynamics $(QED)_4$:

$$\mathcal{L} = \bar{\psi}(i\not{\partial} - m)\psi - \frac{1}{4}F_{\mu\nu}F^{\mu\nu} - e\bar{\psi}\not{A}\psi$$

In (3+1) dimensions, i.e., with 3 spatial dimensions, the coupling constant is dimensionless: $e^2 = 4\pi\alpha$, $\alpha = 1/137$.
Let us now prove that the dimension of e^2, in $(QED)_{n+1}$ with n spatial dimensions, is given by $[M]^{3-n}$.
First note that in a system of units in which $c = 1 = \hbar$, we have

$$[M] = [L]^{-1} = [T]^{-1} \tag{5.1}$$

since $\qquad [c] = [L]\,[T]^{-1} = 1 \Rightarrow [L] = [T]$

and
$$\begin{aligned}[\hbar] &= [\text{energy}]\,[T] = [\text{force}]\,[L]\,[T] \\ &= [M]\,[L]\,[T]^{-2}[L]\,[T] = [M]\,[L]^2[T]^{-1} \\ &\overset{(5.1)}{=} [M]\,[L] = 1 \Rightarrow [M] = [L]^{-1}\end{aligned}$$

In these units, the action operator $W = \int d^{n+1}x\mathcal{L}$ is dimensionless:

$$[W] = 1 = [\mathcal{L}]\,[L]^n\,[T] = [\mathcal{L}]\,[L]^{n+1}$$

$$\Rightarrow \quad [\mathcal{L}] = [M]^{n+1} \tag{5.2}$$

From the expression $\mathcal{L}_0 = \bar{\psi}(i\not{\partial} - m)\psi$, we then conclude

$$[\bar{\psi}\psi] = [M]^n \tag{5.3}$$

The Lagrangian of the free Maxwell field tells us that

$$[F_{\mu\nu}^2] = [(\partial_\mu A_\nu - \partial_\nu A_\mu)^2] = [M]^{n+1}$$

$$\Rightarrow [\partial_\mu A_\nu] = [M]^{\frac{n+1}{2}}$$

$$\Rightarrow [A_\mu] = [M]^{\frac{n+1}{2}-1} = [M]^{\frac{n-1}{2}} \tag{5.4}$$

Finally, the interaction term $\mathcal{L}' = -e\bar{\psi}\not{A}\psi$ yields

$$[M]^{n+1} = [\mathcal{L}'] \underset{(5.4)}{\overset{(5.3)}{=}} [e]\,[M]^n\,[M]^{\frac{n-1}{2}}$$

which proves our original statement

$$[e^2] = [M]^{3-n} \tag{5.5}$$

Examples: $n = 3$, $[e^2] = 1$, $e^2 = 4\pi\alpha$ dimensionless

$n = 2$, $[e^2] = [M]$, $\frac{e^2}{\pi} = m_\gamma^2$ in massless $(QED)_2$

$n = 1$, $[e^2] = [M]^2$ or $[e] = [M]$

Here, then, is the Lagrangian of $(QED)_2$, the Schwinger model in one time, one space dimension:

$$\mathcal{L} = \bar{\psi}(i\not{D})\psi - \frac{1}{4} F_{\mu\nu} F^{\mu\nu} \tag{5.6}$$

with $$D_\mu = \partial_\mu + ieA_\mu \tag{5.7}$$

It should be noted that the mass of the fermion is zero. This is a technical matter-- only then is the model solvable.

In Minkowski (1+1) space, we use the metric

$$g_{\mu\nu} = \begin{matrix} & 0 \quad 1 \\ \begin{matrix} 0 \\ 1 \end{matrix} & \begin{pmatrix} 1 & 0 \\ 0 & -1 \end{pmatrix} \end{matrix} \quad , \mu = \nu = 0,1 \tag{5.8}$$

We also choose $$\{\gamma^\mu, \gamma^\nu\} = 2 g^{\mu\nu} \tag{5.9}$$

from which follows
$$\begin{aligned} (\gamma^0)^2 &= 1 \\ (\gamma^1)^2 &= -1 \\ \{\gamma^0, \gamma^1\} &= 0 \end{aligned} \tag{5.10}$$

γ^0 is taken hermitean, while γ^1 is anti-hermitean:

$$(\gamma^0)^+ = \gamma^0, \ (\gamma^1)^+ = -\gamma^1 \tag{5.11}$$

Furthermore, $$\gamma^5 := \gamma^0\gamma^1 \tag{5.12}$$

which satisfies $$\{\gamma_5, \gamma^\mu\} = \{\gamma^0\gamma^1, \gamma^\mu\} = 0, \tag{5.13}$$

$$(\gamma_5)^+ = (\gamma^0\gamma^1)^+ = -\gamma^1\gamma^0 = \gamma^0\gamma^1 = \gamma_5, \tag{5.14}$$

$$(\gamma_5)^2 = \gamma^0\gamma^1\gamma^0\gamma^1 = -(\gamma^0)^2(\gamma^1)^2 = 1 \tag{5.15}$$

In two dimensions, we need only two γ-matrices and the Pauli matrices suffice. A possible choice is given by

$$\begin{aligned} \gamma^0 &= \sigma^3 \\ \gamma^1 &= i\sigma^1 \\ \gamma_5 = \gamma^0\gamma^1 &= i\sigma^3\sigma^1 = ii\sigma^2 = -\sigma^2 \end{aligned} \tag{5.16}$$

These γ's satisfy the relations (5.9-15). Although we can write down the form of the matrices

$$\gamma^0 = \begin{pmatrix} 1 & 0 \\ 0 & -1 \end{pmatrix}, \quad \gamma^1 = \begin{pmatrix} 0 & i \\ i & 0 \end{pmatrix}, \quad \gamma_5 = \begin{pmatrix} 0 & i \\ -i & 0 \end{pmatrix}$$

we will hardly ever use them in this explicit form.

Let us now return to the Lagrangian (5.6) and use it to derive the equation of motion.

$$\mathcal{L} = \bar{\psi}(i\partial\!\!\!/)\psi - \frac{1}{4}(\partial_\mu A_\nu - \partial_\nu A_\mu)^2 - e\bar{\psi}\gamma^\lambda\psi A_\lambda$$

Employing the Euler-Lagrange equations yields

$$\partial_\mu \frac{\partial \mathcal{L}}{\partial(\partial_\mu A_\lambda)} = \frac{\partial \mathcal{L}}{\partial A_\lambda} = -e\bar{\psi}\gamma^\lambda\psi =: -eJ^\lambda$$

The left-hand side gives $\partial_\mu F^{\lambda\mu}$. Thus the abelian gauge field couples to massless fermions according to

$$\partial_\mu F^{\mu\nu} = eJ^\nu \tag{5.17}$$

$$F^{\mu\nu} = \partial^\mu A^\nu - \partial^\nu A^\mu \tag{5.18}$$

$$J^\nu = \bar{\psi}\gamma^\nu\psi \tag{5.19}$$

The antisymmetry of the field tensor $F^{\mu\nu}$ in (5.17) implies vector current conservatiuon

$$\partial_\nu J^\nu = 0 \tag{5.20}$$

Before we proceed, we introduce the following antisymmetrical ε-tensor:

$$\varepsilon_{\mu\nu} = \begin{array}{c} \\ 0 \\ 1 \end{array}\!\!\overset{\begin{array}{cc} 0 & 1 \end{array}}{\begin{pmatrix} 0 & 1 \\ -1 & 0 \end{pmatrix}}, \quad \mu,\nu = 0,1$$

$$\varepsilon_{\mu\nu} = -\varepsilon_{\nu\mu} \quad \text{, in particular,} \quad \varepsilon_{01} = -\varepsilon_{10} = 1 \tag{5.21}$$

Occasionally we will use

$$\varepsilon^{\mu\nu} = -\varepsilon_{\mu\nu} \tag{5.22}$$

which comes from

$$\varepsilon^{\mu\nu} = g^{\mu\rho} g^{\nu\sigma} \varepsilon_{\rho\sigma} = g^{\mu\rho} \varepsilon_{\rho\sigma} g^{\sigma\nu}$$

$$= \begin{pmatrix} 1 & 0 \\ 0 & -1 \end{pmatrix} \begin{pmatrix} 0 & 1 \\ -1 & 0 \end{pmatrix} \begin{pmatrix} 1 & 0 \\ 0 & -1 \end{pmatrix} = \begin{pmatrix} 0 & -1 \\ 1 & 0 \end{pmatrix}$$

$$= -\varepsilon_{\mu\nu}$$

Two other useful relations are given by

$$\varepsilon^{\mu\nu} \varepsilon_{\nu\lambda} = \begin{pmatrix} 0 & -1 \\ 1 & 0 \end{pmatrix} \begin{pmatrix} 0 & 1 \\ -1 & 0 \end{pmatrix} = \begin{pmatrix} 1 & 0 \\ 0 & 1 \end{pmatrix} = \delta^{\mu}_{\lambda} \tag{5.23}$$

$$\varepsilon_{\mu\nu} \varepsilon^{\mu\nu} = -\varepsilon_{\nu\mu} \varepsilon^{\mu\nu} = -\sum \delta^{\nu}_{\nu} = -2 \tag{5.24}$$

With the ε-tensor, we can prove the following important property for Dirac matrices in two dimensions:

$$\gamma^{\mu} \gamma_5 = \varepsilon^{\mu\nu} \gamma_{\nu} \tag{5.25}$$

which follows from

$$\gamma^{\mu} \gamma_5 = \gamma^{\mu} \gamma^0 \gamma^1$$

$$= \begin{cases} (\gamma^0)^2 \gamma^1 = \gamma^1 = -\gamma_1 = \varepsilon^{01} \gamma_1 = \varepsilon^{0\nu} \gamma_{\nu} & \text{for } \mu = 0 \\ \gamma^1 \gamma^0 \gamma^1 = -\gamma^0 (\gamma^1)^2 = \gamma^0 = \gamma_0 = \varepsilon^{10} \gamma_0 = \varepsilon^{1\nu} \gamma_{\nu} & \text{for } \mu = 1 \end{cases}$$

It is the equation (5.25) of the two-dimensional Dirac algebra which allows us to express the axial vector current as the dual of the vector current:

$$J_5^\mu := \bar{\psi}\gamma^\mu\gamma_5\psi = \varepsilon^{\mu\nu}\,\bar{\psi}\gamma_\nu\psi = \varepsilon^{\mu\nu} J_\nu =: {}^*J^\mu \tag{5.26}$$

The field strength tensor can also be written in ε -notation:

$$F_{\mu\nu} = \varepsilon_{\mu\nu}\, F_{01} \tag{5.27}$$

since
$$F_{00} = F_{11} = 0$$
$$F_{01} = \varepsilon_{01} F_{01} = F_{01},$$
$$F_{10} = \varepsilon_{10} F_{01} = -F_{01},$$

which checks the antisymmetry of $F_{\mu\nu}$.

The dual of $F_{\mu\nu}$ is a scalar:

$$\begin{aligned} {}^*F &= \tfrac{1}{2}\,\varepsilon^{\mu\nu} F_{\mu\nu} \\ &= \tfrac{1}{2}\,\varepsilon^{\mu\nu}\varepsilon_{\mu\nu} F_{01} = \tfrac{1}{2}(-2)\,F_{01} \\ &= -F_{01} \end{aligned} \tag{5.28}$$

and, with (5.27), we obtain

$$F_{\mu\nu} = -\varepsilon_{\mu\nu}\,{}^*F \tag{5.29}$$

In terms of *F, equation (5.17) reads

$$\partial_\mu F^{\mu\nu} = -\partial_\mu \varepsilon^{\mu\nu}\,{}^*F = e\,J^\nu = e\,\varepsilon^{\nu\mu}\,{}^*J_\mu$$

or

$$\partial_\mu {}^*F = e\,{}^*J_\mu \tag{5.30}$$

with

$${}^*J^\mu = \varepsilon^{\mu\nu} J_\nu = J_5^\mu \,.$$

Upon taking the divergence of (5.30), we obtain

$$\begin{aligned} \partial^2\, {}^*F &= e\, \partial_\mu {}^*J^\mu \\ &= e\, \partial_\mu J_5^\mu \end{aligned} \tag{5.31}$$

With massless fermions, one expects J_5^μ to be conserved. However, we will show later that there is an anomaly of the axial vector current:

$$\partial_\mu J_5^\mu = -\frac{e}{\pi}\, \frac{1}{2} \varepsilon_{\mu\nu} F^{\mu\nu} \tag{5.32}$$

or

$$\partial_\mu {}^*J^\mu = -\frac{e}{\pi}\, {}^*F$$

Using equations (5.28) and (5.32), we can rewrite (5.31) in the form

$$\partial^2\, {}^*F = -\partial^2 F_{01} = e\left(-\frac{e}{\pi}\, {}^*F\right) = -\frac{e^2}{\pi}\, {}^*F = \frac{e^2}{\pi} F_{01}$$

or

$$\left(\partial^2 + \frac{e^2}{\pi}\right) F_{01} = 0 = \left(\partial^2 + \frac{e^2}{\pi}\right) {}^*F \tag{5.33}$$

which shows the "photon" to be massive

$$m_\gamma^2 = \frac{e^2}{\pi} \tag{5.34}$$

Note that (5.31) may be written as

$$\begin{aligned} \partial^2 \left(\frac{1}{2} \varepsilon^{\mu\nu} F_{\mu\nu}\right) &= e\, \partial_\mu \left(\varepsilon^{\mu\nu} J_\nu\right) \\ &= e\, \frac{1}{2} \left(\partial_\mu \varepsilon^{\mu\nu} J_\nu + \partial_\nu \varepsilon^{\nu\mu} J_\mu\right) \\ &= e\, \frac{1}{2} \varepsilon^{\mu\nu} \left(\partial_\mu J_\nu - \partial_\nu J_\mu\right) \end{aligned}$$

which means $$\partial^2 F_{\mu\nu} = e(\partial_\mu J_\nu - \partial_\nu J_\mu) \tag{5.35}$$

This formula demonstrates that a mass is generated if the vector current satisfies the London Ansatz:

$$J_\mu = -\frac{e}{\pi} A_\mu \quad \text{+ other possible operators} \tag{5.36}$$

While this Ansatz is consistent with (5.32) which we rewrite as

$$\partial_\mu J_\nu - \partial_\nu J_\mu = -\frac{e}{\pi} F_{\mu\nu} \tag{5.37}$$

since $\partial_\mu {}^*J^\mu = -\frac{e}{\pi} {}^*F$ implies

$$\partial_\mu \varepsilon^{\mu\nu} J_\nu = \frac{1}{2}(\partial_\mu \varepsilon^{\mu\nu} J_\nu + \partial_\nu \varepsilon^{\nu\mu} J_\mu) = \frac{1}{2}\varepsilon^{\mu\nu}(\partial_\mu J_\nu - \partial_\nu J_\mu)$$
$$= -\frac{e}{\pi}\,\frac{1}{2}\varepsilon^{\mu\nu} F_{\mu\nu}$$

we never actually used the London Ansatz. Our basic input was relation (5.32) or (5.37), which is a direct consequence of the axial vector anomaly in the Schwinger model. Later on, we will emphasize its topological origin. But first, let us return to the Green's function $G[A]$ of our theory

$$[i \partial\!\!\!/_x - e A\!\!\!/(x)]\, G_+(x,y|A) = i\,\delta^{(2)}(x-y) \tag{5.38}$$

or $$\gamma^\mu(\partial_\mu^x + ie A_\mu(x))\, G_+(x,y) = \delta(x-y) \tag{5.39}$$

We expect to obtain a solution in terms of $G_+ = G_+[A = 0]$, so that

$$A_\mu = 0: \quad \gamma^\mu \partial_\mu^x G_+(x-y) = \delta(x-y) \tag{5.40}$$

Setting $G_+(x-y) = -\gamma^\nu \partial_\nu^x D_+(x-y)$ (5.41) in equation (5.40), we have

$$-\underbrace{\gamma^\mu \gamma^\nu}_{\substack{=\frac{1}{2}\{\gamma^\mu,\gamma^\nu\}+\cdots \\ = g^{\mu\nu}+\cdots}} \partial_\mu^x \partial_\nu^x D_+(x-y) = \delta(x-y)$$

which means $$-\partial_x^2 D_+(x) = \delta(x) \qquad (5.42)$$

with $$\partial^2 = \partial_0^2 - \partial_1^2$$

Equation (5.42) is satisfied by the causal solution

$$D_+(x) = \int \frac{d^2k}{(2\pi)^2} \frac{e^{ikx}}{k^2+i\varepsilon} \qquad (5.43)$$

since

$$-\partial_x^2 D_+(x) = \int \frac{d^2k}{(2\pi)^2} \frac{\{-(-ik)^2\} e^{ikx}}{k^2+i\varepsilon}$$

$$= \int \frac{d^2k}{(2\pi)^4} e^{ikx}$$

$$= \delta(x)$$

According to usual procedure, we compute

$$D_+(x) = \int \frac{d^2k}{(2\pi)^2} e^{ikx} \frac{1}{k^2+i\varepsilon} = \int \frac{d^2k}{(2\pi)^2} e^{ikx} (-i) \int_0^\infty dt\, e^{it(k^2+i\varepsilon)}$$

$$= -i(2\pi)^{-2} \int_0^\infty dt\, e^{it(i\varepsilon)} \underbrace{\int d^2k\, e^{i(kx+tk^2)}}_{= e^{-i\frac{x^2}{4t}} \cdot e^{i\frac{\pi}{4}}\sqrt{\frac{\pi}{t}} \cdot e^{-i\frac{\pi}{4}}\sqrt{\frac{\pi}{t}}}$$

$$= \frac{-i}{4\pi^2} \pi \int_0^\infty \frac{dt}{t} e^{-i\left(\frac{x^2}{4t} - i\varepsilon t\right)} , \quad \varepsilon \to 0$$

Note that in four dimensions we would have

$$\text{4-dim:} \quad \int d^4k \, e^{itk^2} = \frac{1}{i}\frac{\pi^2}{t^2}$$

So far we have found in two dimensions

$$D_+(x) = -\frac{i}{4\pi}\int_0^\infty \frac{dt}{t} e^{-i\frac{x^2}{4t}} \tag{5.44}$$

For large t, the right-hand side of (5.44) will be logarithmically divergent.

Setting $s = \frac{1}{t}$ so that $ds = -\frac{1}{t^2}dt = -\frac{1}{t}\frac{dt}{t}$

or $\frac{dt}{t} = -t\,ds = -\frac{ds}{s}$ we obtain

$$D_+(x) = -\frac{i}{4\pi}\int_\infty^0 \left(-\frac{ds}{s}\right) e^{-is\frac{x^2}{4}}$$

or

$$D_+(x) = -\frac{i}{4\pi}\int_0^\infty \frac{ds}{s} e^{-is\frac{x^2}{4}} \tag{5.45}$$

and the divergence now appears at the lower limit of integration. In the literature, one usually sees the subtracted form $D_+(x)-D_+(x_o)$, where x_o denotes some fixed point, and this difference is free from divergence. Here, we are only interested in

$$G_+(x) = -\gamma^\mu \partial_\mu D_+(x) = \gamma^\mu \partial_\mu \frac{i}{4\pi}\int_0^\infty \frac{ds}{s} e^{-is\frac{x^2}{4}}$$

$$= \gamma^\mu \frac{i}{4\pi}\int_0^\infty \frac{ds}{s}\left(-is\frac{x_\mu}{2}\right) e^{-is\frac{x^2}{4}}$$

$$= \frac{\gamma^\mu x_\mu}{8\pi}\int_0^\infty ds\, e^{-is\frac{x^2}{4}} = -\frac{i}{2\pi}\frac{\gamma\cdot x}{x^2}$$

For the last integral to exist for large s, we have to replace x^2 by $x^2 - i\varepsilon$. This also shows up in our final form for $G_+(x)$:

$$G_+(x) = -\frac{i}{2\pi}\,\frac{\gamma\cdot x}{x^2 - i\varepsilon} \tag{5.46}$$

To find a solution for equation (5.39) with $A^\mu \neq 0$, let us try the Ansatz

$$G_+(x,y|A) = e^{-ie[\phi(x)-\phi(y)]}\,G_+(x-y) \tag{5.47}$$

Substitution in (5.39) shows that the 2x2 matrix ϕ has to satisfy

$$(\gamma\cdot\partial)\,\phi(x) = \gamma\cdot A(x) \tag{5.48}$$

The latter condition means that

$$\not\partial\not\partial\phi = \partial^2\phi = \not\partial\not A$$

or

$$\phi = \frac{1}{\partial^2}\not\partial\not A\,, \qquad D_+ = -\frac{1}{\partial^2}$$

and, explicitly,

$$\phi(x) = -\int d^2x'\, D_+(x-x')\,\underbrace{\not\partial_{x'}\not A(x')}$$

$$= \gamma^\mu\gamma^\nu\,\partial'_\mu A_\nu(x')$$

$$= \Big(\underbrace{\tfrac{1}{2}\{\gamma^\mu,\gamma^\nu\}}_{=g^{\mu\nu}} + \tfrac{1}{2}[\gamma^\mu,\gamma^\nu]\Big)\,\partial'_\mu A_\nu(x')$$

$$= -\int d^2x'\, D_+(x-x')\Big[\partial'\cdot A(x') + \tfrac{1}{2}[\gamma^\mu,\gamma^\nu]\,\partial'_\mu A_\nu(x')\Big]$$

The simple Dirac algebra allows us to rewrite the last term as

$$\tfrac{1}{2}[\gamma^\mu,\gamma^\nu]\,\partial_\mu A_\nu = \tfrac{1}{2}[\gamma^0,\gamma^1]\,\partial_0 A_1 + \tfrac{1}{2}[\gamma^1,\gamma^0]\,\partial_1 A_0$$

$$= \frac{1}{2}[\gamma^0, \gamma^1](\partial_0 A_1 - \partial_1 A_0) = \frac{1}{2}(\gamma^0\gamma^1 - \gamma^1\gamma^0)(\partial_0 A_1 - \partial_1 A_0)$$

$$= \gamma^0\gamma^1(\partial_0 A_1 - \partial_1 A_0)$$

$$= \gamma_5 (\partial_0 A_1 - \partial_1 A_0)$$

Thus, $\phi(x)$ takes the form

$$\phi(x) = -\int d^2x' \, D_+(x-x')\left[I_2 \, \partial'\cdot A(x') + \gamma_5(\partial_0' A_1(x') - \partial_1' A_0(x'))\right] \qquad (5.49)$$

This is only a solution when $[\partial_\mu \phi(x), \phi(y)] = 0$, in order to be able to differentiate the exponential and obtain the simple desired result

$$-ie\gamma^\mu \partial_\mu \phi(x) \, \exp\{-ie[\phi(x)-\phi(y)]\} \, G_+(x-y) .$$

This, in turn, means that we have a solution only if

$$[(\gamma\cdot\partial')(\gamma\cdot A(x')), (\gamma\cdot\partial'')(\gamma\cdot A(x''))] = 0$$

or

$$[\gamma^\mu\gamma^\nu, \gamma^\lambda\gamma^\sigma] = 0 \qquad (5.50)$$

Now, because we are in two dimensions where we have only γ^0 and γ^1, equation (5.50) is true; i.e., $\gamma^\mu\gamma^\nu$ can be either $(\gamma^0)^2 = 1$ or $\gamma^0\gamma^1$ or $\gamma^1\gamma^0 = -\gamma^0\gamma^1 = -\gamma_5$. Thus, $\phi = \phi\{I_2, \gamma_5\}$ and $[\partial_\mu \phi(x), \phi(y)] = 0$. The matrices I_2 and γ_5 are precisely the ones which show up in (5.49).

After these comments, let us verify that (5.47) is indeed the solution of our Green's function equation (5.39).

$$\gamma^\mu \partial_\mu^x G_+(x,y|A) = \gamma^\mu \partial_\mu^x \left[e^{-ie[\phi(x)-\phi(y)]} G_+(x-y)\right]$$

$$= \gamma^\mu(-ie)(\partial_\mu^x \phi(x)) \underbrace{e^{-ie[\phi(x)-\phi(y)]} G_+(x-y)}_{= G_+(x,y|A)}$$

$$+ \gamma^\mu e^{-ie[\phi(x)-\phi(y)]} \partial_\mu^x G_+(x-y)$$

$$\overset{(5.48)}{\underset{(5.49)}{=}} -ie\gamma\cdot A(x)\,G_+(x,y|A)$$
$$+\gamma^\mu \exp\left(ie\int d^2x'\{D_+(x-x')-D_+(y-x')\}[\partial' A(x')+\gamma_5(\partial_0' A_1'-\partial_1' A_0')]\right)\partial_\mu^x G_+(x-y)$$

$$= -ie\not{A}(x)G_+(x,y|A)+\exp\left(ie\int d^2x'\{D_+(x-x')+D_+(y-x')\}[\partial' A(x')-\gamma_5(\partial_0' A_1'+\partial_1' A_0')]\right)\cdot$$
$$\cdot\underbrace{\gamma^\mu\partial_\mu G_+(x-y)}_{=\delta(x-y)}$$

$$= -ie\not{A}(x)G_+(x,y|A)+\delta^{(2)}(x-y)$$

which yields indeed

$$(\not\partial+ie\not{A}(x))\,G(x,y|A)=\delta^{(2)}(x-y)$$

In the sequel, it is convenient to go to the Lorentz gauge

$$\partial_\mu A^\mu(x)=0 \tag{5.51}$$

Then we are left with

$$\phi(x) = -\gamma_5\int d^2x'\underbrace{D_+(x-x')}_{-\frac{1}{\partial^2}\delta(x-x')}(\partial_0' A_1(x')-\partial_1' A_0(x')) \tag{5.52}$$

Defining

$$\partial^2 B := \partial_0 A_1-\partial_1 A_0 \equiv F_{01}$$

or

$$B = \frac{1}{\partial^2}(\partial_0 A_1-\partial_1 A_0) \equiv \frac{1}{\partial^2}F_{01}$$

we obtain (5.53)

$$\phi(x) = \gamma_5 B(x) =: \gamma_5\frac{\alpha(x)}{e}$$

We can satisfy (5.52) if

$$A_0(x) = \partial_1 B(x)$$
$$A_1(x) = \partial_0 B(x)$$

since

$$\partial_0 A_1-\partial_1 A_0 = \partial_0^2 B-\partial_1^2 B = (\partial_0^2-\partial_1^2)B \equiv \partial^2 B$$

So we have accomplished our task, finding an explicit expression for G[A] :

$$G_+(x,y|A) = e^{-ie[\phi(x)-\phi(y)]}\, G_+(x-y)$$
$$= e^{-ie\gamma_5[B(x)-B(y)]}\, G_+(x-y)$$
$$= e^{-i\gamma_5[\alpha(x)-\alpha(y)]}\, G_+(x-y)$$
$$= e^{-i\gamma_5\alpha(x)}\, G_+(x-y)\, e^{-i\gamma_5\alpha(y)} \tag{5.54}$$

$G_+ \sim x_\mu \gamma^\mu$

with (5.55)

$$\alpha = e\frac{1}{\partial^2}(\partial_0 A_1 - \partial_1 A_0) = e\frac{1}{\partial^2}F_{01}$$

Equation (5.54) tells us that the interaction is merely contained in the phase factor and that it can be transformed away. This fact can also be observed, starting with the Lagrangian

$$\mathcal{L} = \bar\psi[i\partial\!\!\!/ - eA\!\!\!/]\psi$$
$$= \bar\psi[i\partial\!\!\!/ - e(\partial\!\!\!/\phi)]\psi$$
$$= \bar\psi[i\partial\!\!\!/ - (\partial\!\!\!/\gamma_5\alpha)]\psi$$
$$= \bar\psi(i\partial\!\!\!/)\psi - (\partial^\mu\alpha)\,\bar\psi\gamma_\mu\gamma_5\psi$$
$$= \bar\psi e^{i\alpha\gamma_5}(i\partial\!\!\!/)e^{i\alpha\gamma_5}\psi \tag{5.56}$$

or

$$\mathcal{L} = \bar\psi\, G_+[A]^{-1}\psi$$
$$= \bar\psi[e^{-i\alpha\gamma_5}G_+ e^{-i\alpha\gamma_5}]^{-1}\psi$$
$$= \bar\psi e^{i\alpha\gamma_5}G_+^{-1}e^{i\alpha\gamma_5}\psi$$
$$= \bar\psi e^{i\alpha\gamma_5}(i\partial\!\!\!/)e^{i\alpha\gamma_5}\psi$$

Equation (5.56) suggests that the gauge field can be decoupled from the fermion field by a simple phase transformation

$$\psi'(x) = e^{i\alpha(x)\gamma_5}\,\psi(x)$$
$$\bar{\psi}'(x) = \bar{\psi}(x)\,e^{i\alpha(x)\gamma_5}$$

so that the fermionic contribution to the action reduces to the field-free case:

$$\mathcal{L}_\psi \longrightarrow \mathcal{L}_{\psi'} = \bar{\psi}'(i\not{\partial})\psi'$$

This, together with the free photon part, would yield an action

$$W \overset{?}{=} \int d^2x \left\{ \bar{\psi}'(i\not{\partial})\psi' - \tfrac{1}{4}F_{\mu\nu}F^{\mu\nu} + \text{gauge fixing term} \right\}$$

Thus, the Schwinger model would appear to be the theory of massless fermions and photons in two dimensions. However, we know from Schwinger's work that the photon becomes massive, $m_\gamma^2 = e^2/\pi$. Where is the resolution of this paradox and where is the anomaly?

In order to set the stage for a simple path-integral derivation of the axial-vector anonaly, we first rewrite some of our results in euclidean form:

$$\begin{aligned} x^0 &= -i x^4 \,, \quad \partial_0 = i\partial_4 \,, \quad p_0 - i p_4 \\ A_0 &= i A_4 \,, \quad D_0 = i D_4 \end{aligned} \tag{5.57}$$

$$\gamma^4 := i\gamma^0 \Longrightarrow (\gamma^4)^+ = -\gamma^4 \,, \; (\gamma^4)^2 = -1 \tag{5.58}$$

so that $(\gamma^\mu)^+ = -\gamma^\mu$ and $(\gamma^\mu)^2 = -1 \,;\; \mu,\nu = 4,1 .$ (5.59)

Furthermore, $\gamma_5 \equiv \gamma^0\gamma^1 = -i\gamma^4\gamma^1 = i\gamma^1\gamma^4$ (5.60) and $(\gamma_5)^+ = \gamma_5 , \ (\gamma_5)^2 = 1 , \ \{\gamma^\mu, \gamma_5\} = 0$ (5.61)

The metric is changed according to

$$g^{\mu\nu} = \begin{matrix} & 0 & 1 \\ 0 & & \\ 1 & & \end{matrix} \begin{pmatrix} 1 & 0 \\ 0 & -1 \end{pmatrix} \longrightarrow g^{\mu\nu} = \begin{matrix} & 4 & 1 \\ 4 & & \\ 1 & & \end{matrix} \begin{pmatrix} -1 & 0 \\ 0 & -1 \end{pmatrix} = -\delta^{\mu\nu} \tag{5.62}$$

and $\{\gamma^\mu, \gamma^\nu\} = 2 g^{\mu\nu} = 2 \begin{pmatrix} -1 & 0 \\ 0 & -1 \end{pmatrix} = -2\delta^{\mu\nu}$ (5.63)

Using $D_\mu^+ = -D_\mu$ and $(\gamma^\mu)^+ = -\gamma^\mu$, we learn that the product of the two anti-hermitean operators is hermitean $(\not{D}^+ = \not{D})$ with respect to the scalar product

$$(\psi_1, \psi_2) = \int d^2x \ \psi_{1\alpha}^*(x) \psi_{2\alpha}(x) \tag{5.64}$$

where α is a spinor index.

The ε -tensor in its euclidean version is given by

$$\varepsilon_{\mu\nu} = \begin{matrix} & 4 & 1 \\ 4 & & \\ 1 & & \end{matrix} \begin{pmatrix} 0 & 1 \\ -1 & 0 \end{pmatrix} \qquad \mu, \nu = 4, 1 \tag{5.65}$$

i.e., $\varepsilon_{41} = -\varepsilon_{14} = 1$

Then it follows that

$$\varepsilon^{\mu\nu} = g^{\mu\lambda} g^{\nu\sigma} \varepsilon_{\lambda\sigma} = (-\delta^{\mu\lambda})(-\delta^{\nu\sigma}) \varepsilon_{\lambda\sigma}$$

$$= \varepsilon_{\mu\nu} = \begin{pmatrix} 0 & 1 \\ -1 & 0 \end{pmatrix} \tag{5.66}$$

We also need some traces in euclidean space:

$$\operatorname{tr} \gamma^\mu \gamma^\nu = \operatorname{tr} \tfrac{1}{2} \{\gamma^\mu, \gamma^\nu\} + \tfrac{1}{2} \operatorname{tr} (\gamma^\mu \gamma^\nu - \gamma^\nu \gamma^\mu)$$

$$= tr\, g^{\mu\nu} + 0 = -2\,\delta^{\mu\nu} \,, \tag{5.67}$$

$$tr\,\gamma_5 = i\, tr(\gamma^1\gamma^4) = 0 \tag{5.68}$$

It can be shown easily that

$$tr(\gamma_5\,\gamma^\mu\gamma^\nu) = 2i\,\varepsilon^{\mu\nu} \tag{5.69}$$

Clearly, from (5.57)

$$F^M_{01} \equiv \partial_0 A_1 - \partial_1 A_0 = i(\partial_4 A_1 - \partial_1 A_4) = i\, F^E_{41}$$

(M and E stand for Minkowski and Euclidean space, respectively)

so that $F^M_{\mu\nu} = i\, F^E_{\mu\nu}$ (5.70)

Also, $d^2x = dx^0 dx^1 = -i\, dx^4 dx^1 = -i\, d^2x_E$ (5.71)

If we pick the gauge $\partial{\cdot}A = 0$, we have

$$\phi(x) = -\gamma_5 \int d^2x'\, D_+(x-x')\underbrace{(\partial_0' A_1 - \partial_1' A_0)(x')}_{=\,F^M_{01}(x')}$$

$$= -\gamma_5 \int d^2x'_E\; D_+(x-x')\; F^E_{41}(x') \tag{5.72}$$

and, from equation (5.52)

$$\partial^2 B = F^M_{01} = i\, F^E_{41} =: \partial^2 i\, B^E$$

so that $B \equiv B^M = i\,B^E$ (5.73)

and $F^E_{41} = \partial^2 B^E$ (5.74)

Finally, we have the replacements when going to euclidean space:

$$A_0 = \partial_1 B^M \longrightarrow i A_4 \rightarrow \partial_1 B^M$$

or (5.75)

$$A_4 = -i\,\partial_1 B^M = \partial_1 B^E$$

and

$$A_1 = \partial_0 B^M \longrightarrow A_1 = i\,\partial_4 B^M$$

or

$$A_1 = i\,\partial_4 B^M = -\,\partial_4 B^E$$

Equation (5.53) goes into

$$\begin{aligned} \Phi(x) &\equiv \gamma_5 B^M(x) \equiv \gamma_5 \frac{\alpha(x)}{e} \\ &= i\gamma_5 B^E(x) \end{aligned} \tag{5.76}$$

$$\Rightarrow \alpha(x) = ie\,B^E(x) \tag{5.77}$$

Hence, $\alpha(x)$ turns out to be purely imaginary! Note that $\alpha(x)$ does not change since, using equation (5.55), we have

$$\begin{aligned} \alpha(x) &= e^2 \frac{1}{\partial^2}(\partial_0 A_1 - \partial_1 A_0) \\ &= e^2 \frac{1}{\partial^2} F_{01}^M \\ &= e\,\frac{1}{\partial^2}\, i F_{41}^E \\ &= e\,\frac{1}{\partial^2}\, i\,\partial^2 B^E = ie B^E(x) = e B^M(x) \end{aligned}$$

After these intermediate steps, we return to the heart of our investigation: where is the axial vector anomaly, and where is the photon mass? To answer these questions, let us start from the gauge invariant Lagrangian again:

$$\mathcal{L} = \bar\psi(i\not{D} - m)\psi - \frac{1}{4}F_{\mu\nu}F^{\mu\nu} \tag{5.78}$$

Here, we kept the mass of the fermion since the following procedure is not limited to the massless case.

After a Wick rotation to euclidean space-time, we obtain for the generating functional of the Schwinger model:

$$Z[Q^{\mu},K] = \int [d\psi d\bar{\psi} dA]\, e^{\int d^2x \{\bar{\psi}(i\not{D} - m - i\not{Q}\gamma_5 - K\gamma_5)\psi - \frac{1}{4}F_{\mu\nu}F^{\mu\nu}\}} \tag{5.79}$$

where the external sources Q(x) and K(x) are real c-number functions and $[dA]$ is supposed to contain the gauge-fixing Lagrangian.

Now we are going to study the response of the Grassmann integration measure $[d\psi d\bar{\psi}]$ under infinitesimal chiral transformations:

$$\begin{aligned} \psi(x) &\longrightarrow \psi'(x) = e^{i\alpha(x)\gamma_5}\psi(x) = (1 + i\alpha(x)\gamma_5 + O(\alpha^2))\psi(x) \\ \bar{\psi}(x) &\longrightarrow \bar{\psi}'(x) = \bar{\psi}(x)e^{i\alpha(x)\gamma_5} = \bar{\psi}(x)(1 + i\alpha(x)\gamma_5 + O(\alpha^2)) \end{aligned} \tag{5.80}$$

This means that we have to compute the Jacobian $J[\alpha]$ defined by

$$[d\psi d\bar{\psi}] \longrightarrow [d\psi' d\bar{\psi}'] = J[\alpha]\,[d\psi d\bar{\psi}] \tag{5.81}$$

The fermionic part of the Lagrangian in (5.79) changes according to

$$\begin{aligned} \bar{\psi}'(i\not{D} - m)\psi' = \bar{\psi}(i\not{D} - m)\psi &- (\partial^{\mu}\alpha(x))\,\bar{\psi}\gamma_{\mu}\gamma_5\psi \\ &- 2im\alpha(x)\,\bar{\psi}\gamma_5\psi + O(\alpha^2) \end{aligned} \tag{5.82}$$

For our purpose, it suffices to consider only the fermionic part of the path integral (5.79) for vanishing sources. So we obtain from (5.81, 82)

$$\int [d\psi' d\bar{\psi}']\, e^{\int d^2x\, \bar{\psi}'(i\not{D} - m)\psi'} =$$

$$= J[\alpha]\int[d\psi d\bar{\psi}]\, e^{\int d^2x\, \bar{\psi}(i\not{D}-m-\partial^\mu\alpha\,\gamma_\mu\gamma_5-2im\alpha\gamma_5)\psi} \tag{5.83}$$

On both sides we have a standard Gauss-type Grassmann integral leading to

$$\det(i\not{D}-m) = J[\alpha]\det\left(i\not{D}-m-\partial^\mu\alpha\,\gamma_\mu\gamma_5-2im\alpha\gamma_5\right) \tag{5.84}$$

so that $J[\alpha]$ is given as the ratio of two determinants. The value of $J[\alpha]$ for infinitesimal chiral transformation α can be obtained in a number of ways, e.g., via the ζ-function regularization method. The result turns out to be

$$\ln J[\alpha] = -i\int d^2x\, \alpha(x)\, A(x) + O(\alpha^2) \tag{5.85}$$

where

$$A(x) = 2\left.\frac{d}{ds}\right|_0 s\sum_n \lambda_n^{-s}\varphi_n^{+}\gamma_5\varphi_n \tag{5.86}$$

with a complete set of orthonormal eigenfunctions defined by

$$(\not{D}^2+m^2)\,\varphi_n = \lambda_n\varphi_n \tag{5.87}$$

The final answer reads for the Schwinger model

$$A(x) = -\frac{e}{2\pi}\,\varepsilon_{\mu\nu}F^{\mu\nu}(x) \tag{5.88}$$

and

$$J[\alpha] = e^{i\int d^2x\, \alpha(x)\frac{e}{2\pi}\varepsilon_{\mu\nu}F^{\mu\nu}(x)} + O(\alpha^2) \tag{5.89}$$

Incidentally, this form stays intact when going to Minkowski space, since

$$\int d^2x_E \, F_E^{\mu\nu} \ldots = i \int d^2x_M \, (-i) \, F_M^{\mu\nu} \ldots = \int d^2x_M \, F_M^{\mu\nu} \ldots$$

Going back to (5.83), we observe that

$$\begin{aligned} \int [d\psi' d\bar{\psi}'] \, e^{\int \bar{\psi}'(i\not{D}-m)\psi'} &= e^{-i\int \alpha(x) A(x) \, d^2x} \\ &\cdot \int [d\psi d\bar{\psi}] \, e^{\int d^2x \, \bar{\psi}(i\not{D}-m-(\partial_\mu\alpha)\gamma_\mu\gamma_5 - 2im\alpha\gamma_5)\psi} \\ &= \int [d\psi d\bar{\psi}] \, e^{\int d^2x \, \bar{\psi}(i\not{D}-m)\psi} \, . \\ &\cdot e^{-\int d^2x \{ i\alpha(x) A(x) + \underbrace{(\partial\gamma_\mu\alpha)\bar{\psi}\gamma_\mu\gamma_5\psi}_{=-\alpha\,\partial^\mu(\bar{\psi}\gamma_\mu\gamma_5\psi)} + 2im\alpha\bar{\psi}\gamma_5\psi \}} \end{aligned} \tag{5.90}$$

which implies

$$\partial^\mu (\bar{\psi}\gamma_\mu\gamma_5\psi) = 2im\,\bar{\psi}\gamma_5\psi + i\,A \tag{5.91}$$

and, for the particular case $m = 0$:

$$\partial^\mu (\bar{\psi}\gamma_\mu\gamma_5\psi) = -[i]\frac{e}{2\pi}\,\varepsilon_{\mu\nu}\,F^{\mu\nu} \tag{5.92}$$

(The factor i in front of the right-hand side of (5.92) is absent in Minkowski space.)

In order to identify the mass of the photon correctly, we have to determine the Jacobian for finite α .

If the fermion mass is taken to be zero, we obtain as response under the finite chiral transformation

$$\psi' = e^{i\alpha\gamma_5}\psi \, , \quad \bar{\psi}' = \bar{\psi} e^{i\alpha\gamma_5} \; ; \tag{5.93}$$

$$\begin{aligned} \int [d\psi' d\bar{\psi}'] \, e^{\int \bar{\psi}'(i\not{D})\psi'} &= \det[i\not{D}] \\ \parallel \qquad & \\ J[\alpha] \cdot \int [d\psi d\bar{\psi}] \, e^{\int \bar{\psi}(i\not{D} - \partial^{\mu}\alpha\gamma_{\mu}\gamma_5)\psi} & \\ = J[\alpha] \; \det[i\not{D} - \partial^{\mu}\alpha\gamma_{\mu}\gamma_5] & \end{aligned} \tag{5.94}$$

which means

$$\begin{aligned} J[\alpha] &= \frac{\det[i\not{D}]}{\det[i\not{D} - \partial^{\mu}\alpha\gamma_{\mu}\gamma_5]} \\ &= \frac{\det[i\not{\partial} - e\not{A}]}{\det[i\not{\partial} - e\not{A} - \partial^{\mu}\alpha\gamma_{\mu}\gamma_5]} \end{aligned} \tag{5.95}$$

Now, in euclidean space we have

$$\gamma^{\mu}\gamma_5 = -i\,\varepsilon^{\mu\nu}\gamma_{\nu} \, , \qquad g = (-,-) \tag{5.96}$$

which follows from

$$\gamma^\mu\gamma_5 = i\,\gamma^\mu\gamma^1\gamma^4 = \begin{cases} i\gamma^4\gamma^1\gamma^4 = -i\gamma^1(\gamma^4)^2 = i\gamma^1 = -i\gamma_1 = -i\,\varepsilon^{4\nu}\gamma_\nu & \text{for } \mu=4 \\ i\gamma^1\gamma^1\gamma^4 = -i\,\gamma^4 = -i\,\varepsilon^{14}\gamma_4 = -i\,\varepsilon^{1\nu}\gamma_\nu & \text{for } \mu=1 \end{cases}$$

$$= -i\,\varepsilon^{\mu\nu}\gamma_\nu$$

Hence we obtain in (5.95)

$$-(\partial^\mu\alpha)\gamma_\mu\gamma_5 = i\,\varepsilon_{\mu\nu}\gamma^\nu(\partial^\mu\alpha) = -i\,\gamma^\nu\varepsilon_{\nu\mu}(\partial^\mu\alpha) \tag{5.97}$$

so that $$J[\alpha] = \frac{\det[i\partial\!\!\!/ - eA\!\!\!/]}{\det[i\partial\!\!\!/ - e\gamma^\nu(A_\nu + \frac{i}{e}\varepsilon_{\nu\mu}\partial^\mu\alpha)]}$$

Before we continue our calculation, let us discuss the kind of mapping that is induced by our chiral transformation:

$$e^{i\alpha(\cdot)\gamma_5} : S^2 \longrightarrow U(1), \ (\alpha \text{ real}) \tag{5.98}$$
$$x \longmapsto e^{i\alpha(x)\gamma_5}$$

Note that E^2 has been compactified to S^2, a two-sphere.

The set of all mappings is denoted by

$$\mathcal{D} := \left\{ e^{i\alpha(\cdot)\gamma_5} : S^2 \rightarrow U(1) \right\} \tag{5.99}$$

Now, the mathematicians inform us that the second homotopy group of U(1) is zero:

$$\pi_2(U_1) = 0 \tag{5.100}$$

which means that $\mathcal{D}$ is connected, i.e., any two points in $\mathcal{D}$, e.g., $\mathbb{1}$ and $e^{i\alpha(\cdot)\gamma_5}$ can be connected by a path which lies entirely in $\mathcal{D}$.

The various paths in $\mathcal{D}$ are specified by a parameter τ :

$$\text{path in } \mathcal{D} : \quad e^{i\alpha_{(\cdot)}(\cdot)\gamma_5} : S^2 \times [0,1] \longrightarrow U(1), \quad (x,\tau) \longmapsto e^{i\alpha_\tau(x)\gamma_5} \tag{5.101}$$

example:

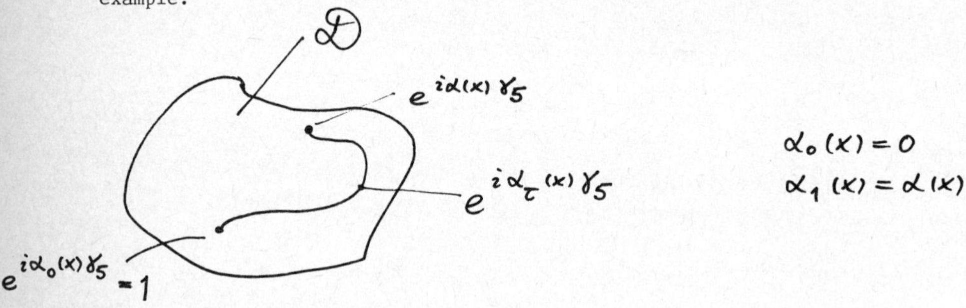

$\mathcal{D}$ is even simply connected, i.e., does not contain any "holes". To understand this fact, we can think of $S^2 \times [0,1]$ as a cylinder which can be deformed continuously into the upper hemisphere of S^3:

$$[0,1] \qquad e^{i\alpha_{(\cdot)}(\cdot)\gamma_5} : S^3 \longrightarrow U(1) \tag{5.102}$$

S^2

However, $\pi_3(U_1) = 0$, i.e., all paths in $\mathcal{D}$ are homotopic to each other. Consequently, $\mathcal{D}$ is simply connected so that we can always choose a "straight" path $e^{i\tau\alpha(x)\gamma_5}$ between two elements, e.g., between 1 and $e^{i\alpha(x)\gamma_5}$.

If we choose α to be purely imaginary, then we have to replace

$$e^{i\alpha(x)\gamma_5} \longrightarrow e^{\beta(x)\gamma_5}, \quad \beta(x) \in R$$
$$U(1) \longrightarrow R \tag{5.103}$$

and

$$\mathcal{D} = \{ e^{\beta(\cdot)\gamma_5} \mid S^2 \rightarrow R \} \tag{5.104}$$

Again, $\pi_2(\mathcal{R}) = 0$ and therefore, $\mathcal{D}$ is connected. Moreover, it is simply connected, since $\pi_3(\mathcal{R}) = 0$. Thus, we can always choose a "straight" path in $\mathcal{D}$.

After these topological remarks, let us return to our Jacobian (5.97) which we are going to write as

$$J[\alpha] = \lim_{N\to\infty} \prod_{n=0}^{N-1} \frac{\det[i\partial\!\!\!/ - e\gamma^\nu(A_\nu + \frac{i}{e}\varepsilon_{\nu\mu}\partial^\mu\alpha\frac{n}{N})]}{\det[i\partial\!\!\!/ - e\gamma^\nu(A_\nu + \frac{i}{e}\varepsilon_{\nu\mu}\partial^\mu\alpha\frac{n+1}{N})]} \tag{5.105}$$

Note that we perform a "straight" path calculation $\tau\alpha(x) = \frac{n}{N}\alpha(x)$ and not $\alpha_\tau(x) = \alpha_{n/N}(x)$.

Introducing (5.106)

$$B_\nu^n := A_\nu + \frac{i}{e}\varepsilon_{\nu\mu}\partial^\mu\alpha\frac{n}{N}$$

we get in the denominator of (5.105)

$$-e\gamma^\nu\left(A_\nu + \frac{i}{e}\varepsilon_{\nu\mu}\partial^\mu\alpha\frac{n+1}{N}\right) = -e\,B\!\!\!/^n - \underbrace{i\gamma^\nu\varepsilon_{\nu\mu}}_{= -\varepsilon_{\mu\nu}\gamma^\nu = -i\gamma_\mu\gamma_5}\partial^\mu\left(\frac{\alpha}{N}\right)$$

$$= -e\,B\!\!\!/^n - \gamma_\mu\gamma_5\,\partial^\mu\left(\frac{\alpha}{N}\right)$$

The Jacobian then reads

$$J[\alpha] = \lim_{N\to\infty} \prod_{n=0}^{N-1} \frac{\det[i\partial\!\!\!/ - eB\!\!\!/^n]}{\det[i\partial\!\!\!/ - eB\!\!\!/^n - \partial^\mu(\frac{\alpha}{N})\gamma_\mu\gamma_5]} \tag{5.107}$$

whereby $\frac{\alpha}{N}$ is infinitesimal.

Taking the logarithm of (5.107) yields

$$\ln J[\alpha] = \lim_{N\to\infty} \ln \prod_{n=0}^{N-1} \frac{\det[i\partial\!\!\!/ - eB\!\!\!/^n]}{\det[i\partial\!\!\!/ - eB\!\!\!/^n - \partial^\mu(\frac{\alpha}{N})\gamma_\mu\gamma_5]}$$

$$= \lim_{N\to\infty} \sum_{n=0}^{N-1} \ln \frac{\det[i\partial\!\!\!/ - e B\!\!\!/^n]}{\det[i\partial\!\!\!/ - e B\!\!\!/^n - \partial^\mu(\frac{\alpha}{N})\gamma_\mu\gamma_5]}$$

$$\underset{(5.84)}{=} \lim_{N\to\infty} \sum_{n=0}^{N-1} \ln J^{\text{inf.}}\left[\frac{\alpha}{N}\right]\Big|_{A_\mu\to B_\mu^n}$$

$$\underset{(5.89)}{=} \lim_{N\to\infty} \sum_{n=0}^{N-1} \frac{ie}{2\pi}\int d^2x\, \frac{\alpha(x)}{N}\, \varepsilon_{\mu\nu} F^{\mu\nu}(B^n(x)) \tag{5.108}$$

The product of $\varepsilon_{\mu\nu} F^{\mu\nu}$ gives

$$\varepsilon_{\mu\nu} F^{\mu\nu}(B^n) = \varepsilon_{\mu\nu}\left[\partial^\mu\left(A^\nu + \frac{i}{e}\varepsilon^{\nu\rho}\partial_\rho \frac{\alpha n}{N}\right) - \partial^\nu\left(A^\mu + \frac{i}{e}\varepsilon^{\mu\rho}\partial_\rho \frac{\alpha n}{N}\right)\right]$$

$$= \varepsilon_{\mu\nu}(\partial^\mu A^\nu - \partial^\nu A^\mu) + \frac{i}{e}\Big(\underbrace{\varepsilon_{\mu\nu}\varepsilon^{\nu\rho}}_{=-\delta_\mu^\rho}\partial^\mu - \underbrace{\varepsilon_{\mu\nu}\varepsilon^{\mu\rho}}_{=\delta_\nu^\rho}\partial^\nu\Big)\partial_\rho \alpha \frac{n}{N}$$

$$= \varepsilon_{\mu\nu} F^{\mu\nu} - \frac{2i}{e}\partial^\mu\partial_\mu \alpha \frac{n}{N} = \varepsilon_{\mu\nu} F^{\mu\nu} - \frac{2i}{e}\partial^2\alpha \frac{n}{N}$$

This changes equation (5.108) into

$$\ln J[\alpha] = \frac{ie}{2\pi}\int d^2x\, \alpha(x) \lim_{N\to\infty} \frac{1}{N}\sum_{n=0}^{N-1}\left\{\varepsilon_{\mu\nu}F^{\mu\nu} - \frac{2i}{e}\frac{n}{N}\partial^2\alpha\right\}$$

We need

$$\lim_{N\to\infty}\sum_{n=0}^{N-1}\frac{1}{N} = \lim_{N\to\infty}\frac{N}{N} = 1$$

$$\lim_{N\to\infty}\frac{1}{N^2}\sum_{n=0}^{N-1} n = \lim_{N\to\infty}\frac{N(N-1)}{2N^2} = \frac{1}{2}$$

Here, then, is our final result for the Jacobian for finite α :

$$\ln J[\alpha] = \frac{ie}{2\pi} \int d^2x \, \alpha(x) \left[\varepsilon_{\mu\nu} F^{\mu\nu} - \frac{i}{e} \partial^2 \alpha \right] \tag{5.109}$$

or

$$J[\alpha] = \exp\left\{ \frac{ie}{2\pi} \int d^2x \, \alpha(x) \left[\varepsilon_{\mu\nu} F^{\mu\nu} - \frac{i}{e} \partial^2 \alpha \right] \right\} \tag{5.110}$$

To further investigate the structure of the exponential, let us go back to equation (5.95) and set $F_{\mu\nu} = 0$:

$$J[\alpha] = \frac{\det[i\partial\!\!\!/]}{\det[i\partial\!\!\!/ - (\partial^\mu \alpha)\gamma_\mu \gamma_5]}$$

Using the relation $\gamma_\mu \gamma_5 = -i \, \varepsilon_{\mu\nu} \gamma^\nu$ again leads to

$$i\partial\!\!\!/ + i \, \varepsilon_{\mu\nu} \gamma^\nu (\partial^\mu \alpha) = i\partial\!\!\!/ - e \left\{ -\frac{i}{e} \varepsilon_{\mu\nu} \partial^\mu \alpha \right\} \gamma^\nu$$

$$= i\partial\!\!\!/ - e \gamma^\nu \left\{ \frac{i}{e} \varepsilon_{\nu\mu} \partial^\mu \alpha \right\}$$

which is then simply a gauge transformation from $A_\mu = 0$ to $\hat{A}_\mu$ in the Lorentz gauge,

where $$\hat{A}_\nu := \frac{i}{e} \varepsilon_{\nu\mu} \partial^\mu \alpha \tag{5.111}$$

with $$\partial_\nu \hat{A}^\nu = 0 \tag{5.112}$$

So, we end up with

$$\det[i\partial\!\!\!/ - e\hat{A\!\!\!/}] = J[\alpha]^{-1} \Big|_{F_{\mu\nu}=0} \cdot \det[i\partial\!\!\!/]$$

$$\overset{(5.110)}{=} \left[e^{\frac{1}{2\pi} \int d^2x \, \alpha \partial^2 \alpha} \right]^{-1} \det[i\partial\!\!\!/]$$

$$= e^{-\frac{1}{2\pi}\int d^2x\, \alpha\, \partial^2 \alpha} \det[i\partial\!\!\!/]$$

$$= e^{\frac{1}{2\pi}\int d^2x\, \partial_\mu \alpha\, \partial^\mu \alpha} \det[i\partial\!\!\!/] \tag{5.113}$$

Since (5.114)

$$\hat{A}_\nu \hat{A}^\nu = \left(\frac{i}{e}\right)^2 \varepsilon_{\nu\mu}\, \varepsilon^{\nu\rho}\, \partial^\mu \alpha\, \partial_\rho \alpha = -\frac{1}{e^2}\, \partial^\mu \alpha\, \partial_\mu \alpha$$

we finally obtain the desired result

$$\det[i\partial\!\!\!/ - e\hat{A\!\!\!/}] = e^{-\frac{e^2}{2\pi}\int d^2x\, \hat{A}_\mu \hat{A}^\mu} \det[i\partial\!\!\!/] \tag{5.115}$$

(The metric is $g = +\delta$).

Going back to (5.93, 94), $D\!\!\!\!/ := i\partial\!\!\!/ - e\hat{A\!\!\!/}$,
we find ($\hat{A} \to A$)

$$\int [d\psi\, d\bar\psi\, dA]\, e^{\int d^2x \{\bar\psi [i\partial\!\!\!/ - eA\!\!\!/]\psi - \frac{1}{4} F_{\mu\nu}F^{\mu\nu}\}}$$

$$= \int [dA] \det[i\partial\!\!\!/ - eA\!\!\!/]\, e^{-\frac{1}{4}\int F_{\mu\nu}F^{\mu\nu}}$$

$$= \int [dA]\, e^{-\int d^2x\, \frac{1}{2}\frac{e^2}{\pi^2} A_\mu A^\mu} \det[i\partial\!\!\!/]\, e^{-\frac{1}{4}\int F_{\mu\nu}F^{\mu\nu}}$$

$$= \int [d\psi\, d\bar\psi\, dA]\, e^{\int d^2x \{\bar\psi (i\partial\!\!\!/)\psi - \frac{1}{2}\frac{e^2}{\pi} A_\mu A^\mu - \frac{1}{4} F_{\mu\nu}F^{\mu\nu}\}} \tag{5.116}$$

which is the generating functional for a free theory of massless fermions and <u>massive</u> vector bosons and, if we remember that we started out with a theory of massless bosons (5.6), we truly have been experiencing a mass generating mechanism.

Additional References to
Schwinger Model $(QED)_2$

J. Schwinger, Phys. Rev. 128, 2425 (1962); Phys. Rev. 125, 397 (1962)

R. Roskies, F. Schaposnik, Phys. Rev. D23, 558 (1981)

R.E. Gamboa Saravi, F. Schaposnik, J.E. Solomin, Nucl. Phys. B185, 239 (1981)

K. Furuya, R.E. Gamboa-Saravi, F. Schaposnik, Nucl. Phys. B208, 159 (1982)

R. Banerjee, Z. Phys. C25, 251 (1984)

R.E. Gamboa Saravi et al., Phys. Lett. 138B, 145 (1984)

R.E. Gamboa Saravi et al., Ann. Phys. 157, 360 (1984)

C. Nash, S. Sen, "Topology and Geometry for Physicists", Academic Press, London (1983)

Chiral Anomalies:

S. Adler, Phys. Rev. 177, 2426 (1969); J. Bell, R. Jackiw, Nuovo Cim. 60A, 47 (1969)

R. Jackiw in: Lectures on current algebra and its applications; ed. D. Gross, R. Jackiw, S. Treiman, Princeton University Press (1972)

S. Adler in: Lectures on elementary particles and quantum field theory, Vol 1, 1970 Brandeis University Summer Institute, MIT Press

K. Huang, "Quarks, Leptons and Gauge Fields", World Scientific, Singapore (1982)

K. Fujikawa, Phys. Rev. Lett. 42, 1195 (1979); Phys. Rev. D21, 2848 (1980)

M. Reuter, Phys. Rev. D31, 1374 (1985)

6. Chiral Anomaly in $(QCD)_2$

In this chapter we want to study quantum chromodynamics in two dimensions with massless fermions, $(QCD)_2$. However, before we go in medias res, let us list some of the most important Jacobians. (The defining equation for $J(\alpha)$ is contained in the last chapter, cf. eqs. (5.81-5.84).)

$(QED)_2$: $\mathcal{L}=\bar{\psi}\, i\not{D}\psi\,,\ D_\mu=\partial_\mu+ieA_\mu\,,$ euclidean

$$\psi'=e^{i\alpha\gamma_5}\psi\,,\quad \bar{\psi}'=\bar{\psi}e^{i\alpha\gamma_5}$$

$$\ln J[\alpha]=-i\int d^2x\,\alpha(x)\,A(x)\,,\quad \alpha\ \text{inf.}$$

$$A(x)=-\frac{e}{2\pi}\varepsilon_{\mu\nu}F^{\mu\nu} \tag{6.1}$$

$$\ln J[\alpha]=\frac{ie}{2\pi}\int d^2x\,\alpha(x)\,\varepsilon_{\mu\nu}F^{\mu\nu}\,,\quad \alpha\ \text{inf.} \tag{6.2}$$

$$\ln J[\alpha]=\frac{ie}{2\pi}\int d^2x\,\alpha(x)\left[\varepsilon_{\mu\nu}F^{\mu\nu}-\frac{i}{e}\,\partial^2\alpha\right]\,,\quad \alpha\ \text{finite} \tag{6.3}$$

$(QED)_4$: $\mathcal{L}=\bar{\psi}(i\not{D}-m)\psi\,,\ D_\mu=\partial_\mu+ieA_\mu$

$$A(x)=\frac{e^2}{8\pi^2}F_{\mu\nu}\,{}^*F^{\mu\nu} \tag{6.4}$$

$$\ln J[\alpha]=-i\int d^4x\,\alpha(x)\,\frac{e^2}{8\pi^2}F_{\mu\nu}\,{}^*F^{\mu\nu}\,,\quad \alpha\ \text{inf.} \tag{6.5}$$

(The details of the derivation for the Jacobian in $(QED)_4$ will be given in the following chapter.)

Non-Abelian $SU(N)_c$-gauge theory in two and four dimensions:

$$\mathcal{L} = \bar{\psi}(i\not{D})\psi\,, \quad D_\mu = \partial_\mu + igA_\mu$$

$$\psi'(x) = e^{i\alpha(x)\gamma_5}\psi(x)\,, \quad \bar{\psi}'(x) = \bar{\psi}(x)\,e^{i\alpha(x)\gamma_5} \tag{6.6}$$

$$\alpha(x) = \alpha^a(x)\,T^a\,, \quad A_\mu(x) = A_\mu^a(x)\,T^a$$

<u>2-dim:</u> $$A^a(x) = -\frac{g}{2\pi}\,\varepsilon_{\mu\nu}\,\mathrm{tr}_c\,(T^a F^{\mu\nu}) \tag{6.7}$$

$$\ln J[\alpha] = \frac{ig}{2\pi}\int d^2x\;\varepsilon_{\mu\nu}\,\mathrm{tr}_c\,[\alpha F^{\mu\nu}]\,, \quad \alpha \text{ inf.} \tag{6.8}$$

The trace operator tr_c is to be taken in color space.

<u>4-dim:</u> $$A^a(x) = \frac{g^2}{8\pi^2}\,\mathrm{tr}_c\,(T^a F_{\mu\nu}{}^*F^{\mu\nu}) \tag{6.9}$$

$$\ln J[\alpha] = -i\int d^4x\;\frac{g^2}{8\pi^2}\,\mathrm{tr}_c\,[\alpha F_{\mu\nu}{}^*F^{\mu\nu}]\,, \quad \alpha \text{ inf.} \tag{6.10}$$

If $\alpha(x)$ is a number, rather than taking values in the Lie algebra of the gauge group, we obtain for <u>$(QCD)_4$ with N_f flavors</u>

$$A(x) = \frac{g^2}{8\pi^2}\,\mathrm{tr}_c\,(F_{\mu\nu}{}^*F^{\mu\nu})\cdot N_f \tag{6.11}$$

$$\ln J[\alpha] = -i\int d^4x\;\alpha(x)\,\frac{N_f\,g^2}{8\pi^2}\,\mathrm{tr}_c\,[F^{\mu\nu}{}^*F_{\mu\nu}]\,, \quad \alpha \text{ inf.} \tag{6.12}$$

Returning to our original problem, $(QCD)_2$, we want to demonstrate how the fermion determinant can be computed using the technique developed in the chapter on the Schwinger model. So, let

$$\psi' = e^{i\alpha\gamma_5}\psi\,, \quad \bar{\psi}' = \bar{\psi}\,e^{i\alpha\gamma_5} \tag{6.13}$$

$$[d\psi d\bar{\psi}]' = J[\alpha]\,[d\psi d\bar{\psi}]\,, \quad \alpha = \alpha^a T^a\,, \qquad \alpha \text{ finite} \tag{6.14}$$

$A_\mu = A_\mu^a T^a$, with $\{T^a \,|\, a = 1,2,3\}$ being the generators of SU(2) in its fundamental representation.

The fermionic part of the generating functional for $(QCD)_2$ with massless fermions reads (in euclidean space)

$$\int [d\psi d\bar\psi]' \, e^{\int d^2x \, \bar\psi'(i\partial\!\!\!/ - eA\!\!\!/)\psi'} = \det(i\partial\!\!\!/ - gA\!\!\!/)$$

$$= J[\alpha] \int [d\psi d\bar\psi] \, e^{\int d^2x \, \bar\psi e^{i\alpha(x)\gamma_5}(i\partial\!\!\!/ - gA\!\!\!/)e^{i\alpha(x)\gamma_5}\psi}$$

$$= J[\alpha] \det\left[e^{i\alpha\gamma_5}(i\partial\!\!\!/ - gA\!\!\!/)e^{i\alpha\gamma_5}\right]$$

which means (for finite, Lie algebra-valued $\alpha(x)$)

$$J[\alpha] = \frac{\det[i\partial\!\!\!/ - gA\!\!\!/]}{\det\left[e^{i\alpha\gamma_5}(i\partial\!\!\!/ - gA\!\!\!/)e^{i\alpha\gamma_5}\right]} \tag{6.15}$$

The denominator of (6.15) can be rewritten using

$$e^{i\alpha\gamma_5}\gamma^\mu(i\partial_\mu - gA_\mu)e^{i\alpha\gamma_5}\psi = e^{i\alpha\gamma_5}\gamma^\mu i\partial_\mu(e^{i\alpha\gamma_5}\psi) - g e^{i\alpha\gamma_5}A\!\!\!/ e^{i\alpha\gamma_5}\psi$$

$$= i\partial\!\!\!/\psi + e^{i\alpha\gamma_5} i(\partial\!\!\!/ e^{i\alpha\gamma_5})\psi - g e^{i\alpha\gamma_5}A\!\!\!/ e^{i\alpha\gamma_5}\psi$$

$$= (i\partial\!\!\!/ - gA\!\!\!/')\psi$$

with

$$A\!\!\!/' := e^{i\alpha\gamma_5}A\!\!\!/ e^{i\alpha\gamma_5} - \frac{i}{g}e^{i\alpha\gamma_5}(\partial\!\!\!/ e^{i\alpha\gamma_5}) \tag{6.16}$$

$$\equiv UA\!\!\!/U - \frac{i}{g}U(\partial\!\!\!/ U), \quad U = e^{i\alpha\gamma_5} \tag{6.17}$$

and

$$U D\!\!\!\!/(A) U = D\!\!\!\!/(A') \tag{6.18}$$

$$J[\alpha] = \frac{\det[i\partial\!\!\!/ - gA\!\!\!/]}{\det[i\partial\!\!\!/ - gA\!\!\!/']} \tag{6.19}$$

Now let us consider paths in the space of the potentials $A^a_\mu(x)$:

$$(x,\tau) \longmapsto A^a_{\mu\tau}(x), \quad \tau \in [0,1], \quad A_{\tau=0} = A, \quad A_{\tau=1} = A' \tag{6.20}$$

$$A\!\!\!/_\tau = U_\tau A\!\!\!/ U_\tau - \frac{i}{g} U_\tau (\partial\!\!\!/ U_\tau), \quad U_\tau := e^{i\alpha_\tau(x)\gamma_5} \tag{6.21}$$

Then we are going to compute (the dot denotes differentiation with respect to τ)

$$\ln J[\alpha] = \ln \frac{\det[i\partial\!\!\!/ - gA\!\!\!/]}{\det[i\partial\!\!\!/ - gA\!\!\!/']} \tag{6.22}$$

$$= \lim_{N\to\infty} \ln \prod_{n=0}^{N-1} \frac{\det[i\partial\!\!\!/ - gA\!\!\!/_{n/N}]}{\det[i\partial\!\!\!/ - gA\!\!\!/_{\frac{n+1}{N}}]}$$

$$= \lim_{N\to\infty} \sum_{n=0}^{N-1} \ln \frac{\det[i\partial\!\!\!/ - gA\!\!\!/_{\frac{n}{N}}]}{\det[i\partial\!\!\!/ - gA\!\!\!/_{\frac{n}{N}+\frac{1}{N}}]} \tag{6.23}$$

$$= \lim_{N\to\infty} N \int_0^1 d\tau \ln \frac{\det[i\partial\!\!\!/ - gA\!\!\!/_\tau]}{\det[i\partial\!\!\!/ - gA\!\!\!/_{\tau+\frac{1}{N}}]}$$

$$= \lim_{N\to\infty} N \int_0^1 d\tau \ln \frac{\det[i\partial\!\!\!/ - gA\!\!\!/_\tau]}{\det[i\partial\!\!\!/ - gA\!\!\!/_\tau - g\dot{A\!\!\!/}_\tau \frac{1}{N}]} \tag{6.24}$$

Since $[\dot{\alpha}_\tau, \alpha_\tau] \neq 0$, $\alpha = \alpha^a T^a$, the calculation will be rather difficult for $(QCD)_2$. Evidently, this complication does not show up in $(QED)_2$ and therefore, in this case, the computation of the Jacobian can be done for an arbitrary path yielding the same result (6.3) as for a "straight" path.

Here, in $(QCD)_2$, we can--like in $(QED)_2$--always pick a "straight" path. (The topology will be discussed at the end of this chapter.)

Thus, let us parametrize the "straight" path according to

$$\alpha_\tau(x) = \tau\,\alpha(x) \tag{6.25}$$

Then

$$\dot{U}_\tau = \frac{d}{d\tau} U_\tau = \frac{d}{d\tau} e^{i\tau\alpha\gamma_5}$$

$$= i\alpha\gamma_5 U_\tau = U_\tau\, i\alpha\gamma_5 \tag{6.26}$$

In equation (6.24), we need $\dot{\not{A}}_\tau$:

$$\dot{\not{A}}_\tau = \frac{d}{d\tau}\left[U_\tau \not{A} U_\tau - \frac{i}{g} U_\tau (\not\partial U_\tau)\right]$$

$$= \dot{U}_\tau \not{A} U_\tau + U_\tau \not{A} \dot{U}_\tau - \frac{i}{g}\dot{U}_\tau(\not\partial U_\tau) - \frac{i}{g} U_\tau (\not\partial \dot{U}_\tau)$$

$$\overset{(6.26)}{=} (i\alpha\gamma_5) U_\tau \not{A} U_\tau + U_\tau \not{A} U_\tau (i\alpha\gamma_5)$$

$$- \frac{i}{g}(i\alpha\gamma_5) U_\tau (\not\partial U_\tau) - \frac{i}{g} U_\tau (\not\partial\, i\alpha\gamma_5 U_\tau)$$

$$= i\gamma_5 U_\tau [\alpha, \not{A}] U_\tau + \frac{1}{g}\gamma_5 \cancel{U_\tau \alpha \not\partial U_\tau}$$

$$- \frac{1}{g}\gamma_5 \cancel{U_\tau \alpha \not\partial U_\tau} - \frac{1}{g}\gamma_5 U_\tau (\not\partial\alpha) U_\tau$$

$$= -\frac{1}{g}\gamma_5 U_\tau \left(\not\partial\alpha + ig[\not{A}, \alpha]\right) U_\tau$$

$$= -\frac{1}{g}\gamma_5\, U_\tau (\not{D}(A)\alpha)\, U_\tau$$

$$= -\frac{1}{g}\gamma_5\, \not{D}(A_\tau)\alpha$$

$$\dot{A}_\tau = -\frac{1}{g}\gamma_5\gamma^\mu T^c D_\mu^{ca}(A_\tau)\alpha^a$$

$$= \frac{1}{g}\gamma^\mu\gamma_5 T^c D_\mu^{ca}(A_\tau)\alpha^a \qquad (6.27)$$

This expression for $\dot{A}_\tau$ allows us to rewrite (6.24) as

$$\ln J[\alpha] = \lim_{N\to\infty} N\int_0^1 d\tau \ln \frac{\det[i\partial\!\!\!/ - g A\!\!\!/_\tau]}{\det[i\partial\!\!\!/ - g A\!\!\!/_\tau - \gamma^\mu\gamma_5 T^c D_\mu^{ca}(A_\tau)\frac{\alpha^a}{N}]}$$

$$= \lim_{N\to\infty} N\int_0^1 d\tau \ \ln J^{\text{infinites.}}\left[\frac{\alpha}{N}\right]\Bigg|_{A^\mu\to A^\mu_\tau}$$

$$= \lim_{N\to\infty} N\int_0^1 d\tau \ \frac{ig}{2\pi}\int d^2x\, \varepsilon_{\mu\nu}\, tr_c\left(\frac{\alpha}{N} F^{\mu\nu}(A_\tau)\right) \qquad F^{\mu\nu} = \partial^\mu A^\nu + ig A^\mu A^\nu - (\mu\leftrightarrow\nu)$$

$$= \frac{ig}{2\pi}\, 2 \int d^2x \int_0^1 d\tau\ \varepsilon_{\mu\nu}\, tr_c[\alpha\,\partial^\mu A^\nu_\tau + ig\alpha A^\mu_\tau A^\nu_\tau]$$

$$= -\frac{ig}{\pi}\int d^2x\int_0^1 d\tau\ \varepsilon_{\mu\nu}\, tr_c[(\partial^\mu\alpha - ig\alpha A^\mu_\tau)A^\nu_\tau]$$

$$= -\frac{ig}{\pi}\int d^2x\int_0^1 d\tau\ \varepsilon_{\mu\nu}\, tr_c[(D^\mu(A_\tau)\alpha)A^\nu_\tau - igA^\mu_\tau\alpha A^\nu_\tau] \qquad (6.28)$$

The last line follows from

$$D_\mu\alpha = \partial_\mu\alpha + ig[A_\mu,\alpha] = \partial_\mu\alpha + igA_\mu\alpha - ig\alpha A_\mu$$

so that

$$\partial_\mu\alpha - ig\alpha A_\mu = D_\mu\alpha - igA_\mu\alpha$$

Now we make use of equation (6.27):

$$\dot{A}_\tau = \gamma_\nu \dot{A}^\nu_\tau = \frac{1}{g}\gamma^\mu\gamma_5\, T^c D_\mu^{ca}(A_\tau)\alpha^a$$

$$= -\frac{i}{g}\, \varepsilon^{\mu\nu}\, \gamma_\nu\, D_\mu \alpha \tag{6.29}$$

which means

$$\begin{aligned} \dot{A}^\nu_\tau &= -\frac{i}{g}\, \varepsilon^{\mu\nu}\, D_\mu(A_\tau)\alpha \\ &= \frac{i}{g}\, \varepsilon^{\nu\mu}\, D_\mu(A_\tau)\alpha \end{aligned} \tag{6.30}$$

or

$$\varepsilon^{\mu\nu} D_\mu(A_\tau)\alpha = ig\, \dot{A}^\nu_\tau$$

This is to be inserted in (6.28):

$$\begin{aligned} \ln J[\alpha] &= -\frac{(ig)^2}{\pi} \int d^2x \int_0^1 d\tau\, tr_c(\dot{A}_{\nu\tau} A^\nu_\tau) \\ &\quad + \frac{(ig)^2}{\pi} \int d^2x \int_0^1 d\tau\, \varepsilon_{\mu\nu}\, tr_c(A^\mu_\tau \alpha A^\nu_\tau) \\ &= \frac{g^2}{\pi} \int d^2x\, tr_c \underbrace{\int_0^1 d\tau\, \frac{1}{2} \frac{d}{d\tau}(A_{\nu\tau} A^\nu_\tau)}_{= \frac{1}{2}(A_{\nu\tau} A^\nu_\tau)\big|_0^1 = \frac{1}{2} A'_\nu A^{\nu\prime} - \frac{1}{2} A_\nu A^\nu} \\ &\quad - \frac{g^2}{\pi} \int d^2x \int_0^1 d\tau\, \varepsilon_{\mu\nu}\, tr_c[A^\mu_\tau \alpha A^\nu_\tau] \end{aligned}$$

A preliminary result for the Jacobian is therefore given by

$$\begin{aligned} \ln J[\alpha] &= \frac{g^2}{2\pi} \int d^2x\, tr_c(A'_\nu A'^\nu - A_\nu A^\nu) \\ &\quad - \frac{g^2}{\pi} \int d^2x\, \varepsilon_{\mu\nu} \int_0^1 d\tau\, tr_c(A^\mu_\tau \alpha A^\nu_\tau) \end{aligned} \tag{6.31}$$

Observe that $A\!\!\!/'$ as it shows up in (6.19) is anti-hermitean

$$A\!\!\!/ = \underbrace{A^a_\mu}_{\text{real}}\; \underbrace{T^a}_{\text{H.}}\; \underbrace{\gamma^\mu}_{\text{A.H.}} = \text{A.H.}\,,\ \text{i.e. } A\!\!\!/^+ = -A\!\!\!/$$

However, $A\!\!\!/'$ is not anti-hermitean, i.e., $A^{a\prime}_\mu$ is not real if we choose $\alpha(x)$ to be real.

Recalling equation (6.16),

$$A\!\!\!/\,' = e^{i\alpha\gamma_5} A\!\!\!/\, e^{i\alpha\gamma_5} - \frac{i}{g} e^{i\alpha\gamma_5}(\partial\!\!\!/ e^{i\alpha\gamma_5}) \; , \; \gamma_5 = \gamma_5^{+}$$

we obtain for the two terms

$$\left[e^{i\alpha\gamma_5} A\!\!\!/\, e^{i\alpha\gamma_5}\right]^{+} = -e^{-i\alpha\gamma_5} A\!\!\!/\, e^{-i\alpha\gamma_5} \; ,$$

$$\left[-\frac{i}{g} e^{i\alpha\gamma_5}(\partial\!\!\!/ e^{i\alpha\gamma_5})\right]^{+} = \left[-\frac{i}{g}\gamma^{\mu} e^{-i\alpha\gamma_5}\partial_{\mu} e^{i\alpha\gamma_5}\right]^{+}$$

$$= \frac{i}{g}\partial_{\mu} e^{-i\alpha\gamma_5} e^{i\alpha\gamma_5} (\gamma^{\mu})^{+}$$

$$= -\frac{i}{g}(-1)\, e^{-i\alpha\gamma_5}(\partial_{\mu} e^{i\alpha\gamma_5})\, e^{-i\alpha\gamma_5} e^{i\alpha\gamma_5}\gamma^{\mu}$$

$$= \frac{i}{g} e^{-i\alpha\gamma_5}\gamma^{\mu}\partial_{\mu} e^{-i\alpha\gamma_5}$$

$$= \frac{i}{g} e^{-i\alpha\gamma_5}\partial\!\!\!/ e^{-i\alpha\gamma_5}$$

So, indeed, $A\!\!\!/$ is not anti-hermitean.

If we then choose α^a to be purely imaginary,

$$\alpha^a = -i\phi^a \; , \; \phi^a \text{ real, i.e., } i\alpha = \phi \tag{6.32}$$

we obtain instead

$$A\!\!\!/\,' = e^{\phi\gamma_5} A\!\!\!/\, e^{\phi\gamma_5} - \frac{i}{g} e^{\phi\gamma_5}(\partial\!\!\!/ e^{\phi\gamma_5}) \tag{6.33}$$

which, however, is anti-hermitean, so that $A_{\mu}'^{a} \in \mathbb{R}$ if $A_{\mu}^{a} \in \mathbb{R}$.

$$[e^{\phi\gamma_5} \not{A} e^{\phi\gamma_5}]^+ = - e^{\phi\gamma_5} \not{A} e^{\phi\gamma_5}$$

$$[-\frac{i}{g}\gamma^\mu e^{-\phi\gamma_5}\partial_\mu e^{\phi\gamma_5}]^+ = \frac{i}{g}\partial_\mu e^{\phi\gamma_5} e^{-\phi\gamma_5}(\gamma^\mu)^+$$

$$= -\frac{i}{g}\gamma^\mu(\partial_\mu e^{-\phi\gamma_5}) e^{\phi\gamma_5}$$

$$= -\frac{i}{g}\gamma^\mu(-1) e^{-\phi\gamma_5}(\partial_\mu e^{\phi\gamma_5}) e^{-\phi\gamma_5} e^{\phi\gamma_5}$$

$$= \frac{i}{g} e^{\phi\gamma_5}(\not{\partial} e^{\phi\gamma_5})$$

A_μ takes values in the Lie algebra: $A_\mu = \underbrace{A^a_\mu}_{\text{real}} T^a$

which gives $tr_c A_\mu = 0$. The same is true for A'_μ :

$$tr_c \not{A}' = -\frac{i}{g}\gamma^\mu tr_c [\underbrace{e^{-\phi\gamma_5}\partial_\mu e^{\phi\gamma_5}} + ig\underbrace{e^{-\phi\gamma_5} A_\mu e^{\phi\gamma_5}}]$$

$$= \int_0^1 dt\, e^{-t\phi\gamma_5}(\partial_\mu\phi\gamma_5) e^{t\phi\gamma_5} \qquad \to tr_c(\ldots) = tr_c(e^{\phi\gamma_5} e^{-\phi\gamma_5} A_\mu)$$

$$= tr_c A_\mu = 0$$

$$= \%$$

$$tr(\%) = \int_0^1 dt\, tr_c (e^{t\phi\gamma_5} e^{-t\phi\gamma_5}\partial_\mu\phi\gamma_5)$$

$$= 0 \quad \text{since} \quad tr_c\phi = tr\,\phi^a T^a = 0$$

where we employed the formula

$$\frac{d}{d\xi} e^{\alpha A(\xi)} = \int_0^\alpha dt\, e^{(\alpha - t)A} \frac{dA(\xi)}{d\xi} e^{tA}$$

Now we rewrite the Jabobian (6.15) in the form

$$J[\phi] = \frac{\det[i\not{\partial} - g\not{A}]}{\det[e^{\phi\gamma_5}(i\not{\partial} - g\not{A})e^{\phi\gamma_5}]} = \frac{\det[i\not{\partial} - g\not{A}]}{\det[i\not{\partial} - g\not{A}']} \qquad (6.34)$$

and equation (6.28) turns into

$$\ln J[\phi] = -\frac{g}{\pi}\int d^2x \int_0^1 d\tau\, \varepsilon_{\mu\nu}\, tr_c\left[(D^\mu(A_\tau)\, i\alpha)\, A^\nu_\tau - ig\, A^\mu_\tau\, i\alpha\, A^\nu_\tau\right]$$

$$= -\frac{g}{\pi}\int d^2x \int_0^1 d\tau\, \varepsilon_{\mu\nu}\, tr_c\left[(D^\mu(A_\tau)\,\phi)\, A^\nu_\tau - ig\, A^\mu_\tau\, \phi\, A^\nu_\tau\right]$$

The relation $\varepsilon^{\mu\nu} D_\mu(A_\tau)\alpha = ig\,\dot{A}^\nu_\tau$ goes into

$$\varepsilon^{\mu\nu} D_\mu(A_\tau)\,\phi = -g\,\dot{A}^\nu_\tau \tag{6.35}$$

which yields, instead of (6.31), the relation

$$\ln J[\phi] = \frac{g^2}{2\pi}\int d^2x\; tr_c\left(A'_\nu A^{\nu\prime} - A_\nu A^\nu\right) + \frac{ig^2}{\pi}\int d^2x\, \varepsilon_{\mu\nu}\int_0^1 d\tau\, tr_c\left(A^\mu_\tau\, \phi\, A^\nu_\tau\right) \tag{6.36}$$

As in the Schwinger model, we can decouple fermions from gauge fields. This is most easily done in the "decoupling gauge" in which every A^a_μ can be represented as

$$A\!\!\!/ = \frac{i}{g}\left(\partial\!\!\!/\, e^{\gamma_5\phi}\right) e^{-\gamma_5\phi} \tag{6.37}$$

which gives, indeed, using (6.33)

$$A\!\!\!/\,' = e^{\phi\gamma_5} A\!\!\!/\, e^{\phi\gamma_5} - \frac{i}{g} e^{\phi\gamma_5}\left(\partial\!\!\!/\, e^{\phi\gamma_5}\right)$$

$$\overset{(6.37)}{=} \frac{i}{g} e^{\phi\gamma_5}\left(\partial\!\!\!/\, e^{\gamma_5\phi}\right) e^{-\phi\gamma_5} e^{\phi\gamma_5} - \frac{i}{g} e^{\phi\gamma_5}\left(\partial\!\!\!/\, e^{\phi\gamma_5}\right)$$

$$= 0$$

Finally, we end up with the Jacobian

$$\ln J[\phi] = \frac{\det[i\partial\!\!\!/ - g A\!\!\!/]}{\det[i\partial\!\!\!/]}$$

or

$$\det[i\not\partial - g\not A] = \det[i\not\partial]\cdot \exp\Big\{-\frac{g^2}{2\pi}\int d^2x\, tr_c\, A_\nu A_\nu$$

$$+ i\frac{g^2}{\pi}\int d^2x\, \varepsilon_{\mu\nu}\int_0^1 d\tau\, tr_c\left[A_\tau^\mu \phi A_\tau^\nu\right]\Big\} \tag{6.38}$$

The first term in the exponential of equation (6.38) shows that the gauge field has obtained a mass m, $m^2 = \frac{g^2}{\pi}$, i.e., the Schwinger mechanism takes place in $(QCD)_2$ as in the abelian $(QED)_2$.

Choosing the decoupling gauge (6.37), we have from (6.21), with $\phi = i\alpha$ and $\alpha_\tau = \tau\alpha$:

$$\not A_\tau = e^{\tau\phi\gamma_5}\not A e^{\tau\phi\gamma_5} - \frac{i}{g} e^{\tau\phi\gamma_5}(\not\partial e^{\tau\phi\gamma_5})$$

$$\overset{(6.37)}{=} \frac{i}{g} e^{\tau\phi\gamma_5}(\not\partial e^{\gamma_5\phi})e^{-\gamma_5\phi}e^{\tau\phi\gamma_5} - \frac{i}{g} e^{\tau\phi\gamma_5}(\not\partial e^{\tau\phi\gamma_5})$$

so that $\not A_{\tau=0} = \frac{i}{g}(\not\partial e^{\gamma_5\phi})e^{-\gamma_5\phi} = \not A$

and $\not A_{\tau=1} = 0 \quad (= \not A')$

Toward the end of this chapter, we will show that the value of $J[\phi]$ is independent of the iteration path. So, let us choose a more convenient parametrization:

$$\not A_\tau = \frac{i}{g}(\not\partial e^{(1-\tau)\gamma_5\phi})e^{-(1-\tau)\gamma_5\phi} \tag{6.39}$$

This choice satisfies our boundary conditions, too:

$$\not A_{\tau=0} = \frac{i}{g}(\not\partial e^{\gamma_5\phi})e^{-\gamma_5\phi} = \not A$$

$$\not A_{\tau=1} = 0$$

Using (6.39), the second term in the exponential of (6.38) can be written as

$$W_2 = \frac{i}{2\pi}\int_0^1 dt \int d^2x \; tr_c \left[(\partial_t \mathcal{U})\mathcal{U}^{-1}(\partial_\mu \mathcal{U})\mathcal{U}^{-1}(\partial_\nu \mathcal{U})\mathcal{U}^{-1}\,\varepsilon_{\mu\nu}\right] \tag{6.40}$$

where $$\mathcal{U} = \mathcal{U}(x,t) = e^{t\phi(x)} \tag{6.41}$$

Since $tr_c\, A_\mu A^\mu$ is proportional to $tr_c\, \partial_\mu \mathcal{U}\, \partial^\mu \mathcal{U}^{-1}$,

we end up with

$$\begin{aligned}\ln J = \ln \frac{\det i\not{D}}{\det i\not{\partial}} &= \frac{1}{\pi}\, tr\Big[\int d^2x\, \partial_\mu \mathcal{U}\, \partial_\mu \mathcal{U}^{-1} + \frac{i}{2}\varepsilon_{\mu\nu}\int_0^1 dt \int d^2x\, (\partial_t \mathcal{U})\mathcal{U}^{-1} \cdot \\ &\quad \cdot (\partial_\mu \mathcal{U})\mathcal{U}^{-1}(\partial_\nu \mathcal{U})\mathcal{U}^{-1}\Big]\end{aligned} \tag{6.42}$$

It is useful to rewrite W_2 in the following form:

$$W_2 = \frac{i}{2\pi}\int_0^1 dt \int d^2x\; \varepsilon^{\mu\nu} \underbrace{tr_c (\partial_t \mathcal{U})\mathcal{U}^{-1}(\partial_\mu \mathcal{U})\mathcal{U}^{-1}(\partial_\nu \mathcal{U})\mathcal{U}^{-1}}_{=:\; T_{t\mu\nu} \quad \text{cyclic in } t,\mu,\nu} \tag{6.43}$$

Let $i = 0,1,2$, i.e., $0 := t$ and $\mu,\nu = 1,2$.

Then it follows that

$$\begin{aligned}\varepsilon^{ijk}\, T_{ijk} &= \varepsilon^{tij}\, T_{tij} + \varepsilon^{\mu ij}\, T_{\mu ij} \\ &= \underbrace{\varepsilon^{t\mu\nu}}_{=:\,\varepsilon^{\mu\nu}} T_{t\mu\nu} + \varepsilon^{\mu\nu j}\, T_{\mu\nu j} + \varepsilon^{\mu tj}\, T_{\mu tj} \\ &= \varepsilon^{\mu\nu} T_{t\mu\nu} + \underbrace{\varepsilon^{\mu\nu\rho}}_{=0} T_{\mu\nu\rho} + \underbrace{\varepsilon^{\mu\nu t}}_{=\varepsilon^{\mu\nu}} T_{\mu\nu t} + \varepsilon^{\mu tj}\, T_{\mu tj} \\ &= 2\,\varepsilon^{\mu\nu}\, T_{t\mu\nu} + \varepsilon^{\mu tj}\, T_{\mu tj} \\ &= 3\,\varepsilon^{\mu\nu}\, T_{t\mu\nu}\end{aligned}$$

Hence, (6.43) takes the form:

$$W_2 = \frac{i}{6\pi} \int d^3x \; e^{ijk} \; tr_c\left[(\partial_i \mathcal{U})\mathcal{U}^{-1}(\partial_j \mathcal{U})\mathcal{U}^{-1}(\partial_k \mathcal{U})\mathcal{U}^{-1}\right] \tag{6.44}$$

Consider for a moment the analytic continuation of U, equation (6.41), to an element U_c of SU(2):

$$\mathcal{U}_c(x,t) = e^{it\phi(x)}, \quad \phi(x) \in \mathbb{R} \tag{6.45}$$

Prior to this, we can define the mapping U(•):

$$\begin{aligned} U_c(\cdot) \; : \quad & S^2 \longrightarrow SU(2)\;, \\ & x \longmapsto \mathcal{U}_c(x) := e^{i\phi(x)} \end{aligned} \tag{6.46}$$

Using $\pi_2(SU(2)) = 0$, we conclude that

$$\mathcal{D} := \left\{ U(\cdot) \; : \; S^2 \rightarrow SU(2) \right\} \tag{6.47}$$

is connected.

Next, consider paths in $\mathcal{D}$:

$$\begin{aligned} U_c(\cdot,\cdot) \; : \quad & S^2 x[0,1] \longrightarrow SU(2)\;, \\ & (x,t) \longmapsto e^{it\phi(x)} \end{aligned} \tag{6.48}$$

Again, we can think of $S^2x[0,1]$ as a disc which can be deformed continuously into the upper hemisphere of S^3. Hence we are considering U(•,•) as a mapping from a disc D in R^3, whose boundary is space-time, into SU(2).

So our analytically continued action W_2 with $\alpha \in \mathbb{R}$ reads

$$W_2^c[\mathcal{U}] = 4\pi i \, \Gamma[\mathcal{U}] \tag{6.49}$$

where Γ is the Wess-Zumino functional

$$\Gamma[\mathcal{U}] \equiv \frac{1}{24\pi^2}\int_D d^3x\, \varepsilon^{ijk}\, tr_c\left[(\partial_i \mathcal{U}_c)\mathcal{U}_c^{-1}(\partial_j \mathcal{U}_c)\mathcal{U}_c^{-1}(\partial_k \mathcal{U}_c)\mathcal{U}_c^{-1}\right] \tag{6.50}$$

$$\mathcal{U}_c \equiv \mathcal{U}_c(x,t)$$

But space-time also bounds (in the opposite direction, i.e., clockwise) a second 3-dimensional disc D' with D+D' = S^3, so that we obtain

$$\Gamma'[\mathcal{U}] = -\frac{1}{24\pi^2}\int_{D'} d^3x\, \varepsilon^{ijk}\, tr_c\left[(\partial_i \mathcal{U}_c)\mathcal{U}_c^{-1}(\partial_j \mathcal{U}_c)\mathcal{U}_c^{-1}(\partial_k \mathcal{U}_c)\mathcal{U}_c^{-1}\right] \tag{6.51}$$

But we have already seen that (cf. chapter 4)

$$n = \frac{1}{24\pi^2}\int_{D\cup D' = S^3} d^3x\, \varepsilon^{ijk}\, tr_c\left[(\partial_i \mathcal{U}_c)\mathcal{U}_c^{-1}(\partial_j \mathcal{U}_c)\mathcal{U}_c^{-1}(\partial_k \mathcal{U}_c)\mathcal{U}_c^{-1}\right] \tag{6.52}$$

gives the homotopy class index for the mappings of $S^3 \rightarrow SU(2)$. It is clear that $W_2^c[U]$ cannot be a single-valued functional; rather, it is determined only up to the homotopy class index with respect to $\pi_3(SU_2) = Z$. Because n is an integer, it holds, however, that

$$e^{i 4\pi \Gamma} = e^{i 4\pi \Gamma'} \tag{6.53}$$

This ambiguity of Γ is thus due to the fact that the mappings of S^2 into SU(2) can be continued in topologically inequivalent ways since, although the space of mappings $U_c(\cdot)$ is connected, $\pi_2(SU(2)) = 0$, it is not simply connected, $\pi_3(SU(2)) = Z$! and therefore, $\ln \mathcal{J}$ is not <u>globally</u> integrable. In other words: paths in $\mathcal{D}$ with identical initial and final points need not be homotopic.

Actually, $(QCD)_2$ is concerned with

$$\mathcal{U}(x,t) = e^{t\phi^a(x)T^a}, \quad \phi^a \in R,$$

so that instead of looking at elements $e^{i\alpha^a T^a} \in SU(2)$, $\alpha^a \in R$,
we are dealing with elements

$$\mathcal{U}(x,t) \in G_5 \subset SL(2,\mathbb{C})$$

forming a subgroup G_5 of $SL(2,\mathbb{C})$ homotopically equivalent to $\mathbb{R}^3$.

Since

$$\pi_2(R^3) = 0 \Rightarrow \mathcal{D} = \{S^2 \to G_5\} \text{ connected}$$

Moreover, $\pi_3(R^3) = 0 \Rightarrow \mathcal{D}$ is simply connected.

This means that W_2 is single-valued and therefore the integration (iteration) path can be chosen arbitrarily.

Additional References to
Chiral Anomaly in $(QCD)_2$

as in $(QED)_2$

R.E. Gamboa Saravi et al., Phys. Rev. D30, 1353 (1984)

R. Roskies in: Festschrift for Feza Gürsey's 60th Birthday (1982), unpublished

7. Fujikawa's Path Integral Method

In this section we describe a few important points of the path integral approach to the problem of chiral anomalies. We thereby consider a multiplet ψ of quantized Dirac fermions interacting with an external non-abelian gauge field A^a_μ. The fermions transform according to the fundamental representation of the gauge group SU(N), the generators of which we denote as T^a, a = 1,..., N^2-1. Thus, the Lagrangian is

$$\begin{aligned} \mathcal{L} &= \overline{\psi}(i\not{D}-m)\psi \\ \not{D} &\equiv \gamma^\mu D_\mu \equiv \gamma^\mu(\partial_\mu + igA_\mu) \\ A_\mu &\equiv A^a_\mu T^a \end{aligned} \tag{7.1}$$

(We use the conventions of Bjorken and Drell).
Now we perform a Wick rotation

$$\begin{aligned} &x^0 \longrightarrow -ix^4 \,, \quad \gamma^0 \longrightarrow -i\gamma^4 \\ &A_0 \longrightarrow iA_4 \end{aligned} \tag{7.2}$$

and g = diag (+1, -1, -1, -1) $\longrightarrow$ g = diag (-1, -1, -1, -1) in euclidean space-time. Thereby all the γ -matrices become anti-hermitian

$$(\gamma^\mu)^+ = -\gamma^\mu$$

and $\not{D}$, with respect to the scalar product,

$$(\phi_1, \phi_2) = \int d^4x\, \phi^*_{1\alpha k}(x)\, \phi_{2\alpha k}(x)$$

becomes hermitian: $\not{D} = \not{D}^+$

α and k denote spinor and color indices, respectively; we do not write out the sum over α = 1,2,...4 and k = 1,2,...N.
Next we consider the change of $\mathcal{L}$ under chiral transformations of the classical fields

$$\begin{aligned} \psi(x) &\longrightarrow \tilde{\psi}(x) = e^{i\alpha(x)\gamma_5}\psi(x) \\ \overline{\psi}(x) &\longrightarrow \overline{\tilde{\psi}}(x) = \overline{\psi}(x)\, e^{i\alpha(x)\gamma_5} \end{aligned} \tag{7.3}$$

with the hermitian γ_5 matrix

$$\gamma_5 = i\gamma^0\gamma^1\gamma^2\gamma^3$$
$$= \gamma^4\gamma^1\gamma^2\gamma^3$$

Because $\{\gamma^\mu, \gamma_5\} = 0$, we obtain to first order in $\alpha(x)$:

$$\tilde{\bar{\psi}}(i\not{D}-m)\tilde{\psi} = \bar{\psi}(i\not{D}-m)\psi - (\partial^\mu\alpha)(\bar{\psi}\gamma_\mu\gamma_5\psi)$$
$$- 2im(\bar{\psi}\gamma_5\psi)\alpha + O(\alpha^2) \tag{7.4}$$

We can now quantize the theory by introducing a generating functional

$$Z[\eta,\bar{\eta}] = \int[d\psi d\bar{\psi}]\, e^{\int d^4x\{\bar{\psi}(i\not{D}-m)\psi + \bar{\eta}\psi + \bar{\psi}\eta\}} \tag{7.5}$$

In order to define the integration measure $[d\psi d\bar{\psi}]$, we consider a complete set of eigenfunctions of the hermitian operator $\not{D}$:

$$\not{D}\varphi_n = \lambda_n\varphi_n$$
$$\int d^4x\, \varphi_n^+(x)\varphi_m(x) = \delta_{nm}$$

and expand the integration variables in this basis:

$$\psi(x) = \sum_n a_n\varphi_n(x)$$
$$\bar{\psi}(x) = \sum_n \varphi_n^+(x)\bar{b}_n$$

Here, a_n and $\bar{b}_n$ are independent elements of the Grassmann algebra. The integration measure is then

$$[d\psi d\bar{\psi}] = \prod_{n,m} da_m\, d\bar{b}_n \tag{7.6}$$

Now Fujikawa's decisive discovery was that this measure is not invariant under the chiral transformations (7.3). To see this, we calculate the expansion coefficients of the chirally rotated field $\tilde{\psi}$:

$$\tilde{\psi}(x) = \sum_m \tilde{a}_m \varphi_m(x) = \sum_m a_m\, e^{i\alpha(x)\gamma_5} \varphi_m(x)$$

Due to the orthonormality of the φ_n's, we get

$$\tilde{a}_n = \sum_m C_{nm} a_m$$

$$C_{nm} = \int d^4x\, \varphi_n^+(x)\, e^{i\alpha(x)\gamma_5} \varphi_m(x)$$

$$= \delta_{nm} + i\int d^4x\, \alpha(x)\, \varphi_n^+(x)\gamma_5 \varphi_m(x) + O(\alpha^2)$$

$$=: \delta_{nm} + \varepsilon_{nm} + O(\alpha^2)$$

and likewise for the $\bar{b}_n$.
The measure then transforms according to

$$\prod_n d\tilde{a}_n = (\det C)^{-1} \prod_n da_n$$

and together with the change of b_n, we end up with

$$[d\tilde{\psi}\, d\tilde{\bar{\psi}}] = (\det C)^{-2} [d\psi\, d\bar{\psi}]$$

(The appearance of inverse functional determinants is caused by the fact that fermiintegration is identical to left-differentiation.)

For infinitesimal α, we obtain, in matrix notation

$$\det C = \det(1+\varepsilon)$$

$$= \exp\{Tr \ln(1+\varepsilon)\}$$

$$= \exp\{Tr\, \varepsilon + O(\alpha^2)\}$$

$$= \exp\{\sum_n \varepsilon_{nn} + O(\alpha^2)\}$$

$$= \exp\{\sum_n i\int d^4x\, \alpha(x)\, \varphi_n^+(x)\gamma_5\varphi_n(x) + O(\alpha^2)\}$$

or $(\det C)^{-1} = \exp\{-i\int d^4x\, \alpha(x) B(x) + O(\alpha^2)\}$

where
$$B(x) = \sum_n \varphi_n^+(x)\gamma_5\varphi_n(x) \tag{7.7}$$

From the completeness of the φ_n's, it follows that B is proportional to

$$\sum_n \varphi_n^+(x)\gamma_5\varphi_n(x) \text{ "=" } tr\,\gamma_5 \cdot \delta(0) \text{ "=" } 0\cdot\infty$$

Hence B(x) makes no sense in the above form (7.7).
Fujikawa solves this problem by regularizing (7.7) by the following cut-off procedure

$$\begin{aligned} B(x) &= \lim_{M\to\infty} \sum_n \varphi_n^+(x)\gamma_5 e^{-(\frac{\lambda_n}{M})^2}\varphi_n(x) \\ &= \lim_{M\to\infty}\sum_n \varphi_n^+(x)\gamma_5 e^{-\frac{\not{D}^2}{M^2}}\varphi_n(x) \end{aligned} \tag{7.8}$$

Again using the completeness of the set $\{\varphi_n\}$, we can write for any function of the Dirac operator $\not{D}^2$

$$\sum_n \varphi_n^+(x) f(\not{D}_x^2)\varphi_n(x) = \int d^4x' \sum_n \varphi_n^+(x)\left(f(\not{D}_x^2)\delta(x-x')\right)\varphi_n(x')$$

$$= \int d^4x' \sum_{a,b}\left(f(\not{D}_x^2)\delta(x-x')\right)_{ab}\sum_n \varphi_{na}^*(x)\varphi_{nb}(x')$$

$$= \int d^4x' \sum_{a,b}\left(f(\not{D}_x^2)\delta(x-x')\right)_{ab}\delta_{ab}\delta(x-x')$$

$$= \lim_{x'\to x} tr\, f(\not{D}_x^2)\,\delta(x-x')$$

This means for our expression (7.8)

$$B(x) = \lim_{M\to\infty} \lim_{x'\to x} \operatorname{tr} \gamma_5 \, e^{-\frac{\not{D}_x^2}{M^2}} \underbrace{\delta(x-x')}_{= \int \frac{d^4k}{(2\pi)^4} e^{ik(x-x')}}$$

$$= \lim_{M\to\infty} \int \frac{d^4k}{(2\pi)^4} \operatorname{tr} \gamma_5 \, e^{-ikx} e^{-\frac{\not{D}_x^2}{M^2}} e^{ikx}$$

$$= \lim_{M\to\infty} \int \frac{d^4k}{(2\pi)^4} \operatorname{tr} \gamma_5 \exp\left\{-\frac{1}{M^2} e^{-ikx} \not{D}_x^2 e^{ikx}\right\}$$

To evaluate the exponential, we need

$$\not{D}^2 = \gamma_\mu \gamma_\nu D^\mu D^\nu$$

$$= \left(\tfrac{1}{2}\{\gamma_\mu, \gamma_\nu\} + \tfrac{1}{2}[\gamma_\mu, \gamma_\nu]\right) D^\mu D^\nu$$

$$= D^\mu D_\mu + \tfrac{i}{2} g \, \gamma_\mu \gamma_\nu F^{\mu\nu}$$

Also note that $\quad [D_\mu, D_\nu] = ig F_{\mu\nu}$

Then we can continue to write

$$B(x) = \lim_{M\to\infty} \int \frac{d^4k}{(2\pi)^4} \operatorname{tr} \gamma_5 \exp\left\{-\frac{1}{M^2}\left[-k^2 + 2ik\cdot D + D^2 + \frac{ig}{2} \gamma_\mu \gamma_\nu F^{\mu\nu}\right]\right\}$$

$$= \lim_{M\to\infty} M^4 \int \frac{d^4k'}{(2\pi)^4} \operatorname{tr} \gamma_5 \exp\left\{-\left[-k'^2 + \frac{2ik'\cdot D}{M} + \frac{D^2 + \frac{ig}{2}\gamma_\mu \gamma_\nu F^{\mu\nu}}{M^2}\right]\right\}$$

This expression is well-defined and can be evaluated by standard methods. Then our result reads

$$B(x) = \frac{g^2}{16\pi^2} \operatorname{tr} \left(F_{\mu\nu} {}^*F^{\mu\nu}\right) \tag{7.9}$$

with $$F_{\mu\nu} \equiv \partial_\mu A_\nu - \partial_\nu A_\mu + ig[A_\mu, A_\nu] \tag{7.10}$$

and $$^*F_{\mu\nu} \equiv \frac{1}{2}\varepsilon_{\mu\nu\alpha\beta}F^{\alpha\beta} \tag{7.11}$$

If one takes into account that in the transformation of $\prod_n d\bar{b}_n$ an identical factor $(\det C)^{-1}$ is created, then it follows that for the total Jacobian of the transformations (7.3) for infinitesimal $\alpha(x)$:

$$[d\tilde{\psi}d\tilde{\bar{\psi}}] = J[\alpha]\ [d\psi d\bar{\psi}]$$

$$J[\alpha] = \exp\left\{-i\frac{g^2}{8\pi^2}\int d^4x\ \alpha(x)\ \mathrm{tr}\left(F_{\mu\nu}{}^*F^{\mu\nu}\right)\right\} \tag{7.12}$$

The analogous result is obtained for the abelian theory (QED) by simply omitting the trace over the algebraic degrees of freedom. If, on the other hand, one wants to consider numerous flavors of fermions, then the exponent is multiplied with its number. Furthermore, (7.12) does not change its form when transformed back into Minkowski space. With the knowledge that the integration measure is not invariant under chiral transformations, we now have a very simple way of characterizing all (anomalous) Ward-Takahashi identities of our model: Since the generating functional does not depend on whether ψ and $\bar{\psi}$ or $\psi(\alpha)$ and $\bar{\psi}(\alpha)$ were used as integration variables, the following relation must be valid for all values of the sources:

$$\frac{\delta}{\delta\alpha(x)}\, Z[\eta,\bar{\eta}]\Big|_{\alpha=0} = 0$$

In order to see what this means, we take as simplest example the calculation of the expectation value of the axial vector (singlet) current. Here it suffices to consider only $Z[0,0]$:

$$\int[d\psi d\bar{\psi}]\ \exp\left\{\int d^4x\ \bar{\psi}(i\not{D}-m)\psi\right\}$$
$$\equiv \int[d\tilde{\psi}d\tilde{\bar{\psi}}]\exp\left\{\int d^4x\ \tilde{\bar{\psi}}(i\not{D}-m)\tilde{\psi}\right\}$$
$$= \int[d\psi d\bar{\psi}]\ \exp\left\{-i\frac{g^2}{8\pi^2}\int d^4x\,\alpha(x)\ \mathrm{tr}\,F_{\mu\nu}{}^*F^{\mu\nu}\right\}\ .$$

$$\cdot \exp\left(\int d^4x \left\{\bar{\psi}(i\not{D}-m)\psi - (\partial^\mu\alpha)(\bar{\psi}\gamma_\mu\gamma_5\psi) - 2im(\bar{\psi}\gamma_5\psi)\alpha + O(\alpha^2)\right\}\right)$$

$$= \int [d\psi d\bar{\psi}] \exp\left\{\int d^4x\, \bar{\psi}(i\not{D}-m)\psi\right\} \cdot \left[1 + \int d^4x\, \alpha(x)\left\{\partial^\mu(\bar{\psi}\gamma_\mu\gamma_5\psi)\right.\right.$$

$$\left.\left. - 2im(\bar{\psi}\gamma_5\psi) - i\frac{g^2}{8\pi^2}\, tr\, F_{\mu\nu}{}^*F^{\mu\nu}\right\} + O(\alpha^2)\right] \tag{7.13}$$

It follows from the equality of the first and last lines of this relation that

$$\partial^\mu \langle\bar{\psi}\gamma_\mu\gamma_5\psi\rangle = 2im\langle\bar{\psi}\gamma_5\psi\rangle + [i]\frac{g^2}{8\pi^2} F_{\mu\nu}{}^*F^{\mu\nu} \tag{7.14}$$

The factor i does not appear in Minkowski space.

The last term in eq. (7.14) is precisely the Adler-Bell-Jackiw anomaly which was found in pertubation theory in lowest approximation by evaluating the triangle graph. Since, however, the above method does not have recourse to perturbation expansion, we not only have produced here the earlier results in a very elegant manner, but also have simultaneously shown that squares, pentagons, etc. do not contribute to the anomaly. This says nothing, however, about possible radiative corrections in presence of a quantized gauge field, i.e., the correctness of the Adler-Bardeen theorem.

Additional References to
Fujikawa's Path Integral Method

K. Fujikawa, Phys. Rev. Lett. 42, 1195 (1979); Phys. Rev. D21, 2848 (1980)

K. Huang, "Quarks, Leptons and Gauge Fields", World Scientific, Singapore (1982)

8. Chiral Anomaly in the Schwinger-Symanzik Formalism

In this chapter we want to demonstrate how chiral anomalies appear in the Schwinger-Symanzik functional formalism of quantum field theory. By way of illustration, we derive the Adler-Bell-Jackiw anomaly of QED. Hence, we start from the Lagrangian

$$\mathcal{L} = \bar{\psi}\,(i \not{D}(A) - m)\psi - \frac{1}{4} F_{\mu\nu} F^{\mu\nu} + \bar{\eta}\psi + \bar{\psi}\eta + j_\mu A^\mu \tag{8.1}$$

with
$$D_\mu(A) = \partial_\mu + ie\,A_\mu \tag{8.2}$$

and where the fermions and photons are coupled to external c-number sources $\bar{\eta}$, η and j_μ, respectively.
To study the response of (8.1) under infinitesimal chiral transformation on the spinor field, we set

$$\begin{aligned} \psi(x) &\longrightarrow \tilde{\psi}(x) = e^{i\alpha(x)\gamma_5}\,\psi(x) \\ \bar{\psi}(x) &\longrightarrow \tilde{\bar{\psi}}(x) = \bar{\psi}(x)\, e^{i\alpha(x)\gamma_5} \end{aligned} \tag{8.3}$$

As a result, we produce a theory containing additional couplings to the external (infinitesimal) pseudoscalar field $\alpha(x)$:

$$\begin{aligned} \mathcal{L}_\alpha &= \bar{\psi}(i\not{D}(A) - m)\psi - \frac{1}{4} F_{\mu\nu} F^{\mu\nu} - (\partial^\mu \alpha)\bar{\psi}\gamma_\mu\gamma_5\psi \\ &\quad - 2im\alpha\,(\bar{\psi}\gamma_5\psi) + \tilde{\bar{\eta}}\psi + \bar{\psi}\tilde{\eta} + j_\mu A^\mu + O(\alpha^2) \end{aligned} \tag{8.4}$$

where, in the source terms, we shifted the chiral transformation from the fields to the sources

$$\begin{aligned} \tilde{\eta}(x) &\equiv e^{i\alpha(x)\gamma_5}\,\eta(x) \\ \tilde{\bar{\eta}}(x) &\equiv \bar{\eta}(x)\, e^{i\alpha(x)\gamma_5} \end{aligned} \tag{8.5}$$

The generating functional for the n-point functions of the theory described by the Lagrangian $\mathcal{L}_\alpha$ is now given by

$$\hat{Z}[\eta,\bar{\eta},j;\alpha] = \langle 0|T\exp\{i\int d^4x(\tilde{\bar{\eta}}\psi+\bar{\psi}\tilde{\eta}+j_\mu A^\mu - (\partial^\mu\alpha)\bar{\psi}\gamma_\mu\gamma_5\psi - 2im\alpha\bar{\psi}\gamma_5\psi)\}|0\rangle \tag{8.6}$$

This generating functional can be used to construct transition matrix elements $\langle a|b\rangle^\alpha$ for in-states $|b\rangle$ and out-states $|a\rangle$ in presence of the external source parameter $\alpha(x)$. To obtain those matrix elements, we first compute the variation of $\langle a|b\rangle^\alpha$ under the change of $\alpha(x)$; using the LSZ reduction formulae, we have, in an obvious notation,

$$\frac{1}{i}\frac{\delta\langle a|b\rangle^\alpha}{\delta\alpha(w)}\bigg|_{\alpha=0} = (\text{const})\cdot\int d^4x \ldots e^{ikx}\ldots\ \bar{u}(i\vec{\partial}\!\!\!/-m)\ldots$$

$$\ldots\ \bar{v}(i\vec{\partial}\!\!\!/-m)\ldots\ \varepsilon\vec{\Box}\ldots\ \frac{1}{i}\frac{\delta}{\delta j^\mu}\cdots\frac{-1}{i}\frac{\delta}{\delta\eta}\cdots\frac{1}{i}\frac{\delta}{\delta\bar{\eta}}\cdots$$

$$\frac{1}{i}\frac{\delta\hat{Z}[\eta,\bar{\eta},j;\alpha]}{\delta\alpha(w)}\bigg|_{\substack{\alpha=0\\ \eta=\bar{\eta}=j=0}}\ (-i\overleftarrow{\partial\!\!\!/}-m)u\ldots(-i\overleftarrow{\partial\!\!\!/}-m)v\ldots\overleftarrow{\Box}\varepsilon\ldots \tag{8.7}$$

To explicitly construct $\hat{Z}$, either Schwinger's action principle or Symanzik's approach can be used.

$$\hat{Z}[\eta,\bar{\eta},j;\alpha] = \exp\left[-\int\frac{\delta}{\delta\tilde{\eta}}(e\gamma^\mu\frac{\delta}{\delta j^\mu})\frac{\delta}{\delta\tilde{\bar{\eta}}}\right]\cdot$$

$$\cdot\exp\left[i\int\frac{\delta}{\delta\tilde{\eta}}\{-(\partial^\mu\alpha)\gamma_\mu\gamma_5-2im\alpha\gamma_5\}\frac{\delta}{\delta\tilde{\bar{\eta}}}\right]Z_0[\tilde{\eta},\tilde{\bar{\eta}},j]\bigg|_{\substack{\tilde{\eta}=e^{i\alpha\gamma_5}\eta\\ \tilde{\bar{\eta}}=\bar{\eta}e^{i\alpha\gamma_5}}} \tag{8.8}$$

with Z_0 the generating functional of the free ($e=0,\alpha=0$) theory contained in

$$Z_0[\tilde{\eta},\tilde{\bar{\eta}},j] = \exp\left[\tfrac{i}{2}jDj\right]\exp\left[-i\tilde{\bar{\eta}}\{i\partial\!\!\!/-m\}^{-1}\tilde{\eta}\right]\det(i\partial\!\!\!/-m) \tag{8.9}$$

where D_+ denotes the photon propagator.
Applying the first exponential operator in (8.8) to Z_o yields the usual QED functional Z_{QED}, which, by employing the formula

$$\exp\left[-i \frac{\delta}{\delta\eta} A \frac{\delta}{\delta\bar{\eta}}\right] \exp\left[i \bar{\eta} B \eta\right]$$

$$= \exp\left[i\bar{\eta}\, B(1+AB)^{-1}\eta\right] \det(1+AB) \tag{8.10}$$

can be written as

$$Z_{QED}[\tilde{\eta},\tilde{\bar{\eta}},j] = \exp\left[-i\tilde{\bar{\eta}}\left\{i\not{D}\left(\tfrac{1}{i}\tfrac{\delta}{\delta j}\right)-m\right\}^{-1}\tilde{\eta}\right]$$

$$\cdot \det\left[i\not{D}\left(\tfrac{1}{i}\tfrac{\delta}{\delta j}\right)-m\right] \exp\left[\tfrac{i}{2} j D_+ j\right] \tag{8.11}$$

More explicitly: $A \equiv e\gamma^\mu \frac{1}{i}\frac{\delta}{\delta j_\mu}$, $B = \frac{-1}{i\not{\partial}-m}$

so that in (8.10)

$$[B(1+AB)^{-1}]^{-1} = (1+AB)B^{-1} = B^{-1}+A = -i\not{\partial}+m+e\gamma\tfrac{1}{i}\tfrac{\delta}{\delta j}$$

$$= -\left[i\not{\partial}-e\gamma\tfrac{1}{i}\tfrac{\delta}{\delta j}-m\right] = -\left[i\not{D}\left(\tfrac{1}{i}\tfrac{\delta}{\delta j}\right)-m\right]$$

and

$$\det(1+AB)\det(i\not{\partial}-m) = \det\left[\left(1+e\gamma\tfrac{1}{i}\tfrac{\delta}{\delta j}\tfrac{(-1)}{i\not{\partial}-m}\right)(i\not{\partial}-m)\right]$$

$$= \det\left[i\not{\partial}-m-e\gamma\tfrac{1}{i}\tfrac{\delta}{\delta j}\right]$$

$$= \det\left[i\not{D}\left(\tfrac{1}{i}\tfrac{\delta}{\delta j}\right)-m\right]$$

A still more useful representation of Z_{QED} is given by

$$Z_{QED}[\tilde{\eta},\tilde{\bar{\eta}},j] = \exp\left[\tfrac{i}{2}\, jD_+ j\right] \exp\left[-\tfrac{i}{2}\frac{\delta}{\delta A} D_+ \frac{\delta}{\delta A}\right]$$

$$\exp\left[-i\tilde{\bar{\eta}}\{i\not{D}(A)-m\}^{-1}\tilde{\eta}\right] \det[i\not{D}(A)-m] \tag{8.12}$$

with A^μ now being defined as $A = D_+ j$.
Inserting (8.12) in (8.8) and applying the second exponential operator with the aid of (8.10), we obtain

$$\hat{Z}[\eta,\bar{\eta},j;\alpha] = \exp\left[\tfrac{i}{2}\, jD_+ j\right] \exp\left[-\tfrac{i}{2}\frac{\delta}{\delta A} D_+ \frac{\delta}{\delta A}\right]$$

$$\cdot \exp\left[-i\tilde{\bar{\eta}}\, e^{-i\alpha\gamma_5}\{i\not{D}(A)-m\}^{-1} e^{-i\alpha\gamma_5}\tilde{\eta}\right]$$

$$\cdot \det[i\not{D}(A)-m-\partial^\mu\alpha\,\gamma_\mu\gamma_5-2im\alpha\gamma_5]\Big|_{\substack{\tilde{\eta}=e^{i\alpha\gamma_5}\eta \\ \tilde{\bar{\eta}}=\bar{\eta}e^{i\alpha\gamma_5}}} \tag{8.13}$$

At this stage, the important point comes in (cf. chapter 7): it holds to first order in α ($^*F^{\mu\nu}$ is the dual of the field strength tensor, $^*F^{\mu\nu} = \frac{1}{2}\varepsilon^{\mu\nu\rho\sigma}F_{\rho\sigma}$)

$$\det[i\not{D}(A)-m-\partial^\mu\alpha\,\gamma_\mu\gamma_5-2im\,\alpha\gamma_5]$$

$$= \exp\left[i\int d^4x\,\alpha(x)\frac{e^2}{8\pi^2}F_{\mu\nu}{}^*F^{\mu\nu}\right]\cdot\det[i\not{D}(A)-m] \tag{8.14}$$

so that we may rewrite $\hat{Z}$:

$$\hat{Z}[\eta,\bar{\eta},j;\alpha] = \exp\left[\tfrac{i}{2}\, jD_+ j\right] \exp\left[-\tfrac{i}{2}\frac{\delta}{\delta A} D_+ \frac{\delta}{\delta A}\right]$$

$$\cdot \exp\left[-i\bar{\eta}\,\{i\not{D}(A)-m\}^{-1}\eta\right]$$

$$\cdot \exp\left[i\int d^4x\,\alpha(x)\frac{e^2}{8\pi^2}(F_{\mu\nu}{}^*F^{\mu\nu})(A(x))\right]\det[i\not{D}(A)-m]$$

$$= \exp\left[i\int d^4x\, \alpha(x) \frac{e^2}{8\pi^2} (F_{\mu\nu}{}^*F^{\mu\nu})\left(\frac{1}{i}\frac{\delta}{\delta j(x)}\right)\right] \hat{Z}[\eta,\bar{\eta},j;\alpha=0] \tag{8.15}$$

We are interested in the functional derivative with respect to α:

$$\frac{1}{i}\frac{\delta \hat{Z}[\eta,\bar{\eta},j;\alpha]}{\delta\alpha(w)}\bigg|_{\alpha=0} = \frac{e^2}{8\pi^2}(F_{\mu\nu}{}^*F^{\mu\nu})\left(\frac{1}{i}\frac{\delta}{\delta j(w)}\right)\hat{Z}[\eta,\bar{\eta},j;\alpha=0] \tag{8.16}$$

Inserting eq. (8.16) in (8.7) yields

$$\frac{1}{i}\frac{\delta \langle a|b\rangle^\alpha}{\delta\alpha(w)}\bigg|_{\alpha=0} = \langle a|\frac{e^2}{8\pi^2}(F_{\mu\nu}{}^*F^{\mu\nu})(A(w))|b\rangle \tag{8.17}$$

with the original meaning of $A_\mu(x)$ as an operator field.
On the other hand, by explicitly differentiating equation (8.6) with respect to α, we see that $\frac{1}{i}\frac{\delta}{\delta\alpha(w)}$ causes an insertion of the operator--after an integration by parts-- $\partial^\mu(\bar{\psi}\gamma_\mu\gamma_5\psi) - 2im(\bar{\psi}\gamma_5\psi)$ in any n-point function calculated by $\hat{Z}$. By LSZ reduction, one then gets

$$\frac{1}{i}\frac{\delta \langle a|b\rangle^\alpha}{\delta\alpha(w)}\bigg|_{\alpha=0} = \langle a|\{\partial^\mu(\bar{\psi}(w)\gamma_\mu\gamma_5\psi(w)) - 2im\,\bar{\psi}(w)\gamma_5\psi(w)\}|b\rangle \tag{8.18}$$

Equating (8.17) and (8.18), we finally end up with the operator relation

$$\partial^\mu(\bar{\psi}\gamma_\mu\gamma_5\psi) = 2im(\bar{\psi}\gamma_5\psi) + \frac{e^2}{8\pi^2} F_{\mu\nu}{}^*F^{\mu\nu} \tag{8.19}$$

which is satisfied between arbitrary states.

Additional References to

Chiral Anomaly in the Schwinger-Symanzik Formalism

H.M. Fried, "Functional Methods and Models in Quantum Field Theory", MIT Press, Cambridge (1972)

J. Schwinger, Proc. Natl. Acad. Sci. U.S. 37, 452, 455 (1951); K. Symanzik, Z. Naturforschg. 9, 809 (1954)

W. Dittrich, M. Reuter, Phys. Rev. D31, 1509 (1985)

M. Reuter, Phys. Rev. D31, 1374 (1985)

9. Anomalous Ward Identities à la Wess-Zumino

To obtain an initial understanding of the Wess-Zumino approach to anomalies of currents, let us look at a simple example, namely, an abelian theory in which Dirac particles are coupled to external fields V_μ and A_μ according to

$$\mathcal{L} = \bar{\psi} i \partial\!\!\!/ \psi - m \bar{\psi}\psi + \bar{\psi}(V\!\!\!/ + A\!\!\!/ \gamma_5)\psi \tag{9.1}$$

By studying the response of this Lagrangian under an infinitesimal local phase transformation--in the spirit of the preceding chapters--

$$\psi(x) \longrightarrow \psi'(x) = (1 + i\alpha(x))\psi(x) \tag{9.2}$$

$$\bar{\psi}(x) \longrightarrow \bar{\psi}'(x) = \bar{\psi}(x)(1 - i\alpha(x)) \tag{9.3}$$

we can generate the vector current.
Since

$$\bar{\psi}' i \partial\!\!\!/ \psi' = -(\partial^\mu \alpha)\bar{\psi}\gamma_\mu\psi + \bar{\psi} i \partial\!\!\!/ \psi + O(\alpha^2) \tag{9.4}$$

and the other terms in (9.1) remain unchanged, we obtain

$$\mathcal{L}_\alpha = \mathcal{L} - (\partial^\mu \alpha(x))\bar{\psi}\gamma_\mu\psi \tag{9.5}$$

which shows that

$$\frac{\delta \mathcal{L}_\alpha}{\delta \alpha} = 0 \tag{9.6}$$

and

$$\frac{\delta \mathcal{L}_\alpha}{\delta(\partial^\mu \alpha)} = -\bar{\psi}\gamma_\mu\psi \tag{9.7}$$

From Schwinger's action principle or the Gell-Mann-Levy equation

$$\partial^\mu \left(\frac{\delta \mathcal{L}_\alpha}{\delta(\partial^\mu \alpha)} \right) = \frac{\delta \mathcal{L}_\alpha}{\delta \alpha} \tag{9.8}$$

we get, using the definition

$$J_\mu = -\frac{\delta \mathcal{L}}{\delta(\partial^\mu \alpha)} \tag{9.9}$$

the current $J_\mu = \bar{\psi}\gamma_\mu\psi$ (9.10)

which is conserved:

$$\partial^\mu J_\mu = \partial^\mu(\bar{\psi}\gamma_\mu\psi) = 0 \tag{9.11}$$

Now we study the variation of (9.1) under the chiral transformation γ_5:

$$\psi'(x) = e^{i\beta(x)\gamma_5}\psi(x) = (1+i\beta(x)\gamma_5)\psi(x) + O(\beta^2) \tag{9.12}$$

$$\bar{\psi}'(x) = \bar{\psi}(x)e^{i\beta(x)\gamma_5} = \bar{\psi}(1+i\beta(x)\gamma_5) + O(\beta^2) \tag{9.13}$$

To understand the positive sign in the exponential of eq. (9.13), we recall that $\{\gamma_\mu, \gamma_5\} = 0$ and $\gamma_5^n\gamma_\mu = (-1)^n\gamma_\mu\gamma_5^n$, which implies

$$\begin{aligned} e^{i\beta(x)\gamma_5}\gamma_\mu &= \sum_{n=0}^{\infty}\frac{(i\beta(x))^n}{n!}\gamma_5^n\gamma_\mu = \gamma_\mu\sum_{n=0}^{\infty}\frac{(i\beta(x))^n}{n!}(-1)^n\gamma_5^n \\ &= \gamma_\mu e^{-i\beta(x)\gamma_5} \end{aligned} \tag{9.14}$$

and therefore,

$$\begin{aligned} \bar{\psi}'(x) = \psi'^{+}\gamma^0 &= \left(e^{i\beta(x)\gamma_5}\psi(x)\right)^{+}\gamma^0 \\ &= \psi^{+}(x)e^{-i\beta(x)\gamma_5^{+}}\gamma^0 \\ &\underset{(9.14)}{\overset{\gamma_5=\gamma_5^{+}}{=}} \psi^{+}(x)\gamma^0 e^{i\beta(x)\gamma_5} = \bar{\psi}(x)e^{i\beta(x)\gamma_5} \end{aligned} \tag{9.15}$$

The transformation of $\mathcal{L}$ under the chiral variation of the fields is then obtained from

$$\bar{\psi}' i\gamma_\mu \partial^\mu \psi' = \bar{\psi} e^{i\beta\gamma_5} i\gamma_\mu \partial^\mu e^{i\beta\gamma_5}\psi$$

$$= \bar{\psi} e^{i\beta\gamma_5} i\gamma_\mu \left[(i\partial^\mu\beta(x)\gamma_5)e^{i\beta\gamma_5}\psi + e^{i\beta\gamma_5}\partial^\mu\psi\right]$$

$$= -(\partial^\mu\beta(x))\bar{\psi} e^{i\beta\gamma_5}\gamma_\mu\gamma_5 e^{i\beta\gamma_5}\psi + \bar{\psi} i \partial\!\!\!/ \psi$$

$$= -(\partial^\mu\beta(x))\,\bar{\psi}\gamma_\mu\gamma_5\psi + \bar{\psi} i \partial\!\!\!/ \psi \tag{9.16}$$

Likewise, $-m\bar{\psi}'\psi' = -m\,\bar{\psi} e^{2i\beta(x)\gamma_5}\psi$

$$= -m\bar{\psi}\psi - 2im\,\bar{\psi}\gamma_5\psi\,\beta(x) + O(\beta^2) \tag{9.17}$$

The terms $\bar{\psi}(\gamma_\mu V^\mu + \gamma_\mu\gamma_5 A^\mu)\psi$ remain unaltered. Hence, the response of $\mathcal{L}$ under an infinitesimal γ_5-phase transformation gives rise to

$$\mathcal{L}_\beta = \mathcal{L} - (\partial_\mu\beta(x))\,\bar{\psi}\gamma_\mu\gamma_5\psi - \beta(x)\,2im\,\bar{\psi}\gamma_5\psi \tag{9.18}$$

Consequently, the current generated by (9.12, 13) follows again from the equations

$$\partial^\mu \frac{\delta\mathcal{L}_\beta}{\delta(\partial^\mu\beta)} = \frac{\delta\mathcal{L}_\beta}{\delta\beta}$$

with
$$-\frac{\delta\mathcal{L}_\beta}{\delta(\partial^\mu\beta)} = \bar{\psi}\gamma_\mu\gamma_5\psi =: J^5_\mu \tag{9.19}$$

and
$$-\frac{\delta\mathcal{L}_\beta}{\delta\beta} = 2im\,\bar{\psi}\gamma_5\psi =: 2im\,J_5 \tag{9.20}$$

The explicit dependence of $\mathcal{L}_\beta$ on $\beta(x)$ in (9.18) yields the non-conserved axial vector current

$$\partial_\mu J^\mu_5 = \partial_\mu(\bar{\psi}\gamma^\mu\gamma_5\psi) = 2im\,\bar{\psi}\gamma_5\psi = 2im\,J_5 \tag{9.21}$$

Had we assumed the fermion to be massless, we would have obtained a conserved axial-vector current, too. However, the "naive" divergence (9.21) is incorrect if the ψ's are treated as operators. Objects like products of operator fields at the same space-time point are highly divergent quantities which must be given precise meaning by assigning a regularization scheme. In the context of the axial vector divergence, this was done in a satisfactory manner for the first time by J. Schwinger in one of the chapters of his superb paper on "gauge invariance and vacuum polarization" (Phys. Rev. 82, 664 (1951)). With the aid of a gauge invariant regularization scheme, we would find instead of (9.21)

$$\partial_\mu(\bar\psi\gamma^\mu\gamma_5\psi) = 2im\bar\psi\gamma_5\psi + \frac{1}{8\pi^2}F_{\mu\nu}{}^*F^{\mu\nu} \tag{9.22}$$

where * denotes the Hodge duality transformation on the field-strength tensor $F_{\mu\nu}$:

$$F_{\mu\nu} = \partial_\mu V_\nu - \partial_\nu V_\mu \tag{9.23}$$

$$^*F_{\mu\nu} = \frac{1}{2}\varepsilon_{\mu\nu\rho\sigma}F^{\rho\sigma} \tag{9.24}$$

If the model were not quantized, equations (9.11, 21) would be correct as a mere consequence of the classical field equations for the spinor field following from the Lagrangian (9.1)

$$(i\partial\!\!\!/ - m)\psi = -(V\!\!\!/ + A\!\!\!/\gamma^5)\psi \tag{9.25}$$

which is to be employed in

$$\begin{aligned}\partial_\mu(\bar\psi\gamma^\mu\gamma^5\psi) &= (\partial_\mu\bar\psi\gamma^\mu)\gamma^5\psi + \bar\psi(\gamma^\mu\gamma^5\partial_\mu\psi)\\ &= (\partial_\mu\bar\psi\gamma^\mu)\gamma^5\psi - \bar\psi\gamma^5(\gamma^\mu\partial_\mu\psi)\end{aligned} \tag{9.26}$$

Now we make use of eq. (9.25) in the form

$$i\gamma^\mu\partial_\mu\psi = m\psi - (V\!\!\!/ + A\!\!\!/\gamma_5)\psi \tag{9.27}$$

or $$\gamma^\mu \partial_\mu \psi = -im\psi + i(\not{V} + \not{A}\gamma_5)\psi \tag{9.28}$$

The adjoint of (9.27) is then given by

$$-i\,\partial_\mu \psi^+ \gamma^{\mu +} = m\psi^+ - \psi^+(\gamma^{\mu +} V_\mu + \gamma^{5+}\gamma^{\mu +} A_\mu)\,,$$

$$-i\,\partial_\mu \psi^+ \gamma^0\gamma^\mu\gamma^0 = m\psi^+ - \psi^+(\gamma^0\gamma^\mu\gamma^0 V_\mu + \gamma^5\gamma^0\gamma^\mu\gamma^0 A_\mu)$$

since in Minkowski space $\gamma^{\mu +} = \gamma^0\gamma^\mu\gamma^0\,,\ \gamma^{5+} = \gamma^5$

Multiplying the last eq. from the r.h.s. by γ^0 and using $\{\gamma^5, \gamma^\mu\} = 0$ again, gives

$$-i\,\partial_\mu \bar{\psi}\gamma^\mu = m\bar{\psi} - \bar{\psi}(\gamma^\mu V_\mu + \gamma^\mu A_\mu \gamma_5)$$

or

$$\partial_\mu \bar{\psi}\gamma^\mu = im\bar{\psi} - i\,\bar{\psi}(\not{V} + \not{A}\gamma_5)$$

so that

$$\begin{aligned}\partial_\mu(\bar{\psi}\gamma^\mu\gamma^5\psi) &= im\bar{\psi}\gamma_5\psi - i\bar{\psi}\,(\not{V} + \not{A}\gamma_5)\,\gamma^5\psi \\ &\quad + im\bar{\psi}\gamma_5\psi - i\bar{\psi}\gamma_5\,(\not{V} + \not{A}\gamma_5)\psi \\ &= 2im\,\bar{\psi}\gamma^5\psi\,, \quad \text{q.e.d.}\end{aligned}$$

The divergence of the vector current is easier to obtain:

$$\begin{aligned}\partial_\mu(\bar{\psi}\gamma^\mu\psi) &= (\partial_\mu\bar{\psi}\gamma^\mu)\psi + \bar{\psi}(\gamma^\mu\partial_\mu\psi) \\ &= im\,\bar{\psi}\psi - i\bar{\psi}\,(\not{V} + \not{A}\gamma^5)\psi \\ &\quad - im\,\bar{\psi}\psi + i\bar{\psi}\,(\not{V} + \not{A}\gamma^5)\psi \\ &= 0\end{aligned}$$

The Lagrangian (9.1) is invariant under the gauge transformation of the second kind:

$$\psi \longrightarrow \psi' = e^{i\alpha(x)}\psi\,, \quad \bar{\psi} \to \bar{\psi}' = \bar{\psi}\, e^{-i\alpha(x)}$$

$$V_\mu \longrightarrow V_\mu' = V_\mu + \partial_\mu \alpha(x)\,, \quad A_\mu \longrightarrow A_\mu' = A_\mu \tag{9.29}$$

Obviously,

$$\begin{aligned}
\mathcal{L}' &= \bar{\psi}' i \partial\!\!\!/ \psi' - m\bar{\psi}'\psi' + \bar{\psi}'(V\!\!\!/' + A\!\!\!/'\gamma_5)\psi' \\
&\overset{(9.4)}{=} -(\partial_\mu \alpha)\bar{\psi}\gamma^\mu\psi + \bar{\psi} i\partial\!\!\!/\psi - m\bar{\psi}\psi \\
&\quad + \bar{\psi} e^{-i\alpha}[(V\!\!\!/ + \partial\!\!\!/\alpha) + A\!\!\!/\gamma_5]e^{i\alpha}\psi \\
&= \bar{\psi} i \partial\!\!\!/ \psi - m\bar{\psi}\psi + \bar{\psi}(V\!\!\!/ + A\!\!\!/\gamma_5)\psi = \mathcal{L}
\end{aligned}$$

The part of $\mathcal{L}$ without the mass term also remains unchanged under the transformation

$$\psi \to \psi' = e^{i\beta(x)\gamma_5}\psi\,, \quad \bar{\psi} \to \bar{\psi}' = \bar{\psi}\, e^{i\beta(x)\gamma_5}$$

$$V_\mu \to V_\mu' = V_\mu\,, \quad A_\mu \to A_\mu' = A_\mu + \partial_\mu \beta(x) \quad : \tag{9.30}$$

$$\begin{aligned}
\mathcal{L}'[m=0] &= \bar{\psi}' i \partial\!\!\!/ \psi' + \bar{\psi}'(V\!\!\!/' + A\!\!\!/'\gamma_5)\psi' \\
&\overset{(9.16)}{=} -(\partial_\mu \beta)\bar{\psi}\gamma^\mu\gamma_5\psi + \bar{\psi} i\partial\!\!\!/\psi + \bar{\psi} e^{i\beta\gamma_5}[V\!\!\!/ + (A\!\!\!/ + \partial\!\!\!/\beta)\gamma_5]e^{i\beta\gamma_5}\psi \\
&= \bar{\psi} i\partial\!\!\!/\psi + \bar{\psi}(V\!\!\!/ + A\!\!\!/\gamma_5)\psi = \mathcal{L}[m=0]
\end{aligned}$$

The mass term transforms according to (9.17) and is therefore the symmetry-breaking term under the chiral transformation.

One can show that the anomalous behavior of the divergence of the axial vector current in (9.22) stems from the fact that it cannot be obtained as a variation of a local fundamental action which depends solely on the fields V_μ and A_μ . So we might ask ourselves whether an action W' exists at all with the property that it is left invariant under the α -transformation (9.29) but responds under the β -transformation (9.30) such that

$$\delta_\alpha W' = 0 \quad , \quad \delta_\beta W' = \int d^4x \, \beta(x) \frac{1}{(4\pi)^2} \varepsilon_{\mu\nu\rho\sigma} F^{\mu\nu} F^{\rho\sigma} \tag{9.31}$$

It turns out that in order to solve equations (9.31) for W', we have to permit a dependence of W' not only on the external fields V_μ and A_μ , but also on a (Goldstone) field $\xi(x)$ that transforms according to

$$\delta_\alpha \xi(x) = 0 \quad , \quad \delta_\beta \xi(x) = \beta(x) \tag{9.32}$$

It is clear that

$$W' = \int d^4x \, \xi(x) \frac{1}{(4\pi)^2} \varepsilon_{\mu\nu\rho\sigma} F^{\mu\nu} F^{\rho\sigma} \tag{9.33}$$

is a solution of (9.31).

From now on, the transformations (9.29) and (9.30) should be supplemented by the variations

(9.29) plus $\quad \xi(x) \longrightarrow \xi'(x) = \xi(x)$

and (9.30) plus $\quad \xi(x) \longrightarrow \xi'(x) = \xi(x) + \beta(x)$

With the ξ -field identified as the pion field π^0, we can regard (9.33) as an effective action for the $\pi^0 \rightarrow 2\gamma$ decay.

In order to study the response of the action under the local symmetries (9.29, 30, 32) in more detail, we are now going to construct operators that generate those transformations. What we have in mind is a way to rewrite the variation of the action with respect to the gauge functions $\alpha(x)$, $\beta(x)$, $\xi(x)$ so as to produce a group theoretical version of the problem expressed in terms of certain generators X(x), Y(x), Z(x) which generate the infinitesimal changes in $\alpha(x)$, $\beta(x)$ and $\xi(x)$, respectively.

Let us look at the functional differentiation

$$\frac{1}{i}\frac{\delta}{\delta\alpha(x)}\,e^{iW[V,A,\xi]} = e^{iW}\,\frac{\delta}{\delta\alpha(x)}W[V,A,\xi]$$

$$= e^{iW}\int d^4x'\,\frac{\delta V_\mu(x')}{\delta\alpha(x)}\,\frac{\delta}{\delta V_\mu(x')}W[V,A,\xi]$$

Using $$\frac{\delta V_\mu(x')}{\delta\alpha(x)} = \frac{\delta}{\delta\alpha(x)}\left[\partial'_\mu\alpha(x')\right] = \partial'_\mu\frac{\delta\alpha(x')}{\delta\alpha(x)} = \partial'_\mu\delta(x'-x)$$

we can continue to write

$$\frac{1}{i}\frac{\delta}{\delta\alpha(x)}\,e^{iW} = e^{iW}\int d^4x'\,\partial'_\mu\delta(x-x')\,\frac{\delta}{\delta V_\mu(x')}\,W$$

$$= e^{iW}\int d^4x'\,\delta(x-x')\left(-\partial'_\mu\frac{\delta}{\delta V_\mu(x')}\right)W$$

$$= e^{iW}\left(-\partial_\mu\frac{\delta}{\delta V_\mu(x)}\right)W \tag{9.34}$$

or $$\frac{1}{i}\frac{\delta}{\delta\alpha}\ln e^{iW} = \frac{\delta W}{\delta\alpha} = \left(-\partial_\mu\frac{\delta}{\delta V_\mu}\right)W \tag{9.35}$$

Similarly, $$\frac{\delta W}{\delta\beta} = \left(-\partial_\mu\frac{\delta}{\delta A_\mu}\right)W \tag{9.36}$$

At this stage, it is useful to introduce the operators

$$X(x) = -\partial_\mu\frac{\delta}{\delta V_\mu(x)} \tag{9.37}$$

$$Y(x) = -\partial_\mu \frac{\delta}{\delta A_\mu(x)} + \frac{\delta}{\delta \xi(x)} =: U(x) + Z(x) \tag{9.38}$$

which allow us to write (α (x), etc. infinitesimal)

$$\delta_\alpha W = \int d^4x\, \alpha(x)\, X(x)\, W =: (\alpha \cdot X) W \tag{9.39}$$

i.e., the role of the variation δ_α has been taken over by the operator$(\alpha \cdot X)$.
Likewise, for

$$\delta_\beta W = \int d^4x\, \beta(x)\, Y(x) W =: (\beta \cdot Y) W \tag{9.40}$$

In this operator language, equations (9.31) read

$$(\alpha \cdot X) W = 0 \tag{9.41}$$

$$(\beta \cdot Y) W = (\beta \cdot G) \tag{9.42}$$

where

$$G(x) = \frac{1}{(4\pi)^2} \varepsilon_{\mu\nu\rho\sigma} F^{\mu\nu} F^{\rho\sigma} \tag{9.43}$$

The (effective) action W is defined by integrating out the fermion fields

$$e^{iW[V,A,\xi]} = \int [d\psi\, d\bar{\psi}]\, e^{iI[\psi,\bar{\psi};V,A,\xi]} \tag{9.44}$$

where the fundamental action I is given by

$$I[\psi,\bar{\psi};V,A,\xi] = \int d^4x\, \mathcal{L}(\psi,\bar{\psi};V,A,\xi) \tag{9.45}$$

where

$$W = W_o + W'$$

with

$$\delta_{\alpha,\beta} W_o\,[m=0] = 0$$

and

$$\delta_\alpha W' = (\alpha \cdot X) W' = 0 \tag{9.46}$$

$$\delta_\beta W' = (\beta \cdot Y) W' = (\beta \cdot G) \tag{9.47}$$

Because the operators X and Y commute, it holds that

$$\begin{aligned} [X(x),\ X(y)] &= 0 \\ [X(x),\ Y(y)] &= 0 \\ [Y(x),\ Y(y)] &= 0 \end{aligned} \tag{9.48}$$

So we are obviously dealing with a group $U(1) \times U(1)$ whose associated algebra is given by (9.48). It is then also true that $(\alpha \cdot X \equiv \int d^4x\, \alpha(x) X(x), \ldots)$

$$\begin{aligned} [(\alpha \cdot X), (\alpha' \cdot X)] &= 0 \\ [(\alpha \cdot X), (\beta \cdot Y)] &= 0 \\ [(\beta \cdot Y), (\beta' \cdot Y)] &= 0 \end{aligned} \tag{9.49}$$

Operating with these commutators on the action W' yields

(1) $$(\alpha \cdot X) \underbrace{(\alpha' \cdot X) W'}_{\underset{(9.46)}{=0}} - (\alpha' \cdot X) \underbrace{(\alpha \cdot X) W'}_{=0} = 0$$

(2) $$(\alpha \cdot X) \underbrace{(\beta \cdot Y) W'}_{\underset{(9.47)}{=} (\beta \cdot G)} - (\beta \cdot Y) \underbrace{(\alpha \cdot X) W'}_{=0} = 0$$

Therefore, $$(\alpha \cdot X)(\beta \cdot G) = 0 \tag{9.50}$$

(3) $$(\beta \cdot Y) \underbrace{(\beta' \cdot Y) W'}_{=(\beta' \cdot G)} - (\beta' \cdot Y) \underbrace{(\beta \cdot Y) W}_{=(\beta \cdot G)} = 0$$

which gives another important relation

$$(\beta\cdot Y)(\beta'\cdot G) = (\beta'\cdot Y)(\beta\cdot G) \tag{9.51}$$

The equations (9.50) and (9.51) are known as the Wess-Zumino integrability conditions.

Here, then, is the important statement we want to prove: Given an arbitrary functional G[V,A], there exists for any G[V,A] which satisfies the W.-Z. integrability conditions (9.50) and (9.51) a solution to the equations

$$(\alpha\cdot X)W' = 0 \tag{9.52}$$

$$(\beta\cdot Y)W' = (\beta\cdot G) \tag{9.53}$$

Here is a way to construct an explicit solution:

$$(\beta\cdot Y)W' = (\beta\cdot G)$$

$$\Rightarrow \quad (\beta\cdot Y)^2 W' = (\beta\cdot Y)(\beta\cdot G)$$

$$= (\beta\cdot \mathcal{U} + \underbrace{\cancel{\beta\cdot Z}}_{\sim \frac{\delta}{\delta\zeta}})(\beta\cdot G) \quad \text{, G independent of } \zeta$$

$$= (\beta\cdot \mathcal{U})(\beta\cdot G)$$

$$\Rightarrow \quad (\beta\cdot Y)^3 W' = (\beta\cdot Y)(\beta\cdot \mathcal{U})(\beta\cdot G)$$

$$= (\beta\cdot U + \beta\cdot Z)(\beta\cdot \mathcal{U})(\beta\cdot G)$$

$$= (\beta\cdot \mathcal{U})^2(\beta\cdot G)$$

$$\vdots$$

$$(\beta\cdot Y)^n W' = (\beta\cdot U)^{n-1}(\beta\cdot G) \tag{9.54}$$

This equation can be employed in the following calculation:

$$e^{(\beta\cdot Y)} W' = [1 + \sum_{n=1}^{\infty} \frac{1}{n!}(\beta\cdot Y)^n] W'$$

$$= W' + \sum_{n=1}^{\infty} \frac{1}{n!} (\beta\cdot U)^{n-1} (\beta\cdot G)$$

$$= W' + (\beta\cdot U)^{-1} [\sum_{n=1}^{\infty} \frac{1}{n!} (\beta\cdot U)^n] (\beta\cdot G)$$

$$= W' + (\beta\cdot U)^{-1} [e^{(\beta\cdot U)} - 1] (\beta\cdot G)$$

$$\Longrightarrow \quad e^{(\beta\cdot Y)} W' = W' + \frac{e^{(\beta\cdot U)} - 1}{(\beta\cdot U)} (\beta\cdot G) \tag{9.55}$$

W' is assumed to be a functional of V , A and ξ which is supposed to vanish for ξ = 0:

$$W'[\xi = 0, V_\mu, A_\mu] = 0 \quad , \quad \forall (V_\mu, A_\mu) \tag{9.56}$$

From the equation of Y, we obtain

$$e^{-(\xi\cdot Y)} W'[\xi, V, A] = e^{-(\xi\cdot U) - (\xi\cdot Z)} W'[\xi, V, A]$$

$$= e^{-(\xi\cdot U)} e^{-(\xi\cdot \frac{\delta}{\delta \xi})} W'[\xi, V, A]$$

$$= \exp\{\int d^4x\, \xi(x)\, \partial_\mu \frac{\delta}{\delta A_\mu(x)}\} W'[0, V, A]$$

$$= \exp\{-\int d^4x\, \partial_\mu \xi \frac{\delta}{\delta A_\mu(x)}\} W'[0, V, A]$$

$$(A'_\mu := A_\mu - \partial_\mu \xi) \qquad = W'[0, V, A']$$

$$= 0 \tag{9.57}$$

the last line being the result of the original assumption $W'[0, V, A] = 0$ for all V and A.

Now let us combine our results found so far:

$$e^{(\beta\cdot Y)} W'[\xi,V,A] \overset{(9.55)}{=} W'[\xi,V,A] + \frac{e^{(\beta\cdot U)}-1}{(\beta\cdot U)}(\beta\cdot G), \quad \forall\beta$$

In particular, for $\beta = -\xi$:

$$e^{-(\xi\cdot Y)} W'[\xi,V,A] \overset{(9.57)}{=} 0 = W'[\xi,V,A] + \frac{e^{-(\xi\cdot U)}-1}{(-\xi\cdot U)}(-\xi\cdot G)$$

which yields the desired result:

$$W'[\xi,V,A] = \frac{1-e^{-(\xi\cdot U)}}{(\xi\cdot U)}(\xi\cdot G) \tag{9.58}$$

$$= \int_0^1 dt\; e^{-t(\xi\cdot U)}(\xi\cdot G) \tag{9.59}$$

Finally, we want to demonstrate that (9.58/59) really solves (9.52, 53). The first equation is rather easily verified:

$$(\alpha\cdot X) W' = (\alpha\cdot X)\int_0^1 dt\; e^{-t(\xi\cdot U)}(\xi\cdot G)$$

$$= \int_0^1 dt\; e^{-t(\xi\cdot U)} \underbrace{(\alpha\cdot X)(\xi\cdot G)}_{\overset{(9.50)}{=}\, 0}$$

$$= 0 \tag{9.60}$$

Before we begin to verify (9.53), we want to prove the following statement:

$$e^{t(\xi\cdot U)}(\beta\cdot Y)e^{-t(\xi\cdot U)} = (\beta\cdot Y) - t(\beta\cdot U) \tag{9.61}$$

Let $F = F[\xi, V, A]$ be an arbitrary functional.

Then

$$[(\beta\cdot Y),(\xi\cdot U)]F = (\beta\cdot Y)(\xi\cdot U)F-(\xi\cdot U)(\beta\cdot Y)F$$

$$= (\beta\cdot U)(\xi\cdot U)F+(\beta\cdot Z)(\xi\cdot U)F$$

$$-(\xi\cdot U)(\beta\cdot U)F-(\xi\cdot U)(\beta\cdot Z)F$$

$$=(\beta\cdot Z)(\xi\cdot U)F-(\xi\cdot U)(\beta\cdot Z)F$$

$$=(\beta\cdot \overline{Z)(\xi}\cdot U)F+(\beta\cdot Z)(\xi\cdot U)F-(\xi\cdot U)(\beta\cdot Z)F$$

$$=(\beta\cdot U)F$$

So we have found

$$[(\beta\cdot Y),(\xi\cdot U)]=(\beta\cdot U) \tag{9.62}$$

Furthermore, from

$$e^{iS}He^{-iS}=H+i[S,H]+\frac{i^2}{2}[S,[S,H]]+\cdots$$

with

$$S=-i(\xi\cdot U)t$$

we obtain

$$e^{t(\xi\cdot U)}(\beta\cdot Y)e^{-t(\xi\cdot U)}=(\beta\cdot Y)+t[(\xi\cdot U),(\beta\cdot Y)]+0$$

$$\underset{(9.62)}{=}(\beta\cdot Y)-(\beta\cdot U)t \tag{9.63}$$

since

$$[S,[S,H]]\sim[(\xi\cdot U),\underbrace{[(\xi\cdot U),(\beta\cdot Y)]}_{\sim(\beta\cdot U)}]=0$$

This brings us back to

$$(\beta\cdot Y)W'=\int_0^1 dt\,(\beta\cdot Y)e^{-t(\xi\cdot U)}(\xi\cdot G)$$

$$\underset{(9.61)}{=}\int_0^1 dt\,e^{-t(\xi\cdot U)}[(\beta\cdot Y)-t(\beta\cdot U)](\xi\cdot G) \tag{9.64}$$

In the following steps, it is necessary to recall that G is a functional of V and A and that

$$Y = U + Z = -\partial \frac{\delta}{\delta A} + \frac{\delta}{\delta \xi}$$

With this in mind, we continue to write

$$\begin{aligned}(\beta\cdot Y)(\xi\cdot G) &= (\beta\cdot Y)(\beta'\cdot G)\big|_{\beta'=\xi} + (\beta\cdot \overline{Y)(\xi}\cdot G) \\ &\underset{(9.51)}{=} (\beta'\cdot Y)(\beta\cdot G)\big|_{\beta'=\xi} + (\beta\cdot G) \\ &= (\xi\cdot Y)(\beta\cdot G) + (\beta\cdot G) = (\xi\cdot U)(\beta\cdot G) + (\beta\cdot G)\end{aligned}$$

which gives us

$$(\beta\cdot Y)(\xi\cdot G) = (\xi\cdot U)(\beta\cdot G) + (\beta\cdot G) \tag{9.65}$$

Similarly,

$$\begin{aligned}(\beta\cdot U)(\xi\cdot G) &= (\beta\cdot Y)(\xi\cdot G) - (\beta\cdot Z)(\xi\cdot G) \\ &\overset{(9.65)}{=} (\xi\cdot U)(\beta\cdot G) + \cancel{(\beta\cdot G)} - \cancel{(\beta\cdot G)} \\ &= (\xi\cdot U)(\beta\cdot G)\end{aligned} \tag{9.66}$$

So we end up with

$$\begin{aligned}(\beta\cdot Y)W' &= \int_0^1 dt\, e^{-t\xi\cdot U}\left[(\xi\cdot U)(\beta\cdot G) + (\beta\cdot G) - t(\xi\cdot U)(\beta\cdot G)\right] \\ &= \int_0^1 dt\, e^{-t\xi\cdot U}\left[1 + (1-t)(\xi\cdot U)\right](\beta\cdot G)\end{aligned} \tag{9.67}$$

The t-integration is very easy to do:

$$\int_0^1 dt\, e^{-t(\xi\cdot U)} = \frac{1-e^{-(\xi\cdot U)}}{(\xi\cdot U)}\,; \qquad x := (\xi\cdot U) \quad \Rightarrow$$

$$\int_0^1 dt\, t\, e^{-tx} = -\frac{d}{dx}\int_0^1 dt\, e^{-tx} = -\frac{d}{dx}(1-e^{-x})x^{-1}$$

$$= -\frac{d}{dx}\left(x^{-1}-x^{-1}e^{-x}\right) = -\left[-x^{-2}+x^{-2}e^{-x}+x^{-1}e^{-x}\right]$$

$$= \frac{1}{x^2}-\frac{1}{x^2}e^{-x}-\frac{1}{x}e^{-x} \equiv \frac{1}{(\xi\cdot U)^2}-\frac{1}{(\xi\cdot U)^2}\,e^{-(\xi\cdot U)}-\frac{1}{(\xi\cdot U)}\,e^{-(\xi\cdot U)}$$

Finally, we obtain

$$(\beta\cdot Y)W' = \Big\{\frac{1-e^{-(\xi\cdot U)}}{(\xi\cdot U)}\,(1+(\xi\cdot U))$$

$$-\left[\frac{1-e^{-(\xi\cdot U)}}{(\xi\cdot U)^{\cancel{2}}}-\frac{1}{\cancel{(\xi\cdot U)}}\,e^{-(\xi\cdot U)}\right]\cancel{(\xi\cdot U)}\Big\}(\beta\cdot G)$$

$$= \left\{\frac{1-e^{-(\xi\cdot U)}}{(\xi\cdot U)}+1-e^{-(\xi\cdot U)}-\frac{1-e^{-(\xi\cdot U)}}{(\xi\cdot U)}+e^{-(\xi\cdot U)}\right\}(\beta\cdot G)$$

$$= (\beta\cdot G) \qquad \text{q.e.d.}$$

This, then, closes our proof that the W.-Z. integrability conditions (9.50, 51) are necessary and sufficient for the system of equations (9.52, 53) to have a solution. The solution itself is given by (9.58/59).

We shall now extend our discussion of Ward identities to SU(3)$\times$SU(3), i.e., we consider a hadronic system in presence of external vector and axial vector fields V^a, A^a, (a = 1, 2, ...8). The Lagrangian is given by

$$\mathcal{L} = \bar{\psi}\, i \not{D}\, \psi = \bar{\psi}(i\not{\partial}+\not{V}+\not{A}\gamma_5)\psi \tag{9.68}$$

where SU(3) indices are omitted; the external fields V and A are matrices in the internal space, e.g., $A_\mu(x) = \frac{1}{2}\lambda^a A^a_\mu(x) \equiv T^a A^a_\mu(x)$.

Again, the Ward identities can be formulated by studying the response of the connected vacuum functional due to infinitesimal vector and chiral gauge transformations. Before we come to that, it is useful to recall the variation of an arbitrary functional $F[V_\mu, A_\mu]$ under V- and A- transformation:

V-transf. (α): $\psi' = e^{i\alpha(x)}\psi , \quad \bar{\psi}' = \bar{\psi} e^{-i\alpha} , \quad \alpha(x) = \alpha^a(x) T^a$ (9.69)

$$\delta \vec{V}_\mu = \vec{\alpha} \times \vec{V}_\mu - \partial_\mu \vec{\alpha} \quad \text{or} \quad \delta V_\mu = -i[\alpha, V_\mu] - \partial_\mu \alpha \tag{9.70}$$

$$\delta \vec{A}_\mu = \vec{\alpha} \times \vec{A} \quad \text{or} \quad \delta A_\mu = -i[\alpha, A_\mu] \tag{9.71}$$

A-transf. (β): $\psi' = e^{i\beta(x)\gamma_5}\psi , \quad \bar{\psi}' = \bar{\psi} e^{i\beta(x)\gamma_5} , \quad \beta(x) = \beta^a(x) T^a$ (9.72)

$$\delta \vec{V}_\mu = \vec{\beta} \times \vec{A}_\mu \quad \text{or} \quad \delta V_\mu = -i[\beta, A_\mu] \tag{9.73}$$

$$\delta \vec{A}_\mu = \vec{\beta} \times \vec{V}_\mu - \partial_\mu \vec{\beta} \quad \text{or} \quad \delta A_\mu = -i[\beta, V_\mu] - \partial_\mu \beta \tag{9.74}$$

with

$$[A, B]^a = A^b B^c [T^b, T^c]^a = i f^{abc} A^b B^c = i(\vec{A} \times \vec{B})^a$$

The Lagrangian (9.68) can be shown to remain invariant under V-as well as A-transformation.

Now the variation of the functional $F[V, A]$ is given by the chain rule (α infinitesimal)

$$\delta_\alpha F[V, A] = \int d^4x \, d^4y \, \alpha^a(y) \left[\frac{\delta V_\mu^b(x)}{\delta \alpha^a(y)} \frac{\delta}{\delta V_\mu^b(x)} + \frac{\delta A_\mu^b(x)}{\delta \alpha^a(y)} \frac{\delta}{\delta A_\mu^b(x)} \right] F[V, A] \tag{9.75}$$

Next, we are going to employ equations (9.70) and (9.71) (α infinitesimal):

$$\frac{\delta V_\mu^b(x)}{\delta \alpha^a(y)} = \frac{\delta}{\delta \alpha^a(y)} \left[f^{bcd} \alpha^c(x) V_\mu^d(x) - \partial_\mu^x \alpha^b(x) \right]$$

$$= f^{bcd} V_\mu^d(x) \delta(x-y) \delta^{ca} - \partial_\mu^x \delta^{ab} \delta(x-y)$$

$$= \left(f^{bad} V_\mu^d(x) - \partial_\mu^x \delta^{ab}\right) \delta(x-y) \tag{9.76}$$

Likewise,

$$\frac{\delta A_\mu^b(x)}{\delta \alpha^a(y)} = \frac{\delta}{\delta \alpha^a(y)} \left[f^{bcd} \alpha^c(x) A_\mu^d(x) \right]$$

$$= f^{bcd} \delta^{ca} \delta(x-y) A_\mu^d(x)$$

$$= f^{bad} A_\mu^d(x) \delta(x-y) \tag{9.77}$$

If we substitute (9.76) and (9.77) into equation (9.75), we obtain

$$\delta_\alpha F[V,A] = \int d^4x\, d^4y\, \alpha^a(y) \Big[\big(\{-\partial_\mu^x \delta^{ab} + f^{bad} V_\mu^d(x)\} \delta(x-y)\big) \frac{\delta}{\delta V_\mu^b(x)} + f^{bad} A_\mu^d(x) \delta(x-y) \frac{\delta}{\delta A_\mu^b(x)} \Big] F$$

integr. by parts

$$= \int d^4x\, d^4y\, \alpha^a(y) \Big[\delta^{ab} \delta(x-y) \partial_\mu^x \frac{\delta}{\delta V_\mu^b(x)} + f^{bad} V_\mu^d(x) \delta(x-y) \frac{\delta}{\delta V_\mu^b(x)} + f^{bad} A_\mu^d(x) \delta(x-y) \frac{\delta}{\delta A_\mu^b(x)} \Big] F$$

$$= \int d^4x\, \alpha^a(x) \Big[\partial_\mu^x \frac{\delta}{\delta V_\mu^a(x)} + f^{adb} V_\mu^d(x) \frac{\delta}{\delta V_\mu^b(x)} + f^{adb} A_\mu^d(x) \frac{\delta}{\delta A_\mu^b(x)} \Big] F$$

$$= - \int d^4x\, \alpha^a(x) X^a(x) F \tag{9.78}$$

whereby

$$-X^a(x) = \partial_\mu^x \frac{\delta}{\delta V_\mu^a(x)} + f^{abc} \left(V_\mu^b(x) \frac{\delta}{\delta V_\mu^c(x)} + A_\mu^b(x) \frac{\delta}{\delta A_\mu^c(x)} \right) \tag{9.79}$$

Similarly, we can study the response of F[V,A] under chiral transformation:

$$\delta_\beta F[V,A] = -\int d^4x\, \beta^a(x)\, Y^a(x)\, F \tag{9.80}$$

with

$$-Y^a(x) = \partial_\mu^x \frac{\delta}{\delta A_\mu^a(x)} + f^{abc}\left(A_\mu^b(x)\frac{\delta}{\delta V_\mu^c} + V_\mu^b(x)\frac{\delta}{\delta A_\mu^c(x)}\right) \tag{9.81}$$

The gauge operators $X^a(x)$ and $Y^a(x)$ satisfy the commutation relations

$$\begin{aligned}
\left[X^a(x), X^b(y)\right] &= f^{abc}\delta(x-y)X^c(x)\\
\left[X^a(x), Y^b(y)\right] &= f^{abc}\delta(x-y)Y^c(x)\\
\left[Y^a(x), Y^b(y)\right] &= f^{abc}\delta(x-y)X^c(x)
\end{aligned} \tag{9.82}$$

which can be verified by using the definition of the operators X^a and Y^a, equations (9.79, 81). Recall that the operators X^a and Y^a are the generators of the infinitesimal variations of the vector and axial vector gauge fields under the $SU(3)_V$ and $SU(3)_A$ gauge group. This can be checked directly:

$$\begin{aligned}
(\alpha\cdot X)\, V_\mu^a(x) &= \int d^4z\, \alpha^b(z)\, X^b(z)\, V_\mu^a(x)\\
&= \int d^4z\, \alpha^b(z)\left[-\partial_\nu^z \frac{\delta}{\delta V_\nu^b(z)} - f^{bcd}\left(V_\nu^c \frac{\delta}{\delta V_\nu^d} + A_\nu^c \frac{\delta}{\delta A_\nu^d}\right)\right] V_\mu^a(x)\\
&= \int d^4z\, \alpha^b(z)\left[-\partial_\nu^z \delta^{\nu\mu}\delta^{ba}\delta(z-x) - f^{bcd}\, V_\nu^c \delta^{\nu\mu}\delta^{da}\delta(z-x)\right]
\end{aligned}$$

integr. by parts

$$= \partial_\mu \alpha^a(x) - f^{abc}\,\alpha^b(x)\, V_\mu^c(x)$$

Altogether:

$$(\alpha\cdot X)\, V_\mu^a(x) = \partial_\mu \alpha^a(x) - f^{abc}\,\alpha^b(x)\, V_\mu^c(x) \tag{9.83}$$

$$(\alpha\cdot X)\, A_\mu^a(x) = - f^{abc}\,\alpha^b(x)\, A_\mu^c(x) \tag{9.84}$$

$$(\beta\cdot Y)\, V_\mu^a(x) = - f^{abc}\,\beta^b(x)\, A_\mu^c(x) \tag{9.85}$$

$$(\beta\cdot Y)\, A_\mu^a(x) = \partial_\mu \beta^a(x) - f^{abc}\,\beta^b(x)\, V_\mu^c(x) \tag{9.86}$$

Now we can proceed as in the abelian case, i.e., the Ward identity involving the divergence of the vector current is given by (cf. (9.41, 42))

$$X^a W = 0 \tag{9.87}$$

and those involving the divergence of the axial vector current can be written as

$$Y^a W = G^a[V_\mu, A_\mu] \tag{9.88}$$

where G^a is the axial vector anomaly.

For SU(3) × SU(3) with quarks, Bardeen has computed the anomalous term:

$$-G^a[V_\mu, A_\mu] = \frac{1}{4\pi^2}\,\varepsilon_{\mu\nu\sigma\tau}\,\mathrm{tr}\Big[\frac{\lambda^a}{2}\Big\{\frac{1}{4}V_{\mu\nu}V_{\sigma\tau} + \frac{1}{12}A_{\mu\nu}A_{\sigma\tau} +$$

$$+ \frac{2}{3}i\left(A_\mu A_\nu V_{\sigma\tau} + A_\mu V_{\nu\sigma}A_\tau + V_{\mu\nu}A_\sigma A_\tau\right) - \frac{8}{3}A_\mu A_\nu A_\sigma A_\tau\Big\}\Big] \tag{9.89}$$

with

$$V_\mu = V_\mu^a \frac{\lambda^a}{2}\,,\quad A_\mu = A_\mu^a \frac{\lambda^a}{2}$$

$$V_{\mu\nu} = \partial_\mu V_\nu - \partial_\nu V_\mu - i\,[V_\mu, V_\nu] - i\,[A_\mu, A_\nu]$$

$$A_{\mu\nu} = \partial_\mu A_\nu - \partial_\nu A_\mu - i\,[V_\mu, A_\nu] - i\,[A_\mu, V_\nu]$$

Again, there is no local functional W depending on the fields V_μ, A_μ alone which would solve equations (9.87, 88).

On the way to the construction of a solution, we encounter certain integrability conditions from equations (9.82); cf. also (9.50, 51):

$$\begin{aligned} X^a(x)\,G^b(y) &= f^{abc}\delta(x-y)\,G^c(x) \\ Y^a(x)\,G^b(y) &= Y^b(x)\,G^a(y) \end{aligned} \tag{9.90}$$

The first of these equations merely states that G^a transforms like an octet under SU(3) transformation. The second equation is a kind of "curl equation" and is more subtle. The integrability equations (9.90) are necessary and sufficient for a solution of (9.87, 88) to exist. In order to find a particular functional which satisfies (9.90), we have to allow for a dependence of W on an octet of pseudoscalar fields $\xi_a(x)$ transforming according to a non-linear realization of SU(3)×SU(3). We are going to use the relation

$$\xi_a(x) = \frac{1}{F_\pi}\pi_a(x), \quad F_\pi \simeq 95\ \text{MeV} \tag{9.91}$$

where the π_a's denote the Goldstone fields.

We shall use the exponential field which transforms under a chiral transformation non-linearly:

$$e^{-i\xi^{a\prime}(x)\lambda^a} = e^{-i\beta^a(x)\frac{\lambda^a}{2}}\, e^{-i\xi^a(x)\lambda^a}\, e^{-i\beta^a(x)\frac{\lambda^a}{2}} \tag{9.92}$$

or, infinitesimal and expanded in ξ:

$$\delta\xi^a\lambda^a = \beta^a(x)\lambda^a + O(\xi^2) \tag{9.93}$$

Equation (9.92) tells us that for $\beta^a(x) = -\xi^a(x)$, $\xi^a(x)$ can be transformed to zero.

Under SU(3)-vector transformation, we simply have

$$\lambda^a\xi^{a\prime}(x) = e^{-i\alpha^a(x)\frac{\lambda^a}{2}}\,\lambda^a\xi^a(x)\, e^{i\alpha^a\frac{\lambda^a}{2}} \tag{9.94}$$

Following the procedure that led us to (9.58), we can construct a particular solution of (9.87, 88) and obtain

$$W'[\xi, V_\mu, A_\mu] = \frac{1-e^{-\xi\cdot U}}{(\xi\cdot U)}\,\xi\cdot G[V,A] \tag{9.95}$$

with $X^aW' = 0$, $Y^aW' = G^a$.

The overall coefficient of G^a is model-dependent and has to be determined by experiment, e.g., by $\pi^0 \to 2\gamma$. For this reason, we are now going to use the above explicit solution (9.95) as an effective Lagrangian to study the $\pi^0 \to 2\gamma$ decay.

$$W'[\xi, V_\mu, A_\mu] = \frac{1-e^{-(\xi\cdot U)}}{(\xi\cdot U)}(\xi\cdot G)$$

$$= \frac{1-1+(\xi\cdot U)+\cdots}{(\xi\cdot U)}(\xi\cdot G)$$

$$= (\xi\cdot G) + \cdots\cdot$$

Using (9.91), we obtain $(\pi^0 \equiv \pi^3)$ with the integral $\int d^4x$ omitted

$$W' = \frac{\lambda}{F_\pi}\pi^a G^a = \frac{\lambda}{F_\pi}\pi^0 G^3 + \cdots\cdot \tag{9.96}$$

where λ is a phenomenological constant. The electromagnetic field is introduced by

$$\mathcal{L} = \bar{\psi}(\cdots\cdot + \gamma^\mu \tfrac{1}{2}\lambda^a V_\mu^a + \cdots)\psi$$
$$= \bar{\psi}(\cdots + \gamma^\mu\{\tfrac{1}{2}\lambda^3 V_\mu^3 + \tfrac{1}{2}\lambda^8 V_\mu^8\} + \cdots\cdot)\psi$$
$$\stackrel{!}{=} e\bar{\psi}\gamma^\mu V_\mu^{EM} Q\psi$$
$$\equiv e\bar{\psi}\gamma^\mu V_\mu^{EM}(\tfrac{1}{2}\lambda^3 + \tfrac{1}{2\sqrt{3}}\lambda^8)\psi \tag{9.97}$$

since an external electromagnetic field is obtained by setting

$A^a_\mu = 0,\ V^{a=3,8}_\mu \neq 0$

and
$$V^3_\mu = e V^{EM}_\mu \,, \quad V^8_\mu = \frac{e}{\sqrt{3}} V^{EM}_\mu$$

so that

$$\tfrac{1}{2}\lambda^3 V^3_\mu + \tfrac{1}{2}\lambda^8 V^8_\mu = e V^{EM}_\mu \left(\tfrac{1}{2}\lambda^3 + \frac{1}{2\sqrt{3}}\lambda^8 \right)$$
$$= e V^{EM}_\mu Q$$

with the charge matrix

$$Q = \tfrac{1}{2}\lambda^3 + \frac{1}{2\sqrt{3}}\lambda^8$$

Furthermore,

$$\begin{aligned} V^{\mu\nu} &= \partial^\mu V^\nu - \partial^\nu V^\mu + 0 \\ &= \partial^\mu(\dots + \tfrac{1}{2}\lambda^3 V^\nu_3 + \tfrac{1}{2}\lambda^8 V^\nu_8 + \dots) - \partial^\nu(\dots + \tfrac{1}{2}\lambda^3 V^\mu_3 + \tfrac{1}{2}\lambda^8 V^\mu_8 + \dots) \\ &= eQ\left(\partial^\mu V^{\nu EM} - \partial^\nu V^{\mu EM}\right) \\ &= eQ F^{\mu\nu EM} \end{aligned} \tag{9.98}$$

$$A^{\mu\nu} = 0$$

Equation (9.89) is then reduced to

$$-G^3 = \frac{1}{4\pi^2}\,\frac{1}{8}\,\varepsilon_{\mu\nu\sigma\tau}\,\mathrm{tr}\left[\lambda^3 V^{\mu\nu} V^{\sigma\tau}\right]$$

and hence, (9.96) is given by

$$W' = \frac{\lambda}{F_\pi} \pi_0 G^3$$

$$= -\frac{1}{(4\pi)^2} \frac{\lambda}{F_\pi} \pi_0 \, \varepsilon_{\mu\nu\sigma\tau} \, \frac{1}{2} tr[\lambda^3(\cdots + \frac{1}{2}\lambda^3 V_3^{\mu\nu} + \frac{1}{2}\lambda^8 V_8^{\mu\nu} + \cdots)$$
$$\cdot (\cdots + \frac{1}{2}\lambda^3 V_3^{\sigma\tau} + \frac{1}{2}\lambda^8 V_8^{\sigma\tau} + \cdots)]$$

$$= -\frac{1}{(4\pi)^2} \frac{\lambda e^2}{F_\pi} \pi_0 \, \varepsilon_{\mu\nu\sigma\tau} \, \frac{1}{2} tr[\lambda^3 Q^2] F_{EM}^{\mu\nu} F_{EM}^{\sigma\tau}$$

Using

$$Q = \begin{pmatrix} 2/3 & & 0 \\ & -1/3 & \\ 0 & & -1/3 \end{pmatrix}, \quad Q^2 = \begin{pmatrix} 4/9 & & 0 \\ & 1/9 & \\ 0 & & 1/9 \end{pmatrix}$$

$$\lambda^3 = \begin{pmatrix} 1 & & 0 \\ & -1 & \\ 0 & & 0 \end{pmatrix}$$

so that

$$tr[\lambda^3 Q^2] = tr\begin{pmatrix} 4/9 & & \\ & -1/9 & \\ & & 0 \end{pmatrix} = \frac{1}{3}$$

we finally obtain for the $\pi^0 \to 2\gamma$ vertex

$$W' = -\frac{1}{(4\pi)^2} \frac{\lambda e^2}{6 F_\pi} \pi_0 \, \varepsilon_{\mu\nu\sigma\tau} F_{EM}^{\mu\nu} F_{EM}^{\sigma\tau} \tag{9.99}$$

Another part of the effective action which describes purely strong interactions among pseudoscalar mesons is contained in the five pseudoscalar vertex which arises from the last term in equation (9.89).

Now we need the fifth power in the expansion

$$e^{-(\xi \cdot U)} = 1 - \frac{1}{1!}(\xi \cdot U) + \cdots - \frac{1}{5!}(\xi \cdot U)^5 + \cdots$$

so that

$$\frac{1 - e^{-(\xi \cdot U)}}{(\xi \cdot U)} \longrightarrow \frac{1}{5!}(\xi \cdot U)^4$$

The relevant term in (9.89) is given by

$$-G^a = \frac{1}{4\pi^2}\,\varepsilon_{\mu\nu\sigma\tau}\,\mathrm{tr}\left[\frac{\lambda^a}{2}\left(-\frac{8}{3}\right)A_\mu A_\nu A_\sigma A_\tau\right]$$

which, when inserted into equation (9.95), yields ($\xi^a = \frac{1}{F_\pi}\pi^a$):

$$W' \longrightarrow \frac{1}{5!}(\xi\cdot U)^4\,\xi^a G^a$$

$$= \frac{1}{5!}\,\frac{1}{4\pi^2}\,\frac{8}{3}\,\frac{1}{F_\pi}\,(\xi\cdot U)^4\,\varepsilon_{\mu\nu\sigma\tau}\,\mathrm{tr}\Big[\underbrace{\pi^a\frac{\lambda^a}{2}}_{=\pi}\;A_\mu\,\underbrace{A_\nu}_{=A^c_\nu\frac{\lambda^c}{2}}\,A_\sigma\,A_\tau\Big]$$

Here it is useful to recall the definition of $U^a(x)$, equations (9.38, 81):

$$-U^a(x) = \partial^x_\mu\frac{\delta}{\delta A^a_\mu(x)} + f^{abc}\Big(\underbrace{A^b_\mu\frac{\delta}{\delta V^c_\mu(x)}}_{\text{acting on }F[A]\,!} + V^b_\mu\frac{\delta}{\delta A^c_\mu(x)}\Big)$$

(the first term inside the parentheses is struck through)

Therefore, we get

$$(\xi^a(x)\cdot U^a(x)) = \int d^4x\left[\xi^a(x)\,\partial^x_\mu\frac{\delta}{\delta A^a_\mu(x)} + f^{abc}\,\xi^a(x)V^b_\mu(x)\frac{\delta}{\delta A^c_\mu(x)}\right]$$

integr. by parts

$$= \int d^4x\left[\frac{1}{F_\pi}\left(-\partial^x_\mu\pi^a(x)\right)\frac{\delta}{\delta A^a_\mu(x)} + \cdots\right]$$

When applied to the expression contained in the trace of W' (note the combinatorial factor 4!), we obtain

$$W' \longrightarrow \frac{4!}{5!}\,\frac{1}{4\pi^2}\,\frac{8}{3}\,\frac{1}{F_\pi^5}\,\varepsilon_{\mu\nu\sigma\tau}\,\mathrm{tr}[\pi\,\partial_\mu\pi\,\partial_\nu\pi\,\partial_\sigma\pi\,\partial_\tau\pi]$$

or

$$W' \to \frac{2}{15}\,\frac{1}{\pi^2}\,\frac{1}{F_\pi^5}\,\varepsilon_{\mu\nu\sigma\tau}\,\mathrm{tr}[\pi\,\partial_\mu\pi\,\partial_\nu\pi\,\partial_\sigma\pi\,\partial_\tau\pi] \qquad (9.100)$$

It can be shown (no go theorem) that this vertex W' cannot be obtained as part of a chiral invariant.

Again, our result (9.100) is only determined up to a phenomenological constant which will be of great interest later on when we study some topological aspects of the W.-Z. action (9.100).

Before we leave this chapter, we want to return to our original Lagrangian (9.1, 68) and assume that we know the explicit form of the anomaly, e.g., by showing how the fermionic measure changes under the chiral transformation β :

$$[d\psi' d\bar{\psi}'] = e^{-i\int d^4x\, \beta(x)\, G(x)} [d\psi\, d\bar{\psi}] \tag{9.101}$$

where G is given by the non-abelian anomaly as presented in (9.89) or by the U(1) abelian anomaly (for N_f flavors)

$$G(x) = \frac{N_f}{8\pi^2} \, tr_c \, {}^*F^{\mu\nu} F_{\mu\nu} \tag{9.102}$$

The question then arises as to whether G(x) satisfies the W.-Z. conditions. For this reason, let us look again at the effective action W contained in (cf. also (9.44, 45))

$$e^{iW[V,A]} = \int [d\psi d\bar{\psi}]\, e^{iI[\psi,\bar{\psi};V,A]}$$

or, taking the logarithm:

$$W[V,A] = -i \ln \int [d\psi d\bar{\psi}]\, e^{iI[\psi,\bar{\psi};V,A]} \tag{9.103}$$

under the axial-vector transformation β which is of interest here (9.72-74), the action changes by

$$\delta_\beta W[V,A] = W[V',A'] - W[V,A]$$
$$= -i\left\{ \ln \int [d\psi d\bar\psi]\, e^{iI[\psi,\bar\psi;V',A']} - \ln \int [d\psi d\bar\psi]\, e^{iI[\psi,\bar\psi;V,A]} \right\} \qquad (9.104)$$

with

$$V_\mu' = V_\mu + \delta V_\mu$$
$$A_\mu' = A_\mu + \delta A_\mu$$

and the variations δV_μ and δA_μ given by equations (9.72–74). The fact that the fundamental action I is vector and axial vector gauge invariant is conveyed by

$$I[\psi,\bar\psi;V,A] = I[\psi',\bar\psi';V',A']$$

because

$$\bar\psi \not{D} \psi = \bar\psi e^{i\beta\gamma_5} e^{-i\beta\gamma_5} \not{D} e^{-i\beta\gamma_5} e^{i\beta\gamma_5} \psi$$
$$= \bar\psi' \not{D}' \psi'$$

So we find

$$\int [d\psi d\bar\psi]\, e^{iI[\psi,\bar\psi;V,A]} = \int [d\psi d\bar\psi]\, e^{iI[\psi',\bar\psi';V',A']} \qquad (9.105)$$

with the same fermionic measure on both sides.

Employing equation (9.101) yields

$$\int [d\psi d\bar\psi]\, e^{iI[\psi,\bar\psi;V,A]} = e^{i\int d^4x\, \beta(x) G(x)} \int [d\psi' d\bar\psi']\, e^{iI[\psi',\bar\psi';V',A']}$$

$$= e^{i\int d^4x\, \beta(x) G(x)} \int [d\psi d\bar\psi]\, e^{iI[\psi,\bar\psi;V',A']}$$

or, inverted,

$$\int[d\psi d\bar\psi]\, e^{iI[\psi,\bar\psi;V',A']} = e^{-i\int d^4x\,\beta(x)G(x)} \int[d\psi d\bar\psi]\, e^{iI[\psi,\bar\psi;V,A]}$$

This equation allows us to rewrite (9.104) in the form

$$\delta_\beta W[V,A] = -i\left\{\ln e^{-i\int d^4x\,\beta(x)G(x)} \int[d\psi d\bar\psi]e^{iI} - \ln\int[d\psi d\bar\psi]e^{iI}\right\}$$

$$= -i\left\{(-i)\int d^4x\,\beta(x)\,G(x) + \ln\int[d\psi d\bar\psi]\,e^{iI} - \ln\int[d\psi d\bar\psi]\,e^{iI}\right\}$$

$$= -\int d^4x\,\beta(x)\,G(x)$$

Thus, we obtain

$$\delta_\beta W[V,A] = -\int d^4x\,\beta^a(x)\,G^a(x) \tag{9.106}$$

Earlier we found for the response of the action with regard to the axial vector transformation β (9.80):

$$\delta_\beta W = -\int d^4x\,\beta^a(x)\,Y^a(x)\,W \tag{9.107}$$

and for the α-response:

$$\delta_\alpha W = -\int d^4x\,\alpha^a(x)\,X^a(x)\,W$$

so that,together with (9.106), we read off:

$$Y^a(x)W = G^a(x) \tag{9.108}$$

and $\quad X^a(x)W = 0$

At this stage, let us make use of the commutation relations (9.82) applied to W:

$$X^a(x)\underbrace{X^b(y)W}_{=0} - X^b(y)\underbrace{X^a(x)W}_{=0} = f^{abc}\delta(x-y)\underbrace{X^c(x)W}_{=0}$$

$$X^a(x)\underbrace{Y^b(y)W}_{=G^b(y)} - Y^b(y)\underbrace{X^a(x)W}_{=0} = f^{abc}\delta(x-y)\underbrace{Y^c(x)W}_{=G^c(x)}$$

$$Y^a(x)\underbrace{Y^b(y)W}_{=G^b(y)} - Y^b(y)\underbrace{Y^a(x)W}_{=G^a(x)} = f^{abc}\delta(x-y)\underbrace{X^c(x)W}_{=0}$$

which results in

$$X^a(x)G^b(y) = f^{abc}\delta(x-y)G^c(x)$$

$$Y^a(x)G^b(y) = Y^b(y)G^a(x)$$

Thus, we have arrived exactly at the W.-Z. integrability conditions (9.90).

Additional References to
<u>Anomalous Ward Identities à la Wess-Zumino</u>

J. Wess, B. Zumino, Phys. Lett. <u>37B</u>, 95 (1971)

J. Wess, Acta Physica Austriaca, Suppl. IX, 494 (1972)

10. Topological Aspects of the Wess-Zumino Effective Action

In our earlier discussion of the W.-Z. effective action, the integrability conditions did not impose any restrictions on the overall scale of the anomalies G^a. In an effective action like $W' = \frac{\lambda}{F_\pi} \pi^a G^a$, cf. equation (9.96), the constant λ had to be adjusted by experiment, e.g., via the $\pi^0 \rightarrow 2\gamma$ decay process.

From our studies of model field theories QED_2 and QCD_2, it should have become clear that global (topological) arguments can further restrict the form of an effective action. In fact, we will demonstrate that the constant λ cannot be arbitrary: it must be an integer.

However, before we start with a discussion of local and global integrability conditions of anomalous Ward identities, let us lay out a similar problem which is simple enough to understand the basic questions involved.

Our example consists of a particle of mass m, constrained to move on a two-dimensional sphere of radius one. The Lagrangian is

$$L = \frac{1}{2} m \dot{\vec{x}}^2 , \quad \vec{x} = (x_1, x_2, x_3) \tag{10.1}$$

and the constraint is

$$\vec{x}^2 = \sum_{i=1}^{3} x_i^2 = 1$$

Defining

$$g(\vec{x}) = \vec{x}^2 - 1$$

$$\Rightarrow \vec{\nabla} g(\vec{x}) = 2\vec{x}$$

we obtain the equation of motion with Lagrangian multiplier μ

$$m\ddot{\vec{x}} - \mu \vec{\nabla} g(\vec{x}) = 0$$

or

$$m\ddot{\vec{x}} - \lambda \vec{x} = 0 , \quad \lambda = 2\mu \tag{10.2}$$

If we take the scalar product of this equation with $\vec{x}$, we get ($\vec{x}^2 = 1$)

$$m\ddot{\vec{x}}\cdot\vec{x} - \lambda = 0 \tag{10.3}$$

Now, $\vec{x}^2 = 1 \Rightarrow 0 = \frac{d}{dt}\vec{x}^2 = 2\vec{x}\cdot\dot{\vec{x}}$

$$\Rightarrow \vec{x}\cdot\dot{\vec{x}} = 0$$

$$0 = \frac{d}{dt}(\dot{\vec{x}}\cdot\vec{x}) = \ddot{\vec{x}}\cdot\vec{x} + \dot{\vec{x}}^2$$

$$\Rightarrow \ddot{\vec{x}}\cdot\vec{x} = -\dot{\vec{x}}^2$$

With this equation, we rewrite (10.3) as

$$\lambda = -m\dot{\vec{x}}^2$$

which allows us to write for equation (10.2)

$$m\ddot{\vec{x}} + m\vec{x}\,\dot{\vec{x}}^2 = 0 \tag{10.4}$$

or

$$m\ddot{x}_i + m x_i\left(\sum_{k=1}^{3}\dot{x}_k^2\right) = 0$$

This system respects the symmetries

(a) $t \rightarrow -t$ (10.5)

and separately (b) $\vec{x} \rightarrow -\vec{x}$

If we want a system that is only invariant under the combined operation

$$\{t \rightarrow -t,\ \vec{x} \rightarrow -\vec{x}\} \tag{10.6}$$

the simplest choice would be

$$\begin{aligned}\left(m\ddot{\vec{x}} + m\,\vec{x}\,\dot{\vec{x}}^2\right)_i &= \alpha\left(\vec{x}\times\dot{\vec{x}}\right)_i \\ &= \alpha\,\varepsilon_{ijk}\,x_j\,\dot{x}_k\end{aligned} \tag{10.7}$$

The question arises whether this equation of motion can be derived from a Lagrangian. An extra term in the Lagrangian whose variation equals the right-hand side would be $\varepsilon_{ijk}\,x_i\,x_j\,\dot{x}_k$; this, however, vanishes by anti-symmetry of ε_{ijk}.

Now, if we recall the Lorentz force for an electric charge in presence of a magnetic monopole at the center of the sphere

$$\vec{F} = e\,\frac{\vec{v}}{c}\times\vec{B}, \quad \vec{B} = g\,\frac{\vec{x}}{|\vec{x}|^3} = \vec{\nabla}\times\vec{A}$$

or

$$F_i = \frac{eg}{c}\,\varepsilon_{ijk}\,\dot{x}_j\,x_k\,\left(\sum x_i^2\right)^{-\frac{3}{2}}$$

we recognize the structure of the right-hand side of equation (10.7). The action for our problem is therefore

$$I = \int dt\,\left\{\frac{m}{2}\,\dot{\vec{x}}^2 + \alpha\,\vec{A}\cdot\dot{\vec{x}}\right\} \tag{10.8}$$

Is this a single-valued action? The answer is no, since $\vec{A}$ contains a Dirac string in which case the closed line integral $\oint \vec{A}\cdot d\vec{x}$ is determined only modulo $2\pi n$, $n\in\mathbb{Z}$. Let us explore our problematical action (10.8) quantum mechanically. Since we want to look at closed orbits of the particle, we consider the Feynman path integral

$$\mathrm{Tr}\, e^{-\beta H} = \int\limits_{\vec{x}(0)=\vec{x}(\beta)} [d\vec{x}(t)]\; e^{-I} \tag{10.9}$$

In e^{-I}, the term of interest is

$$e^{i\alpha\int_\gamma \vec{A}\cdot d\vec{x}} \tag{10.10}$$

The path integral in (10.9) goes over all closed paths, i.e., we are talking about a mapping

$$\vec{x} : S^1 \rightarrow S^2 \ , \quad t \mapsto \vec{x}(t) \tag{10.11}$$

where $S^1 \cong [0, 2\pi]$, with the endpoints identified.

A mapping of a circle γ into S^2 is then given by

$$\gamma := \vec{x}(S^1) = \left\{ \vec{x}(t) \,\middle|\, t \in [0, 2\pi] \right\}$$

The set of all mappings,

$$\mathcal{D} := \left\{ \vec{x} : S^1 \longrightarrow S^2 \right\}$$

is connected, since $\pi_1(S^2) = 0$; i.e., any two mappings $\vec{x}_1$ and $\vec{x}_2$ can be connected continuously by a path that lies entirely in $\mathcal{D}$. (However, $\mathcal{D}$ is not simply connected: $\pi_2(S^2) = Z$.) From $\pi_1(S^2) = 0$, it follows that a mapping γ of a circle into S^2 can be extended to a mapping of a disc D into S^2 whose boundary is γ : $\vec{x}(S^1) = \partial D$.

Using Gauss's law, we rewrite (10.10) in favor of the magnetic field. The closed orbit γ is the boundary of a disc D. In terms of the magnetic flux through D, we have

$$e^{i\alpha \int_{\gamma = \partial D} A_i dx_i} = e^{i\alpha \int_D F_{ij} d\sigma^{ij}} = e^{-i\alpha \int_{D'} F_{ij} d\sigma^{ij}} \tag{10.12}$$

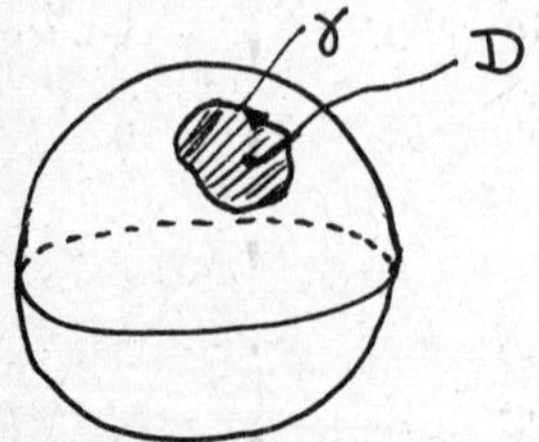

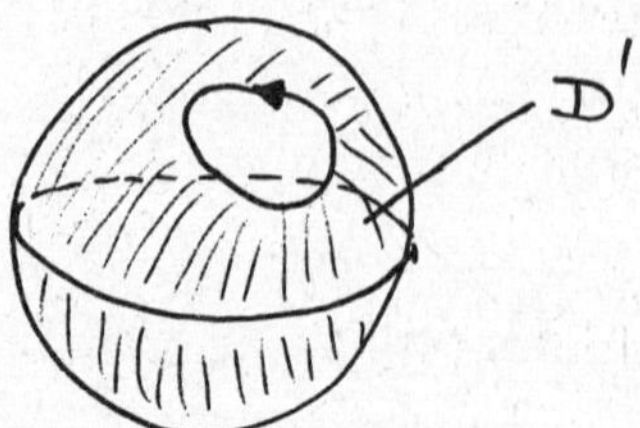

Evidently, D is not unique. There is no consistent way to decide whether to choose D or D' (the curve γ could continuously be looped around the sphere or turned inside out). Either expression on the right-hand side of (10.12) can be used in the Feynman path integral and, since we require that they be equal, we obtain

$$1 = e^{i\alpha \int_{D \cup D'} F_{ij}\, d\sigma^{ij}} = e^{i\alpha \int_{S^2} F_{ij}\, d\sigma^{ij}} \tag{10.13}$$

To evaluate the flux integral over S_2, we write

$$\begin{aligned} \int_{S^2} F_{ij}\, d\sigma^{ij} &= \iint F_{ij}(\vec{x}(u)) \frac{\partial x^i}{\partial u^1} \frac{\partial x^j}{\partial u^2}\, du^1 du^2 \\ &= \iint B_k \varepsilon_{ijk} \frac{\partial x^i}{\partial u^1} \frac{\partial x^j}{\partial u^2}\, du^1 du^2 \\ &= \oint_{S^2} \vec{B} \cdot d\vec{f} \end{aligned}$$

Using $\vec{B} = \frac{1}{R^2} \hat{r} = \vec{\nabla} \times \vec{A}$

and $d\vec{f} = \hat{r}\, df$, we obtain for the integration over a sphere with radius R:

$$\int_{S^2} F_{ij}\, d\sigma^{ij} = \iint \frac{1}{R^2} \hat{r} \cdot \hat{r}\, df = \frac{1}{R^2} 4\pi R^2 = 4\pi$$

Equation (10.13) is then equivalent to

$$1 = e^{i\alpha \int_{S^2} F_{ij}\, d\sigma^{ij}} = e^{i\alpha \cdot 4\pi} \tag{10.14}$$

which tells us that the constant α is not arbitrary, but

$$\alpha = \frac{n}{2}, \quad n \in \mathbb{Z} \tag{10.15}$$

a conclusion that followed entirely from topological arguments.

Equation (10.15) is nothing but Dirac's quantization condition for the product of electric and magnetic charges.

Now we want to show how topological arguments restrict the a priori form of the Wess-Zumino effective action. One starts with the $SU(3)_L \times SU(3)_R$ invariant effective Lagrangian

$$\mathcal{L} = \frac{1}{16} F_\pi^2 \int d^4x \, Tr \left[\partial_\mu U \, \partial_\mu U^{-1} \right] \tag{10.16}$$

Here we describe the low-energy phenomena by introducing a field U(x) that transforms according to the non-linear realization of $SU(3)_L \times SU(3)_R$:

$$\begin{aligned} U(x) &= e^{2i\xi(x)} \\ &= 1 + \frac{2i}{F_\pi} \sum_{a=1}^{8} \lambda^a \pi^a + \cdots , \end{aligned} \tag{10.17}$$

$$\xi(x) = \frac{1}{F_\pi} \sum_a \lambda^a \pi^a(x)$$

The pion decay constant is $F_\pi \simeq$ 190 MeV and the λ^a ($Tr\, \lambda^a \lambda^b = 2\delta^{ab}$) are the SU(3)- flavor generators. Finally, the $\pi^a(x)$ are the Goldstone boson fields.

The Lagrangian (10.16) possesses an extra symmetry that is not a symmetry of QCD. The operation $(-1)^{n_\pi}$ which counts modulo two the number of pseudoscalars is not conserved , i.e., the symmetry $U \rightarrow U^{-1}$ or, equivalently, $\pi^a \rightarrow -\pi^a$, is not respected in the real world. Furthermore, (10.16) is also invariant under the naive parity operation $P_o(\vec{x} \rightarrow -\vec{x},\ t \rightarrow t,\ U \rightarrow U)$. The parity operation in QCD corresponds to the joint transformation $\vec{x} \rightarrow -\vec{x}$, $t \rightarrow t$, $U \rightarrow U^{-1}$, which we call $P = P_o(-1)^n$. QCD is invariant under P but not under P_o and $(-1)^{n_\pi}$ separately. The simplest process that shows this is $K^+K^- \rightarrow \pi^+ \pi^- \pi^o$. Now the question arises again as to whether we can find an additional higher-order term to (10.16) that obeys only the appropriate symmetries.

The Euler-Lagrange equation derived from (10.16) is given by

$$\partial_\mu \left(\frac{1}{8} F_\pi^2 \, U^{-1} \partial_\mu U \right) \tag{10.18}$$

A Lorentz-invariant term which violates $P_o(\vec{x} \to -\vec{x},\ t \to t,\ U \to U)$ must contain the Levi-Civita $\varepsilon_{\mu\nu\alpha\beta}$. There is a unique P_o-violating term with only four derivatives. So let us modify (10.18) according to

$$\partial_\mu\left(\tfrac{1}{8}F_\pi^2 U^{-1}\partial_\mu U\right) + \lambda\varepsilon^{\mu\nu\alpha\beta}(U^{-1}\partial_\mu U)(U^{-1}\partial_\nu U)(U^{-1}\partial_\alpha U)(U^{-1}\partial_\beta U) = 0 \tag{10.19}$$

Although (10.19) violates P_o, it does respect $P = P_o(-1)^{n_\pi}$. However, we cannot obtain (10.19) as a variation of a certain Lagrangian, $\int d^4x\, \mathcal{L}$, since the only pseudoscalar of dimension four vanishes identically by anti-symmetry of $\varepsilon^{\mu\nu\alpha\beta}$ and cyclic symmetry of the trace.

$$\varepsilon^{\mu\nu\alpha\beta}\, tr(U^{-1}\partial_\mu U)(U^{-1}\partial_\nu U)(U^{-1}\partial_\alpha U)(U^{-1}\partial_\beta U) = 0 \tag{10.20}$$

since $\quad tr\,\mu\nu\alpha\beta = tr\,\beta\mu\nu\alpha$, but $\varepsilon^{\mu\nu\alpha\beta} = -\varepsilon^{\beta\mu\nu\alpha}$.

What, then, is the solution to our problem? To demonstrate the existence of an effective action that meets our requirements, we proceed in the same way as for the monopole case. Instead of the one-dimensional compactified world $\mathcal{R}^1 \rightsquigarrow S^1$, we now imagine space-time to be a very large four-sphere $M = S^4$ (compactifiable because $U(x) \to I$ for $|x| \to \infty$).

The field U is a mapping of M into the SU(3) manifold

$U(M)$ = image of space-time $S^4 = M$ in SU(3)

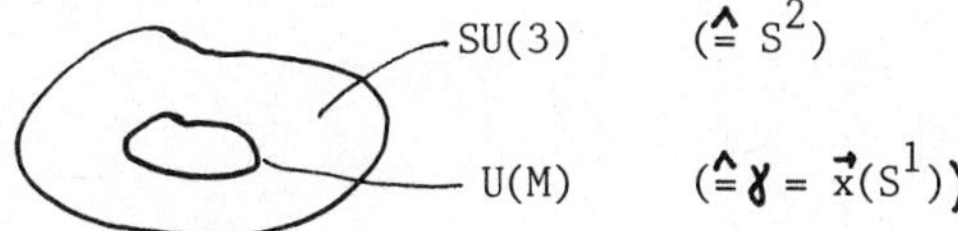

Since $\pi_4(SU(3)) = 0$, we know that $\mathcal{D} := \{U : S^4 \to SU(3)\}$ is connected, however not simply connected. ($\pi_5(SU(3)) = Z$)

We then conclude that the image of the four-sphere in SU(3), U(M), is the boundary of a five-dimensional disc, Q. So, again, we have extended the mapping $U: S^4 = M \to SU(3)$ to $Q \subset SU(3)$, $\partial Q = U(M)$, which corresponds to our earlier monopole discussion where we extended the mapping $\vec{x}: S^1 \to S^2$ to $D \subset S^2$ with $\partial D = \vec{x}(S^1)$.

Now, over ∂Q, with $\dim \partial Q = 4$, we can only integrate 4th-rank antisymmetric tensors (4-forms); there is, however, no 4-form in ∂Q whose variation would provide the additional term in the equations of motion. Likewise, over Q, with dim Q = 5, we can only integrate 5th-rank antisymmetric tensors (5-forms). On the manifold $SU(3) \supset Q$, there is a unique 5th-rank antisymmetric tensor, namely, the antisymmetric part of

$$\mathrm{tr}\left[(U^{-1}\partial_i U)(\ _j)(\ _k)(\ _l)(\ _m)\right] \sim \omega_{ijklm} \tag{10.21}$$

so that with the surface element on Q: $d\Sigma^{ijklm} \sim \varepsilon^{mijkl}$ and the cyclic trace operation: $\omega_{ijklm} = \omega_{mijkl}$, we obtain

$$d\Sigma^{ijklm}\,\omega_{ijklm} \neq 0$$

The image of our four-sphere $S^4 = M = \partial D$, $U(M) = \partial Q$ in SU(3) is the boundary of two five-dimensional discs Q and Q'.

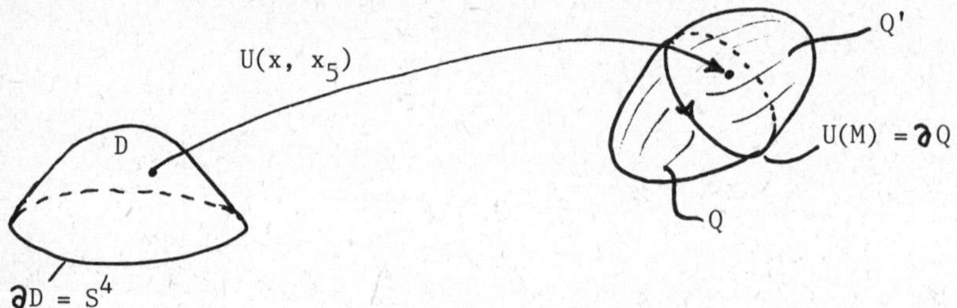

where we introduced a fifth coordinate $x_5 (0 \leq x_5 \leq 1)$ and consider the function $U(x, x_5)$ such that $U(x, 0) = 1$ and $U(x, 1) = U(x)$.

Analogous to the right-hand side of (10.12), we construct

$$\Gamma = \int_Q \omega_{ijklm}\, d\Sigma^{ijklm} =: \int_Q \omega\, d\Sigma \tag{10.22}$$

and

$$\Gamma' = -\int_{Q'} \omega\, d\Sigma$$

We require

$$e^{i\Gamma} = e^{i\int_Q \omega d\Sigma} = e^{-i\int_{Q'} \omega d\Sigma} = e^{i\Gamma'}$$

or $$e^{i\int_{Q\cup Q'} \omega d\Sigma} = 1 \,, \quad Q\cup Q' = U(S^5)$$

Consequently,

$$\int_{U(S^5)} \omega d\Sigma = 2\pi n, \; n\in \mathbb{Z}, \; \forall U: S^5 \rightarrow SU(3) \tag{10.23}$$

From $$\int_{U(S^5)=Q\cup Q'} \omega d\Sigma = \int_Q \omega d\Sigma - \int_{Q'} \omega d\Sigma = 2\pi n \tag{10.24}$$

we find that $\Gamma = \int_Q \omega d\Sigma$ is defined

only up to $$\Gamma \rightarrow \Gamma + 2\pi n, \; n\in \mathbb{Z} \tag{10.25}$$

or $$e^{ic\Gamma} \rightarrow e^{ic(\Gamma+2\pi n)} = e^{ic\Gamma} e^{i2\pi nc} \tag{10.26}$$

which yields $$c \in \mathbb{Z}. \tag{10.27}$$

Hence, we may work with the effective action

$$I = \frac{1}{16} F_\pi^2 \int d^4x \; tr[\partial_\mu U \partial_\mu U^{-1}] + n\Gamma \tag{10.28}$$

where n is an arbitrary integer which is a consequence of the multi-valued action Γ, the reason being: $\mathcal{D} = \{ S^4 \rightarrow SU(3)\}$ is not simply connected $\pi_5(SU(3)) = \mathbb{Z}$. Γ is, in fact, the Wess- Zumino Lagrangian and our earlier constant λ must be chosen an integer.

The mappings U : $S^5 \rightarrow SU(3)$ are divided into an infinite number of homotopy classes following from $\pi_5(SU(3)) = \mathbb{Z}$. If U_o is taken from the "one-soliton sector", we normalize ω so that

$$\int_{U_0(S^5)=S_0} \omega_{ijklm}\, d\Sigma^{ijklm} = 2\pi \tag{10.29}$$

where S_0 denotes a basic 5-sphere in SU(3). Every five sphere in SU(3) is then a multiple of the basic five-sphere, so that

$$\frac{\Gamma[U]}{2\pi} = \int_{U(S^5)} \omega_{ijklm}\, d\Sigma^{ijklm} = n \in \mathbb{Z} \tag{10.30}$$

where n is the homotopy class index of $U : S^5 \to SU(3)$. Clearly, the explicit form of Γ, eq. (10.30), satisfies eq. (10.25).

Writing

$$U = e^{\frac{2i}{F_\pi} A}\ , \quad A = \sum \lambda^a \pi^a$$
$$U^+ = e^{-\frac{2i}{F_\pi} A} \tag{10.31}$$

we find

$$U^{-1}\partial_\mu U = \frac{2i}{F_\pi}\partial_\mu A + O(A^2) \tag{10.32}$$

so that $n\Gamma$ can be rewritten as (cf. (9. 100))

$$n\,\Gamma[A] = n\frac{2}{15\pi^2 F_\pi^5}\int d^4x\, \varepsilon^{\mu\nu\alpha\beta}\, tr\left[A\partial_\mu A\,\partial_\nu A\,\partial_\alpha A\,\partial_\beta A\right] + O(A^6) \tag{10.33}$$

It is useful to recall that A(x) is only dependent on $x \in S^4$. Also note that $\Gamma[A]$ vanishes for SU(2) with $A = A^a\tau^a$. Finally, $\frac{\delta\Gamma}{\delta A}$ provides the desired additional terms in the equation of motion up to $O(A^4)$:

$$\frac{\delta\Gamma}{\delta A} \sim \varepsilon^{\mu\nu\alpha\beta}\,\partial_\mu A\,\partial_\nu A\,\partial_\alpha A\,\partial_\beta A + O(A^4) \tag{10.34}$$

The value of the integer n turns out to be the number of the three colors, $n = N_c = 3$. This can best be seen by coupling the Goldstone bosons to photons. So let us introduce the electromagnetic current $J_{E.M.}$ of the field U(x):

$$\frac{\delta W}{\delta A_\mu^{EM}(x)} = \frac{1}{i} \frac{i\int[d\ldots]\, J^\mu_{EM}(x)\, e^{i\int \%}}{\int[d\ldots]\, e^{i\int \%}} \equiv \frac{\langle 0|J^\mu_{EM}(x)|0\rangle}{\langle 0|0\rangle} \equiv J^\mu_{EM}(U) \tag{10.35}$$

with the action $W = W_o + n\Gamma$ given by the logarithm of

$$e^{iW[U=e^{2i\pi}, A_\mu, \ldots]} = \int[d \begin{smallmatrix}\text{fundam. fields}\\ \text{quarks, gluons, ..}\end{smallmatrix}]\, e^{i\int[\ldots + J^\mu_{EM} A_\mu^{EM} + \ldots]} \quad \text{fctl. of fundamental fields} \tag{10.36}$$

The functional Γ is invariant under global charge rotation $U \to U + i\varepsilon[Q, U]$, with ε = const. Gauging this symmetry, it is necessary to introduce the photon field $A_\mu(x)$:

$$A_\mu^\varepsilon = A_\mu - \partial_\mu \varepsilon$$

From
$$\left.\frac{\delta W[A^\varepsilon]}{\delta \varepsilon(x)}\right|_{\varepsilon=0} = \int d^4y \left.\frac{\delta W[A^\varepsilon]}{\delta A_\mu^\varepsilon(y)}\right|_{\varepsilon=0} \frac{\delta A_\mu^\varepsilon(y)}{\delta \varepsilon(x)}$$

$$= \int d^4y\, J^\mu_{EM}(y) \left(-\partial^y_\mu \delta(y-x)\right)$$

$$= \partial_\mu J^\mu_{EM}(x)$$

we observe
$$\delta_\varepsilon W = \int d^4x\, J_\mu^{EM} \left(-\partial^\mu \varepsilon(x)\right) \tag{10.37}$$

Under a local charge rotation

$$\Lambda(x) = e^{i\varepsilon(x) Q}, \quad Q = \begin{pmatrix} 2/3 & & \\ & -1/3 & \\ & & -\frac{1}{3} \end{pmatrix} \tag{10.38}$$

U transforms according to

$$U \to \Lambda U \Lambda^{+} = (1+i\varepsilon Q)U(1-i\varepsilon Q) + O(\varepsilon^2)$$

$$= U + i\varepsilon[Q,U] + O(\varepsilon^2) \tag{10.39}$$

and, under a local charge rotation, Γ is supplemented by a term

$$\Gamma \to \Gamma - \int d^4x\,(\partial_\mu \varepsilon)\, J^\mu \tag{10.40}$$

with

$$J^\mu = \frac{1}{48\pi^2}\varepsilon^{\mu\nu\alpha\beta}\, tr[Q(\partial_\nu U U^{-1})(\partial_\alpha U U^{-1})(\partial_\beta U U^{-1}) + Q(U^{-1}\partial_\nu U)(U^{-1}\partial_\alpha U)(U^{-1}\partial_\beta U)] \tag{10.41}$$

The expression (10.40) is not gauge invariant because J^μ is not; but one can construct a gauge invariant action which has in it a piece that describes the process $\pi^0 \to 2\gamma$ and this, in turn, fixes $n = N_c = 3$.

Additional References to
Topological Aspects of the Wess-Zumino Effective Action

E. Witten, Nucl. Phys. B223, 422 (1983) and B223, 433 (1983) (1st paper)

11. Topological Baryon Number and the Skyrme Model

The idea that baryons might appear as topological solitons of an effective Lagrangian was first suggested by Skyrme. His soliton approach to hadron physics is based on the non-linear sigma model, i.e., on a $SU(2)\times SU(2)$ invariant chiral theory of pions. Using this dynamical symmetry, one obtains some very interesting features of the chiral solitons (baryons) which are of topological origin. First of all, they carry an exactly conserved homotopic charge. This charge is localized and hence is ideally suited as a candidate for a baryon number. Secondly, they are extended objects and interact strongly. Finally, their masses exceed by far the masses of the fields in the Lagrangian. While the Skyrme model does not admit exact analytical solutions, one can still try to analyze its basic physical features.

Let us recall that all results of current algebra can be derived from an effective Lagrangian. Let $G = SU(N)_R \times SU(N)_L$ be the symmetry group of such a Lagrangian and $H = SU(N)_V$ be the subgroup of G which leaves the vacuum invariant. Now the vacuum of this theory is described by an element U(x) of the coset space G/H, i.e., U(x) belongs to

$$SU(N)_A = \frac{SU(N)_R \times SU(N)_L}{SU(N)_V} \tag{11.1}$$

The low-energy pion physics (N = 2) is described by a space-time dependent choice of U (coordinatization of the G/H group manifold)

$$U: \qquad M_4 \ni x \longrightarrow U(x) \in SU(2) \tag{11.2}$$

A useful parametrization of U(x) is

$$U(x) = e^{\frac{2i}{F_\pi}\pi^a(x)\lambda^a} \tag{11.3}$$

where $\pi^a(x)$ are the Goldstone boson fields and $F \simeq 190$ MeV is the pion decay constant. The effective Lagrangian which describes the low-energy interactions of these Goldstone bosons has the form

$$\mathcal{L} = -\frac{F_\pi^2}{16} \, tr(L^\mu L_\mu) \tag{11.4}$$

with

$$L_\mu(x) := U^{-1}(x)\, \partial_\mu U(x) \tag{11.5}$$

Higher derivative terms in (11.4) are assumed to be unimportant at low energies, but--as Skyrme noticed-- they are essential to make the solitons of the theory stable against collapse into a single point.

Hence we seek finite, static, topological solutions to some chiral model. Using our knowledge of instanton physics in classical Yang-Mills theory, the first step should be the homotopic classification of the field configurations.

Let the dynamics of the field $U(x) \in SU(2)$ be such that finite-energy stationary solutions satisfy

$$U(x) \xrightarrow[|\vec{x}| \to \infty]{} I \tag{11.6}$$

at spatial infinity. The condition (11.6) means that R^3 can be compactified onto S^3 so that we have the mappings

$$U : \quad S^3 \longrightarrow SU(2) \cong S^3$$

or

$$L_i : \quad S^3 \longrightarrow S^3 \tag{11.7}$$

Now $U(x)$ or $L_\mu(x) = U^{-1} \partial_\mu U$ maps the space S^3 into the group space SU(2) and the mappings L_i become decomposed into an infinite set of Chern classes of

$$\Pi_3(SU(2)) = \Pi_3(S^3) = Z \tag{11.8}$$

Static configurations $U(x)$ are then classified by their topological charge

$$B = \int d^3x \; B_0(x) \tag{11.9}$$

where B_μ is the conserved topological current

$$B_\mu = \frac{1}{24\pi^2} \varepsilon_{\mu\nu\rho\sigma} \, tr \, L^\nu L^\rho L^\sigma$$

$$= \frac{1}{24\pi^2} \varepsilon_{\mu\nu\rho\sigma} \, tr \{U^+ \partial^\nu U U^+ \partial^\rho U U^+ \partial^\sigma U\} \qquad (11.10)$$

Note that B_μ is not a total divergence and therefore charge can be localized arbitrarily in space--in contrast to the monopole charge whose detection involves Gauss' law, i.e., the flux through a large sphere at infinity. B_μ is a number-current which means that while electromagnetic charge distribution of a particle can be measured through its form factor, the baryonic charge is not coupled to any long-range field and its measurement is performed through simple counting. Additivity of homotopy classes in $\pi_3(SU(2))$ takes the role of charge combination.

Let us quickly show that B_μ is indeed conserved.

From

$$L_\nu = U^+ \partial_\nu U$$

and

$$\partial_\mu (U^+ U) = \partial_\mu I = 0$$

$$\leadsto \partial_\mu U^+ U + U^+ \partial_\mu U = 0$$

$$\leadsto \partial_\mu U^+ = - U^+ \partial_\mu U U^+ \qquad (11.11)$$

we obtain

$$\partial_\mu L_\nu = \partial_\mu U^+ \partial_\nu U + U^+ \partial_\mu \partial_\nu U$$

$$= - U^+ \partial_\mu U \cdot U^+ \partial_\nu U + U^+ \partial_\mu \partial_\nu U$$

$$= - L_\mu L_\nu + U^+ \partial_\mu \partial_\nu U$$

or $$\partial_\mu L_\nu - \partial_\nu L_\mu = -(L_\mu L_\nu - L_\nu L_\mu) = -[L_\mu, L_\nu] \tag{11.12}$$

This relation enables us to prove $\partial_\mu B^\mu = 0$:

$$\begin{aligned}
\partial^\mu B_\mu &= \frac{1}{24\pi^2} \underline{\varepsilon_{\mu\nu\rho\sigma} \, \partial^\mu \, tr\, L^\nu L^\rho L^\sigma} \\
&= \varepsilon_{\mu\nu\rho\sigma} \, tr\,(\partial^\mu L^\nu L^\rho L^\sigma) + \cdots \\
&\underset{(12)}{=} - \varepsilon_{\mu\nu\rho\sigma} \, tr\,(L^\mu L^\nu L^\rho L^\sigma) + \cdots \\
&= - \varepsilon_{\mu\nu\rho\sigma} \, tr\,(L^\sigma L^\mu L^\nu L^\rho) + \cdots \\
&= - \varepsilon_{\nu\rho\sigma\mu} \, tr\,(L^\mu L^\nu L^\rho L^\sigma) + \cdots \\
&= + \varepsilon_{\mu\nu\rho\sigma} \, tr\,(L^\mu L^\nu L^\rho L^\sigma) + \cdots \qquad = 0
\end{aligned}$$

The topological charge is given by

$$\begin{aligned}
B &= \int d^3x \;\; B_0(x) \\
&= \int d^3x \; \frac{1}{24\pi^2} \underbrace{\varepsilon_{0ijk}}_{= -\varepsilon_{ijk}} \, tr[L^i L^j L^k]
\end{aligned}$$

where we used

$$- \varepsilon_{ijk} = \varepsilon_{0ijk}$$

$$\varepsilon_{0123} = 1 = - \varepsilon^{0123}$$

$$\varepsilon_{123} = -1 = - \varepsilon^{123}$$

Here are some useful equivalent representations of the topological charge:

$$B = - \frac{1}{24\pi^2} \, \varepsilon^{ijk} \int d^3x \;\; tr[L_i L_j L_k] \tag{11.13}$$

$$\equiv -\frac{1}{24\pi^2}\,\varepsilon^{ijk}\int d^3x\ \mathrm{tr}\left[(U^+\partial_i U)(U^+\partial_j U)(U^+\partial_k U)\right] \tag{11.14}$$

$$= \frac{1}{24\pi^2}\,\varepsilon^{ijk}\int d^3x\ \mathrm{tr}\left[(U\partial_i U^+)(U\partial_j U^+)(U\partial_k U^+)\right] \tag{11.15}$$

$$= \frac{1}{24\pi^2}\,\varepsilon^{ijk}\int d^3x\ \mathrm{tr}\left[U^+\partial_i U\ \partial_j U^+\ \partial_k U\right] \tag{11.16}$$

Expression (11.15) is simply a consequence of (11.11):

$$\mathrm{tr}\left[(U\partial_i U^+)(U\partial_j U^+)(U\partial_k U^+)\right]$$

$$\underset{(11)}{=} (-1)^3\ \mathrm{tr}\big[\underbrace{UU^+}_{=1}\ \partial_i U\ \underbrace{U^+U}_{=1}\ U^+\partial_j U\,\underbrace{U^+U}_{=1}U^+\ \partial_k U U^+\big]$$

$$= -\,\mathrm{tr}\left[\partial_i U\, U^+\ \partial_j U\ U^+\partial_k U\, U^+\right]$$

$$= -\,\mathrm{tr}\left[(U^+\partial_i U)(U^+\partial_j U)(U^+\partial_k U)\right]$$

Equation (11.16) reproduces (11.14), since

$$\mathrm{tr}\left[U^+\partial_i U\ \partial_j U^+\partial_k U\right]$$

$$= -\,\mathrm{tr}\left[(U^+\partial_i U)(U^+\partial_j U)(U^+\partial_k U)\right]$$

It might perhaps be instructive to also present a detailed proof of $B\in\mathbb{Z}$, where B(U) is the Chern index of the mapping U.

$$B = -\frac{1}{24\pi^2}\,\varepsilon^{ijk}\int d^3x\ \mathrm{tr}\ L_i L_j L_k$$

$$= -\frac{1}{24\pi^2}\,\varepsilon^{ijk}\,\frac{1}{2}\int d^3x\ \mathrm{tr}\ (L_iL_j - L_jL_i)\,L_k$$

$$= \frac{-1}{48\pi^2}\, \varepsilon^{ijk} \int d^3x \; tr\,[L_i, L_j]\, L_k$$

$$= -\frac{1}{48\pi^2}\, \varepsilon^{ijk} \int_{R^3} d^3x \; tr[U^+\partial_i U, U^+\partial_j U]\, U^+\partial_k U \qquad (11.17)$$

Let U be parametrized by Θ^a, a = 1, 2, 3, so that

$$\partial_i U \equiv \frac{\partial U}{\partial x^i} = \frac{\partial \Theta^a}{\partial x^i}\,\frac{\partial U}{\partial \Theta^a} \equiv \partial_i \Theta^a\, \partial_a U$$

Then we can rewrite B in the form,

$$B = -\frac{1}{48\pi^2} \int_{R^3} d^3x \; \varepsilon^{ijk}\, \partial_i \Theta^a \partial_j \Theta^b \partial_k \Theta^c \cdot \underbrace{tr[U^+\partial_a U, U^+\partial_b U]\, U^+\partial_c U]}$$

$$= tr\, U^+\partial_a U\, U^+\partial_b U\, U^+\partial_c U - (a \leftrightarrow b)$$

$=$ cyclic in (a,b,c) and antisymm. in (a,b)

Here we meet a sum of the type

$$\sum_{a,b,c} \overbrace{A_{abc}}^{\text{totally antisymm.}}\; \underbrace{B_{abc}}_{\substack{\text{cyclic + symm.}\\ \text{in a,b}\\ B_{abc} = -B_{bac}}} = A_{123} B_{123} + \underbrace{A_{132}}_{-A_{123}}\, \underbrace{B_{132}}_{B_{213} = -B_{123}}$$

$$+ \underbrace{A_{213}}_{-A_{123}}\, \underbrace{B_{213}}_{-B_{123}} + \underbrace{A_{231}}_{A_{123}}\, \underbrace{B_{231}}_{B_{123}}$$

$$+ \underbrace{A_{312}}_{A_{123}}\, \underbrace{B_{312}}_{B_{123}} + \underbrace{A_{321}}_{-A_{123}}\, \underbrace{B_{321}}_{-B_{123}}$$

$$= 6\, A_{123}\, B_{123}$$

This result enables us to write for the toplogical charge

$$\mathcal{B} = -\frac{6}{48\pi^2}\int_{R^3} d^3x\, \varepsilon^{ijk}\, \partial_i\Theta^1 \partial_j\Theta^2 \partial_k\Theta^3 \cdot$$

$$\cdot\ tr\,[U^+\partial_1 U, U^+\partial_2 U]\,U^+\partial_3 U$$

$$= -\frac{6}{48\pi^2}\int_{R^3} d^3x \left|\frac{D(\Theta^1,\Theta^2,\Theta^3)}{D(x^1,x^2,x^3)}\right| tr[L_1,L_2]L_3$$

derivative with respect to Θ_3, etc.

$$= -\frac{6}{48\pi^2}\, n \int_{SU(2)} d^3\Theta\ tr[L_1,L_2]L_3 \qquad (11.18)$$

where n is the number of points of R^3 which are mapped into the same element $U(x) \in SU(2)$.

To further evalauate (11.18), we parametrize U by the Euler angles Θ^a:

$$U = e^{\frac{i}{2}\Theta^1\tau_3}\, e^{\frac{i}{2}\Theta^2\tau_2}\, e^{\frac{i}{2}\Theta^3\tau_3} \qquad (11.19)$$

Using the well-known relation

$$e^{\frac{i}{2}\Theta\tau_a} = \cos\frac{\Theta}{2} + i\,\tau_a\, \sin\frac{\Theta}{2} \qquad (11.20)$$

we immediately obtain

$$\frac{d}{d\Theta}\, e^{\frac{i}{2}\Theta\tau_a} = \frac{i}{2}\tau_a\, e^{\frac{i}{2}\Theta\tau_a} = \frac{i}{2}\, e^{\frac{i}{2}\Theta\tau_a}\,\tau_a \quad \text{(no summation)} \qquad (11.21)$$

It is also convenient to reexpress the trace in (11.18):

$$\begin{aligned} tr\,([L_1, L_2]\,L_3) &= tr\,(L_1L_2L_3 - L_2L_1L_3) \\ &= tr\,(L_3L_1L_2 - L_1L_3L_2) \\ &= tr\,([L_3, L_1]\,L_2) \end{aligned} \qquad (11.22)$$

With the aid of

$$\partial_1 U = \frac{i}{2} \tau_3 U$$

and

$$U^{-1} = e^{-\frac{i}{2}\theta^3\tau_3} e^{-\frac{i}{2}\theta^2\tau_2} e^{-\frac{i}{2}\theta^1\tau_3}$$

we obtain

$$\begin{aligned} L_1 &\equiv U^{-1}\partial_1 U \\ &= \frac{i}{2} e^{-\frac{i}{2}\theta^3\tau_3} e^{-\frac{i}{2}\theta^2\tau_2} \underbrace{e^{-\frac{i}{2}\theta^1\tau_3} \tau_3 e^{\frac{i}{2}\theta^1\tau_3}}_{=\tau_3} e^{\frac{i}{2}\theta^2\tau_2} e^{\frac{i}{2}\theta^3\tau_3} \\ &\underset{\{\tau_2,\tau_3\}=0}{=} \frac{i}{2} \tau_3 e^{-\frac{i}{2}\theta^3\tau_3} e^{i\theta^2\tau_2} e^{\frac{i}{2}\theta^3\tau_3} \end{aligned} \tag{11.23}$$

Likewise, it follows that

$$\begin{aligned} \partial_3 U &= \frac{i}{2} U \tau_3 \\ L_3 &\equiv U^{-1}\partial_3 U = \frac{i}{2}\tau_3 \end{aligned} \tag{11.24}$$

We need the commutator of $[L_3, L_1]$:

$$\begin{aligned} [L_3, L_1] &= -\frac{1}{4}\left[\tau_3, \tau_3 e^{-\frac{i}{2}\theta^3\tau_3} e^{i\theta^2\tau_2} e^{\frac{i}{2}\theta^3\tau_3}\right] \\ &= -\frac{1}{4}\tau_3 e^{-\frac{i}{2}\theta^3\tau_3} \underline{\left[\tau_3, e^{i\theta^2\tau_2}\right]} e^{\frac{i}{2}\theta^3\tau_3} \\ &\qquad = [\tau_3, \cos\theta^2 + i\tau_2 \sin\theta^2] \\ &\qquad = i[\tau_3, \tau_2]\sin\theta^2 \\ &\qquad = 2\tau_1 \sin\theta^2 \end{aligned}$$

$$= -\tfrac{1}{2}\tau_3\, e^{-\frac{i}{2}\theta^3\tau_3}\,\tau_1\, e^{\frac{i}{2}\theta^3\tau_3}\,\sin\theta^2$$

$$\underset{\{\tau_3,\tau_1\}=0}{=} -\tfrac{1}{2}\tau_3\tau_1\, e^{i\theta^3\tau_3}\,\sin\theta^2$$

$$= -\tfrac{i}{2}\tau_2\,\sin\theta^2\, e^{i\theta^3\tau_3} \tag{11.25}$$

We also need L_2, which follows from

$$\partial_2 U = \tfrac{i}{2}\, e^{\frac{i}{2}\theta^1\tau_3}\, e^{\frac{i}{2}\theta^2\tau_2}\,\tau_2\, e^{\frac{i}{2}\theta^3\tau_3}$$

$$= \tfrac{i}{2}\, e^{\frac{i}{2}\theta^1\tau_3}\, e^{\frac{i}{2}\theta^2\tau_2}\, e^{-\frac{i}{2}\theta^3\tau_3}\,\tau_2$$

$$\Rightarrow L_2 \equiv U^{-1}\partial_2 U$$

$$= \tfrac{i}{2}\, e^{-\frac{i}{2}\theta^3\tau_3}\,\underbrace{e^{-\frac{i}{2}\theta^2\tau_2}\, e^{-\frac{i}{2}\theta^1\tau_3}\, e^{\frac{i}{2}\theta^1\tau_3}\, e^{\frac{i}{2}\theta^2\tau_2}}_{=1}\, e^{-\frac{i}{2}\theta^3\tau_3}\,\tau_2$$

$$= \tfrac{i}{2}\, e^{-i\theta^3\tau_3}\,\tau_2 \tag{11.26}$$

Finally, we can compute the commutator

$$[L_3, L_1]L_2 = (-\tfrac{i}{2})\,\tfrac{i}{2}\,\tau_2\,\sin\theta^2\, e^{i\theta^3\tau_3}\, e^{-i\theta^3\tau_3}\,\tau_2$$

$$= \tfrac{1}{4}\, I_2\,\sin\theta^2$$

so that $$\mathrm{tr}\,[L_3, L_1]L_2 = \tfrac{1}{2}\sin\theta^2 \tag{11.27}$$

Altogether we obtain for the topological charge

$$B = -\frac{6}{48\pi^2}\, n \cdot 2 \cdot \tfrac{1}{2}\,\underbrace{\int_0^{2\pi} d\theta^1 \int_0^{2\pi} d\theta^3}_{=(2\pi)^2}\,\underbrace{\int_0^{\pi} d\theta^2\,\sin\theta^2}_{=-\cos\theta^2|_0^\pi = 2}$$

where the extra factor of 2 reminds us that the Euler angles parametrize proper rotations only.

Here, then, is the end of the proof:

$$B = (-n)\,\frac{6}{48}\,2\cdot\frac{1}{2}\cdot\frac{1}{\pi^2}\,4\pi^2\cdot 2 = -n \in \mathbb{Z} \qquad \text{q.e.d.} \tag{11.28}$$

We leave it as an exercise to show that

$$B[U^+] = -B[U] \tag{11.29}$$

Now we come to Skyrme's chiral Lagrangian. It is composed of two parts

$$\mathcal{L} = -\frac{F_\pi^2}{16}\,tr\,(L^\mu L_\mu) + \frac{1}{32e^2}\,tr\,[L_\mu, L_\nu]^2 \tag{11.30}$$

e is a dimensionless coupling

Again, we introduced

$$L_\mu \equiv U^+ \partial_\mu U$$

so that $tr\,(L^\mu L_\mu) = tr\,(U^+ \partial^\mu U\, U^+ \partial_\mu U)$

$$= tr\,(\partial^\mu U \underbrace{U^+ \partial_\mu U\, U^+}_{-\partial_\mu U^+}) = -\,tr\,(\partial_\mu U\, \partial^\mu U^+) \tag{11.31}$$

There exist several other equivalent forms for $\mathcal{L}$ which can be computed rather easily with the aid of (11.31):

$$\mathcal{L} = -\frac{F_\pi^2}{16} tr[(U^+\partial_\mu U)(U^+\partial^\mu U)] + \frac{1}{32e^2} tr[U^+\partial_\mu U, U^+\partial_\nu U]^2$$

$$= +\frac{F_\pi^2}{16} tr[\partial_\mu U \partial^\mu U^+] - \frac{1}{16e^2} tr(\partial^\mu U^+ \partial_\mu U \partial^\nu U^+ \partial_\nu U - \partial^\mu U^+ \partial^\nu U \partial_\mu U^+ \partial_\nu U) \quad (11.32)$$

(11.33)

$$= +\frac{F_\pi^2}{16} tr[\partial_\mu U \partial^\mu U^+] + \frac{1}{32e^2} tr[(\partial_\mu U)U^+, (\partial_\nu U)U^+]^2$$

(11.34)

For static fields, $\partial_0 U(x) \equiv \partial_0 U(\vec{x}) = 0$, eq. (11.32) reduces to

$$\mathcal{L} = -\frac{F_\pi^2}{16} tr(L^i L_i) + \frac{1}{32e^2} tr[L_i, L_j]^2$$

$$= +\frac{F_\pi^2}{16} tr(L_i L_i) + \frac{1}{32e^2} tr[L_i, L_j]^2$$

$g = (+---)$ (11.35)

Recalling that the static limit of H = $\sum p_i \dot{q}_i - L$ is given by H = -L, we obtain for the static Hamiltonian

$$\mathcal{H} = -\frac{F_\pi^2}{16} tr(L_i L_i) - \frac{1}{32e^2} tr[L_i, L_j]^2 \quad (11.36)$$

and for the soliton mass

$$M = \int d^3x \, \mathcal{H} \quad (11.37)$$

Beside various other reasons for the choice of the commutator quartic term in (11.36), we will now demonstrate that for the particular form of $L_\mu = U^+\partial_\mu U$, the soliton mass is bounded from below by the topological charge B.

Starting with $L_\mu = U^+\partial_\mu U$, we obtain for the adjoint

$$L_\mu^+ = (U^+\partial_\mu U)^+ = \partial_\mu U^+ U = -U^+ \partial_\mu U\, U^+ U$$
$$= -U^+\partial_\mu U = -L_\mu \tag{11.38}$$

which says that L_μ is antihermitean.

Let us recall that $L_\mu = U^{-1}\partial_\mu U$ is a 2×2 matrix. Hence, expanding L in $i\tau^a$ gives (the 2×2 unit matrix can be excluded from the basis, since tr $L_\mu = 0$)

$$L_\mu = L_\mu^a \, i\tau^a \tag{11.39}$$

with
$$L_\mu^a = \frac{1}{2i}\, tr\,(\tau^a L_\mu)$$

$$= \frac{1}{2i}\, tr\,(\tau^a U^+\partial_\mu U) \in \mathbb{R} \tag{11.40}$$

which checks:

$$\frac{1}{2i}\, tr(\tau^b L_\mu) = \frac{1}{2i}\, iL_\mu^a \underbrace{tr(\tau^b\tau^a)}_{=2\delta^{ba}} = L_\mu^b$$

Next we want to reexpress (11.36) in terms of the components L_i^a.

$$tr(L_iL_i) = i^2 L_i^a L_i^b\, tr(\tau^a\tau^b) = -2 L_i^a L_i^a \tag{11.41}$$

$$tr\,[L_i,L_j]^2 = tr\left[iL_i^a\tau^a,\, iL_j^b\tau^b\right]^2$$
$$= (-i)^2\, tr\Big(L_i^a L_j^b \underbrace{[\tau^a,\tau^b]}_{=2i\varepsilon^{abc}\tau^c}\Big)^2$$

$$= -4 L_i^a L_j^b L_i^{\bar a} L_j^{\bar b} \underbrace{\varepsilon^{abc} \varepsilon^{\bar a \bar b \bar c}}_{= \delta^{a\bar a}\delta^{b\bar b} - \delta^{a\bar b}\delta^{b\bar a}} \underbrace{tr(\tau^c \tau^{\bar c})}_{= 2\delta^{c\bar c}}$$

$$= -8 (L_i^a L_i^a L_j^b L_j^b - L_i^a L_j^a L_i^b L_j^b)$$

$$= -4 \, {}^*L_i^a \, {}^*L_i^a \tag{11.42}$$

with

$${}^*L_i^a := \varepsilon_{ijk} \varepsilon^{abc} L_j^b L_k^c \tag{11.43}$$

With the results (11.41) and (11.42), we can recast (11.37) into

$$M = \int d^3x \, \mathcal{H}$$

$$= \int d^3x \left[\frac{F_\pi^2}{8} L_i^a L_i^a + \frac{1}{8e^2} {}^*L_i^a \, {}^*L_i^a \right] (\geq 0)$$

$$= \underbrace{\int d^3x \left(\frac{F_\pi}{\sqrt{8}} L_i^a \mp \frac{1}{e\sqrt{8}} {}^*L_i^a \right)^2}_{\geq 0} \pm \frac{F_\pi}{4e} \int d^3x \, L_i^a \, {}^*L_i^a \tag{11.44}$$

Consequently,

$$M \geq \pm \frac{F_\pi}{4e} \int d^3x \, L_i^a \, {}^*L_i^a$$

or

$$M \geq \frac{F_\pi}{4e} \left| \int d^3x \, L_i^a \, {}^*L_i^a \right| \tag{11.45}$$

A relation of M with respect to the topological charge B follows from rewriting B in the form

$$B \equiv -\frac{1}{24\pi^2} \varepsilon^{ijk} \int d^3x \, tr(L_i L_j L_k)$$

$$= -\frac{1}{24\pi^2} \int d^3x \, \varepsilon^{ijk} \, tr(L_i^a L_j^b L_k^c \, i\tau^a \, i\tau^b \, i\tau^c)$$

$$= -\frac{i^3}{24\pi^2} \int d^3x\, \varepsilon^{ijk} L_i^a L_j^b L_k^c \; \underbrace{tr(\underbrace{\tau^a \tau^b}_{= \delta^{ab} + i\varepsilon^{abd}\tau^d} \tau^c)}_{= i\varepsilon^{abd} \underbrace{tr(\tau^d \tau^c)}_{2\delta^{dc}} = 2i\,\varepsilon^{abc}}$$

$$= -\frac{2i^4}{24\pi^2} \int d^3x\, L_i^a \underbrace{\varepsilon^{ijk} \varepsilon^{abc} L_j^b L_k^c}_{= {}^*L^{ia} = -{}^*L_i^a}$$

Therefore, we obtain

$$B = \frac{1}{12\pi^2} \int d^3x\, L_i^a\, {}^*L_i^a \tag{11.46}$$

which, together with eq. (11.45), yields

$$M \geq \frac{F_\pi}{4e}\, 12\pi^2 |B| = \frac{3\pi^2 F_\pi}{e} |B|$$

or, finally,

$$M[U] \geq 3\pi^2 \frac{F_\pi}{e} |B[U]| \tag{11.47}$$

Indeed, then, eq. (11.47) yields a lower bound for the soliton mass in each homotopy class.

Additional References to
<u>Topological Baryon Number and the Skyrme Model</u>

N. Pak, H.C. Tze, Ann. Phys. <u>117</u>, 164 (1979)

T. H. R. Skyrme, Proc. Roy. Soc. <u>A260</u>, 127 (1960)

T.H. R. Skyrme, J. Math. Phys. <u>12</u>, 1735 (1971)

G.S. Adkins, C.R. Nappi, E. Witten, Nucl. Phys. <u>B228</u>, 552 (1983)

12. The Atiyah-Singer Index Theorem (for the Euclidean Dirac Operator)

First we give a "physicist's proof" and then a more rigorous derivation of an important theorem that provides an understanding of the axial vector anomaly in purely topological terms.

Let $A^a_\mu(x)$ be a background field. Furthermore, let $n_\pm[A^a_\mu]$ be the number of eigenvectors of the Euclidean Dirac operator $\not{D}(A^a)$ with vanishing eigenvalues of positive and negative chirality, respectively, i.e., with eigenvalues $\gamma_5' = \pm 1$. The Pontryagin index ν is denoted by

$$\nu[A^a_\mu] = \frac{1}{16\pi^2}\int d^4x \; tr\,({}^*F_{\mu\nu}F^{\mu\nu}) \tag{12.1}$$

Then it holds that

$$n_+ - n_- = \nu , \tag{12.2}$$

i.e., the number of positive chirality minus the number of negative chirality zero-modes equals the topological charge.

To give a heuristic proof of this statement, let us start with

$$\begin{aligned} &\not{D}\varphi_n(x) = \lambda_n \varphi_n(x) \\ &\not{D}(\gamma_5\varphi_n)(x) = -\lambda_n(\gamma_5\varphi_n)(x) \quad \text{since} \quad \{\gamma_\mu, \gamma_5\} = 0 \end{aligned} \tag{12.3}$$

and consider the two different possibilities

(a) $\underline{\lambda_n \neq 0}$: In this case, (12.3) tells us that with λ_n also $-\lambda_n$ is an eigenvalue of $\not{D}$; and since $\not{D}^+ = \not{D}$, the associated eigenfunctions are thus orthogonal: $\varphi_n \perp \gamma_5\varphi_n$.

(b) $\underline{\lambda_n = 0}$: Here, (12.3) reduces to

$$\not{D}\varphi_n^0 = 0 , \quad \not{D}(\gamma_5\varphi_n^0) = 0 \tag{12.4}$$

so that

$$\not{D}\,(1\pm\gamma_5)\,\varphi_n^0 = 0 \tag{12.5}$$

We know that zero mode eigenfunctions are also γ_5 - eigenfunctions:

$$\gamma_5\,\varphi_n^0 = \chi_n\,\varphi_n^0 \tag{12.6}$$

with

$$\chi_n = \pm 1$$

since (12.7)

$$\gamma_5\,\varphi_\pm \equiv \gamma_5\,\tfrac{1}{2}(1\pm\gamma_5)\varphi = \tfrac{1}{2}(\gamma_5\pm 1)\varphi = \pm\tfrac{1}{2}(1\pm\gamma_5)\,\varphi = \pm\,\varphi_\pm$$

If we now look at (cf. eq. (7.7))

$$B(x) \equiv \sum_n \varphi_n^+(x)\,\gamma_5\,\varphi_n(x) = \frac{1}{16\pi^2}\,tr\left({}^*F_{\mu\nu}\,F^{\mu\nu}\right) \tag{12.8}$$

and integrate over $\int d^4x$, we obtain

$$\nu[A] = \frac{1}{16\pi^2}\int d^4x\; tr\left({}^*F_{\mu\nu}\,F^{\mu\nu}\right)$$

$$= \sum_n \int d^4x\; \underbrace{\varphi_n^+(x)}\;\underbrace{\gamma_5\,\varphi_n(x)}_{\text{orthogonal for } \lambda_n \neq 0}$$

$$= \sum_{\substack{n \\ \lambda_n=0}} \int d^4x\; \varphi_n^+(x)\; \underbrace{\gamma_5\,\varphi_n(x)}_{=\chi_n\varphi_n(x)}$$

$$= \sum_{\substack{n \\ \lambda_n=0}} \chi_n\; \underbrace{\int d^4x\; \varphi_n^+(x)\,\varphi_n(x)}_{=1}$$

This leaves us with

$$\nu[A] = n_+[A] - n_-[A] \tag{12.9}$$

Now we turn to a more rigorous proof and start out with a few definitions.

In general, the index of an operator A is defined by

$$\text{index } A = \dim \ker A - \dim \ker A^{+} \equiv \ell(A) - \ell(A^{+}) \tag{12.10}$$

where the kernel of A is the space of solution of equation $Af = 0$.

So we have

$$\begin{aligned} \ker A &= \{ f \mid Af = 0 \} \\ \ker A^{+} &= \{ g \mid A^{+} g = 0 \} \end{aligned} \tag{12.11}$$

$\ell(A)$ denotes the dimension of the kernel of A, i.e., the number of zero-modes of the operator A. Obviously, (12.11) implies

$$A^{+} A f = 0 \,, \quad A A^{+} g = 0 \tag{12.12}$$

and it can be shown that

$$\ell(A) = \ell(A^{+} A) \tag{12.13}$$

and

$$\ell(A^{+}) = \ell(A A^{+}) \tag{12.14}$$

One can easily show that the operators $A^{+}A$ and AA^{+} are positive semi-definite (spectrum $A^{+}A$, $AA^{+} \geqslant 0$) and, furthermore, that--for non-zero modes--they consist of identical spectra: spectrum$[A^{+}A]$ = spectrum $[AA^{+}]$.

In terms of (12.13, 14), we can rewrite equation (12.10) as

$$\text{index } A = \ell(A^{+}A) - \ell(AA^{+}) \tag{12.15}$$

Now it becomes absolutely necessary to specify the domain in which the various operators act in order to set up an eigenvalue problem.

Let $S = \{ \text{spinor fields } \psi \text{ on } E^4 \}$ (E^4 = Euclidean four-space) (12.16)

We may then split the space of the Dirac spinors into two eigenspaces of chirality $\pm$:

$$\gamma_5 \psi_{\pm} = \pm \psi_{\pm}$$

$$S_{\pm} = \{ \psi \in S \mid P_{\pm} \psi = \psi \} \,, \quad P_{\pm} = \tfrac{1}{2}(1 \pm \gamma_5) \tag{12.17}$$

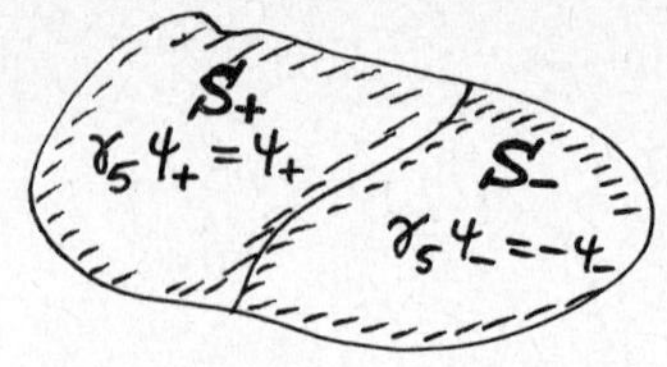

Note, however, that the Dirac operator $\not{D}$ acts on the spinor space S_+ of positive chirality, while the image lies in the spinor space S_- of negative chirality:

$$\not{D}\psi_\pm \in S_\mp \tag{12.18}$$

since

$$P_\pm(\not{D}\psi_\pm) = \not{D} P_\mp \psi_\pm = 0 \tag{12.19}$$

$$P_\mp(\not{D}\psi_\pm) = \not{D} P_\pm \psi_\pm = \not{D}\psi_\pm \tag{12.20}$$

If we then denote with $A := \not{D}|_{S_+}$, the operator $\not{D}$ acting on spinors $\psi \in S_+$, we have

$$A := \not{D}|_{S_+} \quad : S_+ \rightarrow S_-, \ \psi_+ \mapsto A\psi_+ \tag{12.21}$$

$$A^+ = \not{D}|_{S_-} \quad : S_- \rightarrow S_+, \ \psi_- \mapsto A\psi_- \tag{12.22}$$

Hence, A and A^+ do not have eigenvalue problems, but A^+A and AA^+ do:

$$A^+A = \not{D}^2 P_+ \quad : \quad S_+ \longrightarrow S_+ \tag{12.23}$$

$$A\,A^+ = \not{D}^2 P_- \quad : \quad S_- \longrightarrow S_- \tag{12.24}$$

The following picture should help to illustrate the various domains on which the operators act:

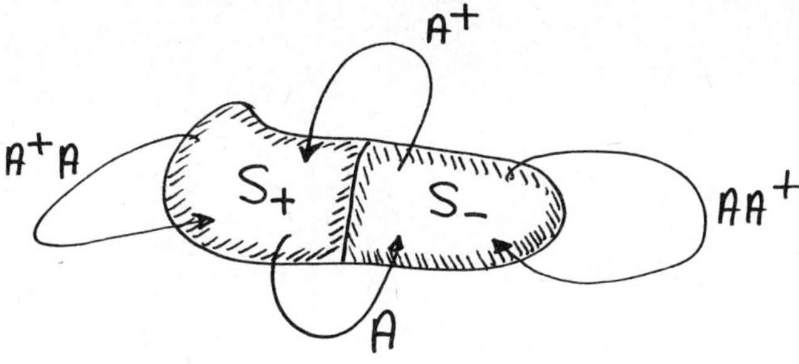

The eigenvalue problem for $A^{+}A$ is given by

$$A^{+}A\varphi_n = \lambda_n \varphi_n \;, \quad \lambda_n \neq 0 \tag{12.25}$$

$$A^{+}A\varphi_n^{o} = 0 \;, \quad \varphi_n^{o} \in \mathrm{Ker}\, A$$

and likewise for AA^{+}:

$$AA^{+}\phi_n = \lambda_n \phi_n \;, \quad \lambda_n \neq 0 \tag{12.26}$$

$$AA^{+}\phi_n^{o} = 0 \;, \quad \phi_n^{o} \in \mathrm{Ker}\, A^{+}$$

The number of non-zero mode eigenvalues λ_n in (12.25) and (12.26) is identical.

Now it is important to realize that the following traces are taken in different spaces:

$$\mathrm{Tr}\,[e^{-tA^{+}A}] - \mathrm{Tr}\,[e^{-tAA^{+}}] \tag{12.27}$$

$$= \sum_n \langle \varphi_n | e^{-tA^{+}A} | \varphi_n \rangle - \sum_n \langle \phi_n | e^{-tAA^{+}} | \phi_n \rangle$$

$$= \sum_{\substack{n \\ \lambda_n \neq 0}} e^{-t\lambda_n} + \underbrace{\sum_{\substack{n=1 \\ \lambda_n = 0}}^{\ell(A^{+}A)} \langle \varphi_n^{o} | e^{-tA^{+}A} | \varphi_n^{o} \rangle}_{= \sum \langle \varphi_n^{o} | \varphi_n^{o} \rangle = \sum 1 = \ell(A^{+}A)}$$

$$- \sum_{\substack{n \\ \lambda_n \neq 0}} e^{-t\lambda_n} - \underbrace{\sum_{\substack{n=1 \\ \lambda_n = 0}}^{\ell(AA^{+})} \langle \phi_n^{o} | e^{-tAA^{+}} | \phi_n^{o} \rangle}_{= \ell(A^{+}A)}$$

$$= \ell(A^+A) - \ell(AA^+) \tag{12.28}$$

So we have found

$$\text{index } A = \mathrm{Tr}\, e^{-tA^+A} - \mathrm{Tr}\, e^{-tAA^+}, \quad \forall t \neq 0 \tag{12.29}$$

The statement (12.29) holds also true if we replace exp $\to f$ (t) with $f(0)=1$ f (t) analytical.

Now we want to repeat everything for our special Dirac operator $\not{D}$:

$$\text{index } \not{D}|_{S_+} = \mathrm{Tr}_{S_+}[e^{-t\not{D}^2}] - \mathrm{Tr}_{S_-}[e^{-t\not{D}^2}] \tag{12.30}$$

where Tr_{S_+} denotes summation over eigenvalues of eigenspinors with positive chirality only. We continue to write

$$\begin{aligned}\text{index } \not{D}|_{S_+} &= \mathrm{Tr}_{S=S_+\cup S_-}[e^{-t\not{D}^2}P_+ - e^{-t\not{D}^2}P_-] \\ &= \mathrm{Tr}_S[(P_+ - P_-)e^{-t\not{D}^2}] \\ &= \mathrm{Tr}_S[\gamma_5 e^{-t\not{D}^2}]\end{aligned} \tag{12.31}$$

which again is true if we replace exp $\to f$ with $f(0) = 1$.

The choice f = exp and $t \to 0$ yields Fujikawa's result:

$$\begin{aligned}\text{index } \not{D}|_{S_+} &= \lim_{t\to 0} \mathrm{Tr}[\gamma_5 e^{-t\not{D}^2}] \\ &= \lim_{t\to 0} \sum_n \int d^4x\, \varphi_n^+ \gamma_5 e^{-t\not{D}^2} \underbrace{\varphi_n}_{\text{arb. orthon. system on } E^4}\end{aligned}$$

$$\equiv \lim_{M\to\infty} \sum_n \int d^4x \, \varphi_n^+ \gamma_5 \, e^{-\frac{\not{D}^2}{M^2}} \varphi_n$$

following Fujikawa (chapter 7)

$$= q[A_\mu^a] = \text{Pontryagin index of } A_\mu^a(x) \tag{12.32}$$

On the other hand, we have found

$$\begin{aligned} \dim\ker A^+A &= \dim\ker \not{D}^2|_{S_+} \\ &= n_+[A_\mu^a] \\ &= \#\text{ of zero-modes with positive chirality} \end{aligned} \tag{12.33}$$

$$\begin{aligned} \dim\ker AA^+ &= \dim\ker \not{D}^2|_{S_-} \\ &= n_-[A_\mu^a] \\ &= \#\text{ of zero-modes with negative chirality} \end{aligned} \tag{12.34}$$

so that our final result reads

$$\text{index } \not{D}|_{S_+} = n_+[A_\mu^a] - n_-[A_\mu^a] = q[A_\mu^a] \tag{12.35}$$

where the essence of the equation is contained in the equal sign on the right-hand side.

Additional References to

The Atiyah-Singer Index Theorem (for the Euclidean Dirac Operator)

M. Atiyah, I. Singer, Ann. Math. 87, 484 (1968)

A.S. Schwarz, Commun. Math. Phys. 64, 233 (1979)

J.S. Dowker, J. Phys. A11, 347 (1978)

T. Eguchi, P. Gilkey, A. Hanson, Phys. Rep. 66, 213 (1980)

K. Fujikawa, Phys. Rev. Lett. 42, 1195 (1979); Phys. Rev. D21, 2848 (1980)

H. Huang, "Quarks, Leptons and Gauge Fields", World Scientific, Singapore (1982)

ζ-Function Method:

M. Reuter, Phys. Rev. D31, 1374 (1985)

13. Chiral Anomalies and Differential Geometry

The purpose of this chapter is to demonstrate how chiral anomalies, abelian and non-abelian, can be determined by differential geometric methods. In particular, we are going to set up integral representations of abelian and non-abelian anomalies and show that they are always total divergences. To carry out the differential geometric procedure, we will make substantial use of the language of differential forms. The gauge field in Yang-Mills theory is a 1-form

$$\overset{1}{A} = A_\mu dx^\mu \tag{13.1}$$

Since $A_\mu = -iA^a_\mu \lambda^a$, the gauge field is also a matrix. The field strength is a 2-form

$$\overset{2}{F} = \underset{\substack{\uparrow \\ \text{Maxwell}}}{\overset{\substack{\text{ext. derivative} \\ \downarrow}}{dA}} + \underset{\substack{\uparrow \\ \text{non-linear} \\ \text{Yang-Mills}}}{A^2} \tag{13.2}$$

Explicitly:

$$\begin{aligned} d\overset{1}{A} &= d(A_\nu dx^\nu) \\ &= \partial_\mu A_\nu \, dx^\mu \wedge dx^\nu \\ &= \tfrac{1}{2} \partial_\mu A_\nu \, dx^\mu \wedge dx^\nu + \tfrac{1}{2} \partial_\nu A_\mu \, dx^\nu \wedge dx^\mu \\ &= \tfrac{1}{2} (\partial_\mu A_\nu - \partial_\nu A_\mu) \, dx^\mu \wedge dx^\nu \end{aligned} \tag{13.3}$$

$$\begin{aligned} A^2 &= A \wedge A = A_\mu dx^\mu \wedge A_\nu dx^\nu \\ &= A_\mu A_\nu \, dx^\mu \wedge dx^\nu \\ &= \tfrac{1}{2} (A_\mu A_\nu - A_\nu A_\mu) \, dx^\mu \wedge dx^\nu \\ &\equiv \tfrac{1}{2} [A_\mu, A_\nu] \, dx^\mu \wedge dx^\nu \end{aligned} \tag{13.4}$$

Hence, writing F out in longhand, we find instead of (13.2)

$$\mathcal{F} = (\partial_\mu A_\nu - \partial_\nu A_\mu + [A_\mu, A_\nu])\,\tfrac{1}{2}\,dx^\mu \wedge dx^\nu$$

$$\equiv \tfrac{1}{2}\,\mathcal{F}_{\mu\nu}\,dx^\mu \wedge dx^\nu \qquad (13.5)$$

Note again that $F_{\mu\nu}$ is a matrix-valued function: $F_{\mu\nu} = -iF^a_{\mu\nu}\lambda^a$.

As another example of the use of forms, we derive Bianchi's identity:

From

$$\begin{aligned} dF &= d\,(dA + A^2) \\ &= d(dA) + d(A^2) = \underbrace{d^2A}_{\overset{(+)}{=}0} + \underbrace{d(AA)}_{\substack{= dA\,A + (-1)^1 A\,dA \\ = dA\,A - A\,dA}} \end{aligned}$$

$$\begin{aligned} (+):\ d^2A &= d\,(\partial_\mu A_\nu - \partial_\nu A_\mu)\,\tfrac{1}{2}\,dx^\mu \wedge dx^\nu \\ &= \underbrace{\partial_\lambda(\partial_\mu A_\nu - \partial_\nu A_\mu)}_{\text{symm.}}\,\tfrac{1}{2}\,\underbrace{dx^\lambda \wedge dx^\mu \wedge dx^\nu}_{\text{ant-symm.}} = 0 \end{aligned}$$

we obtain $\qquad dF = dAA - AdA \qquad (13.6)$

The two terms on the right-hand side of (13.6) can also be written as

$$\begin{aligned} -[A, \mathcal{F}] &= -[A, dA + A^2] \\ &= -\,AdA - \cancel{AA^2} + dA\,A + \cancel{A^2A} \end{aligned}$$

so that we have $\qquad dF = -[A, F]$

or

$$DF \equiv dF + [A, F] = 0 \qquad (13.7)$$

Often we will suppress wedge symbols, e.g.,

$$\begin{aligned} dA\,A \equiv dA \wedge A &= \partial_\mu A_\nu\,dx^\mu \wedge dx^\nu \wedge A_\lambda\,dx^\lambda \\ &\equiv \partial_\mu A_\nu A_\lambda\,dx^\mu dx^\nu dx^\lambda \end{aligned}$$

The Bianchi-identity is essentially a consequence of

$$d^2 = 0 \tag{13.8}$$

and the fact that d anticommutes with the one-form A. The operation D in (13.7) is the covariant differential.

Another simple exercise is given by

$$d\{\mathrm{tr}\, F^2\} = \mathrm{tr}\left(dF\,F + (-1)^2 F\,dF\right) = 2\,\mathrm{tr}\,(dF\,F)$$

$$\underset{(7)}{=} -2\,\mathrm{tr}\left([A,F]\,F\right)$$

$$= -2\,\mathrm{tr}\left(AF\,F - FAF\right)$$

$$= 0 \tag{13.9}$$

This relation holds in any dimension. In four dimensions, equation (13.9) is evident: F is a 2-form; therefore, F^2 is a 4-form and so is tr F^2. However, the exterior derivative of a 4-form in 4 dimensions is zero:

$$\omega_4 := \frac{1}{4!}\,\omega_{\mu_1 \dots \mu_4}\, dx^{\mu_1} \wedge \dots \wedge dx^{\mu_4}$$

$$d\omega_4 = \frac{1}{4!}\,\partial_\lambda \omega_{\mu_1 \dots \mu_4}\, dx^\lambda \wedge dx^{\mu_1} \wedge \dots \wedge dx^{\mu_4}$$

since the wedge product contains two differentials which are equal.

Let us consider an n-dimensional space. If a p-form is given by

$$\omega = \frac{1}{p!}\, F_{\mu_1 \mu_2 \dots \mu_p}\, dx^{\mu_1} \wedge \dots \wedge dx^{\mu_p} \tag{13.10}$$

then we define its conjugate (or dual) as

$$*\omega = \frac{1}{(n-p)!} \tilde{F}_{\nu_1 \dots \nu_{n-p}}(x)\, dx^{\nu_1} \wedge \dots \wedge dx^{\nu_{n-p}} \tag{13.11}$$

where (13.12)

$$\tilde{F}_{\nu_1 \dots \nu_{n-p}} = \frac{1}{p!} \varepsilon_{\nu_1 \dots \nu_{n-p} \mu_1 \dots \mu_p} F^{\mu_1 \dots \mu_p}$$

and

$$\varepsilon_{\mu_1 \dots \mu_p} = \sqrt{g(x)}\, \eta_{\mu_1 \dots \mu_p}$$

$$F^{\mu_1 \dots \mu_p} = g^{\mu_1 \nu_1} \dots g^{\mu_p \nu_p} F_{\nu_1 \nu_2 \dots \nu_p}$$

In order to show that these definitions make sense, let us have a closer look at the dual of the current (n = 4)

$$\overset{1}{J} = J_\mu\, dx^\mu$$

$$* \overset{1}{J} = J_\mu * dx^\mu \tag{13.13}$$

The formal definition of the dual in (13.11) is a consequence of the $*$-operator being defined as a mapping (Λ^p = set of p-forms)

$$*: \Lambda^p \to \Lambda^{n-p}, \quad v \to *v, \quad v \wedge w = (*v, w)e \quad \forall w \in \Lambda^{n-p} \tag{13.14}$$

where the scalar product for $v = v_1 \wedge v_2 \dots \wedge v_p$ and $w = w_1 \wedge w_2 \wedge \dots \wedge w_p$ is given by

$$(v, w) = (v_1 \wedge v_2 \wedge \dots \wedge v_p, w_1 \wedge \dots \wedge w_p) = \det\{(v_i, w_j)\}$$

and we have set (13.15)

$$e := dx^1 \wedge dx^2 \wedge \dots \wedge dx^n \in \Lambda^n$$

In equation (13.13), we need $*dx^\mu$, i.e., the dual of dx^μ:

$$dx^\mu \wedge \omega = (*dx^\mu, \omega)\, e \tag{13.16}$$

Let us write the 3-form ω as

$$\omega = \omega_{\alpha\beta\gamma}\, dx^\alpha dx^\beta dx^\gamma \tag{13.17}$$

Then (13.16) is given by

$$\underline{dx^\mu \wedge dx^\alpha \wedge dx^\beta \wedge dx^\gamma}\, \omega_{\alpha\beta\gamma} = \omega_{\alpha\beta\gamma}\, (*dx^\mu, dx^\alpha \wedge dx^\beta \wedge dx^\gamma)\, e$$

$$= \varepsilon^{\mu\alpha\beta\gamma}\, dx^1 \wedge dx^2 \wedge dx^3 \wedge dx^4$$

$$= \varepsilon^{\mu\alpha\beta\gamma}\, e$$

so that $$\varepsilon^{\mu\alpha\beta\gamma} = (*dx^\mu, dx^\alpha \wedge dx^\beta \wedge dx^\gamma) \tag{13.18}$$

The solution with respect to $*dx^\mu$ is given by

$$*dx^\mu = \frac{1}{3!}\, \varepsilon^\mu{}_{\rho\sigma\tau}\, dx^\rho dx^\sigma dx^\tau \tag{13.19}$$

Let us prove (13.19) by verifying (13.18).

$$(*dx^\mu, dx^\alpha dx^\beta dx^\gamma) = \frac{1}{3!}\, \varepsilon^\mu{}_{\rho\sigma\tau}\, (dx^\rho \wedge dx^\sigma \wedge dx^\tau, dx^\alpha \wedge dx^\beta \wedge dx^\gamma)$$

$$= \frac{1}{3!}\, \varepsilon^\mu{}_{\rho\sigma\tau} \begin{vmatrix} (dx^\rho, dx^\alpha) & & \\ & \ddots & \\ & & (dx^\tau, dx^\gamma) \end{vmatrix} = \frac{1}{6}\, \varepsilon^\mu{}_{\rho\sigma\tau} \begin{vmatrix} g^{\rho\alpha} & g^{\rho\beta} & g^{\rho\gamma} \\ g^{\sigma\alpha} & g^{\sigma\beta} & g^{\sigma\gamma} \\ g^{\tau\alpha} & g^{\tau\beta} & g^{\tau\gamma} \end{vmatrix}$$

$$= \frac{1}{6}\, \varepsilon^\mu{}_{\rho\sigma\tau} \big(g^{\rho\alpha} g^{\sigma\beta} g^{\tau\gamma} - g^{\rho\alpha} g^{\sigma\gamma} g^{\tau\beta} - g^{\sigma\alpha} g^{\rho\beta} g^{\tau\gamma} + g^{\sigma\alpha} g^{\rho\gamma} g^{\tau\beta}$$
$$+ g^{\tau\alpha} g^{\rho\beta} g^{\sigma\gamma} - g^{\tau\alpha} g^{\rho\gamma} g^{\sigma\beta} \big)$$

$$= \frac{1}{6}\left(\varepsilon^{\mu\alpha\beta\gamma} - \varepsilon^{\mu\alpha\gamma\beta} - \varepsilon^{\mu\beta\alpha\gamma} + \varepsilon^{\mu\gamma\alpha\beta} + \varepsilon^{\mu\beta\gamma\alpha} - \varepsilon^{\mu\gamma\beta\alpha}\right)$$

$$= \varepsilon^{\mu\alpha\beta\gamma} \quad \checkmark$$

Returning to equation (13.13), we find for the dual of J (cf. equations (13.11, 13.12):

$$* J \underset{(19)}{=} J_\mu \frac{1}{3!} \varepsilon^\mu{}_{\rho\sigma\tau} \, dx^\rho dx^\sigma dx^\tau$$

$$= \frac{1}{3!} J^\mu \varepsilon_{\mu\rho\sigma\tau} \, dx^\rho dx^\sigma dx^\tau \qquad (13.20)$$

The components of the 3-form $*J$ are $(*J)_{\rho\sigma\tau} = \frac{1}{3!} J^\mu \varepsilon_{\mu\rho\sigma\tau}$ (13.21)

Taking the exterior derivative of (13.20), we obtain

$$d*J = d\left(\frac{1}{3!} J^\mu \varepsilon_{\mu\rho\sigma\tau} \, dx^\rho dx^\sigma dx^\tau\right)$$

$$= \frac{1}{3!} \partial_\alpha J^\mu \, \varepsilon_{\mu\rho\sigma\tau} \underbrace{dx^\alpha dx^\rho dx^\sigma dx^\tau}_{= \varepsilon^{\alpha\rho\sigma\tau} dx^1 dx^2 dx^3 dx^4 = \varepsilon^{\alpha\rho\sigma\tau} e}$$

$$= \frac{1}{3!} \partial_\alpha J^\mu \underbrace{\varepsilon_{\mu\rho\sigma\tau} \varepsilon^{\alpha\rho\sigma\tau}}_{= -6 g^\alpha_\mu} e$$

$$d*J = -\partial_\mu J^\mu \, e \qquad (13.22)$$

The 4-form e is given by

$$e = -\frac{1}{4!}\,\varepsilon_{\mu\nu\rho\sigma}\,dx^\mu dx^\nu dx^\rho dx^\sigma \tag{13.23}$$

To verify this statement, we continue to write

$$e \overset{?}{=} -\underbrace{\frac{1}{4!}\varepsilon_{\mu\nu\rho\sigma}\,\varepsilon^{\mu\nu\rho\sigma}}_{=-24}\ \underbrace{dx^1dx^2dx^3dx^4}_{=e}$$

$$\overset{!}{=} e$$

Employing (13.23) in (13.22), we end up with

$$d{*}J = (\partial_\lambda J^\lambda)\,\frac{1}{4!}\,\varepsilon_{\mu\nu\rho\sigma}\,dx^\mu dx^\nu dx^\rho dx^\sigma \tag{13.24}$$

In D dimensions this reads

$$d{*}J = d\left(\varepsilon_{\nu\mu_1\cdots\mu_{D-1}}\,J^\nu\,\frac{1}{(D-1)!}\,dx^{\mu_1}\cdots dx^{\mu_{D-1}}\right)$$

$$= (\partial_\lambda J^\lambda)\,\frac{1}{D!}\,\varepsilon_{\mu_1\cdots\mu_D}\,dx^{\mu_1}\wedge\ldots\wedge dx^{\mu_D} \tag{13.25}$$

In four-dimensional space-time, the $U_A(1)$ abelian anomaly is given by

$$\partial^\lambda J_\lambda^5 = -\frac{1}{16\pi^2}\,\varepsilon^{\mu\nu\rho\sigma}\,\mathrm{tr}\,(F_{\mu\nu}F_{\rho\sigma}) \tag{13.26}$$

When substituted in (13.22), we obtain for the chiral anomaly in the language of forms

$$d{*}J_5 \sim \underbrace{\varepsilon^{\mu\nu\rho\sigma}\, e}_{=dx^\mu dx^\nu dx^\rho dx^\sigma} \operatorname{tr}(F_{\mu\nu} F_{\rho\sigma})$$

$$\sim \operatorname{tr}\left(\tfrac{1}{2} F_{\mu\nu}\, dx^\mu dx^\nu \; \tfrac{1}{2} F_{\rho\sigma}\, dx^\rho dx^\sigma\right)$$

$$= \operatorname{tr} F^2 = \operatorname{tr}\left[d\left(A\,dA + \tfrac{2}{3} A^3\right)\right] \qquad (13.27)$$

where the last term makes use of the fact that the right-hand side of (13.26) is a total divergence

$$\partial^\lambda J_\lambda^5 = -\frac{1}{4\pi^2} \operatorname{tr}\left[\partial^\mu \varepsilon_{\mu\nu\rho\sigma}\left(A^\nu \partial^\rho A^\sigma + \tfrac{2}{3} A^\nu A^\rho A^\sigma\right)\right] \qquad (13.28)$$

The non-abelian anomalies are given by the right-hand side of the covariant derivative of $J^H_{\lambda a} = \bar{\psi}_H \gamma_\lambda \lambda_a \psi_H$, $H = L, R$, $\eta_L = -\eta_R = -1$

$$D^\mu_H J^H_{\mu a} = \eta_H \frac{1}{24\pi^2} \varepsilon^{\mu\nu\rho\sigma} \operatorname{tr}\left[\lambda_a \partial_\mu \left(A^H_\nu \partial_\rho A^H_\sigma + \tfrac{1}{2} A^H_\nu A^H_\rho A^H_\sigma\right)\right] \qquad (13.29)$$

Notice the numerical coefficient $\frac{2}{3}$ in equation (13.28) and $\frac{1}{2}$ in (13.29). It is then clear that the non-abelian anomaly cannot be rewritten in terms of Yang-Mill curls.

To reformulate (13.29) in terms of forms, we use equation (13.22) once more:

$H=L$:

$$(D{*}J_5)_a \sim \underbrace{\varepsilon^{\mu\nu\rho\sigma}\, e}_{=dx^\mu dx^\nu dx^\rho dx^\sigma} \operatorname{tr}\left[\lambda_a \partial_\mu \left(A_\nu \partial_\rho A_\sigma + \tfrac{1}{2} A_\nu A_\rho A_\sigma\right)\right]$$

$$= \operatorname{tr}\left[\lambda_a \underline{dx^\mu \partial_\mu}\left(\underline{A_\nu dx^\nu}\; \underline{dx^\rho \partial_\rho}\; \underline{A_\sigma dx^\sigma} + \tfrac{1}{2} \underline{A_\nu dx^\nu}\; \underline{A_\rho dx^\rho}\; \underline{A_\sigma dx^\sigma}\right)\right]$$

$$= tr[\lambda_a d(AdA + \tfrac{1}{2}A^3)]$$

$$\sim - G_a(A) \qquad (13.30)$$

Equation (13.9) says that tr F^2 is closed:

$$d\,\mathrm{tr}F^2 = 0 \qquad (13.31)$$

By Poincaré's lemma, there exists locally a 3-form (not necessarily globally) such that

$$\mathrm{tr}F^2 = d\omega_3^o \qquad (13.32)$$

Remark: Topological quantization always involves a closed form which is not globally exact.

Now we want to integrate equation (13.32), i.e., we want to find the form ω_3^o of which tr F^2 is the derivative. To this end, we study the response of tr F^2 under the variation $A \to A + \delta A$.

$$F = dA + A^2 \qquad (13.33)$$

$$\delta F = d\delta A + \delta AA + A\delta A = D(\delta A) \qquad (13.34)$$

The last equality in (13.34) follows from

$$D_\mu \delta A_\nu = \partial_\mu \delta A_\nu + [A_\mu, \delta A_\nu]$$

which we multiply by $dx^\mu dx^\nu$:

$$\underline{D_\mu \delta A_\nu dx^\mu dx^\nu} = dx^\mu \partial_\mu \delta A_\nu dx^\nu + (A_\mu \delta A_\nu - \delta A_\nu A_\mu) dx^\mu dx^\nu$$

$$\parallel$$

$$D\delta A \qquad = d\delta A + A\delta A - \underline{\delta A_\nu A_\mu dx^\mu dx^\nu}$$

$$= -\delta A_\nu dx^\nu A_\mu dx^\mu$$

$$= d\delta A + A\delta A + \delta A\ A$$

On our way to evaluating $\delta \mathrm{tr}\, F^2$, we need for $F_1, F_2 \in \Lambda^2$

$$D(F_1 F_2) = d(F_1 F_2) + [A, F_1 F_2]$$

$$= dF_1\, F_2 + F_1\, dF_2 + F_1 [A, F_2] + [A, F_1] F_2$$

$$= F_1 \{dF_2 + [A, F_2]\} + \{dF_1 + [A, F_1]\} F_2$$

$$= F_1\, DF_2 + DF_1\, F_2 \tag{13.35}$$

For $W \in \Lambda^{2n}$ we obtain (W : $DW = dW + [A, W]$) (W^a_μ: octet)

$$\mathrm{tr}(DW) = \mathrm{tr}(dW + [A, W]) = \mathrm{tr}(dW + AW - WA)$$

$$= d\, \mathrm{tr}\, W \tag{13.36}$$

Finally, we can compute ($F \in \Lambda^2$)

$$\delta\, \mathrm{tr}\, F^2 = \mathrm{tr}(\delta F\, F + F\, \delta F)$$

$$= 2\, \mathrm{tr}(\delta F\, F)$$

$$\underset{(34)}{=} 2\, \mathrm{tr}[D(\delta A) F]$$

$$= 2\, \mathrm{tr}[D(\delta A F)] - 2\, \mathrm{tr}[\delta A \underbrace{DF}_{=0}]$$

$$\underset{(36)}{=} 2d\, \mathrm{tr}[\delta A F] \tag{13.37}$$

We now introduce the variation of A via a parameter t :

$$A_t = tA \qquad \delta A = \delta t A \quad , \quad 0 \leq t \leq 1 \tag{13.38}$$

$$F_t = t\,dA + t^2A^2 \equiv t\,dA + tA^2 + t^2A^2 - tA^2$$

$$= tF + (t^2 - t)A^2 \tag{13.39}$$

Then we can rewrite equation (13.37) as

$$\delta(\mathrm{tr}\, F_t^2) = 2\,d\,\mathrm{tr}[\delta A F_t] = 2\,d\,\mathrm{tr}[A F_t]\,\delta t \tag{13.40}$$

This equation is equivalent to

$$\frac{d}{dt}\,\mathrm{tr}\, F_t^{\,2} = 2d\;\mathrm{tr}\,[A F_t]$$

$$\underset{(39)}{=} 2d\;\mathrm{tr}\,[t\,AF + (t^2 - t)A^3] \tag{13.41}$$

The integration of (13.41) is easily done:

$$\mathrm{tr}\, F^2 = 2d \int_0^1 dt\;\mathrm{tr}\,[t\,AF + (t^2 - t)A^3]$$

$$= 2d\;\mathrm{tr}\,[\tfrac{1}{2}AF + (\tfrac{1}{3} - \tfrac{1}{2})A^3]$$

$$\mathrm{tr}\, F^2 = d\;\mathrm{tr}\,[AF - \tfrac{1}{3}A^3] =: d\omega_3^0 \,, \quad \omega_3^0 \in \Lambda^3 \tag{13.42}$$

with

$$\omega_3^0 = \mathrm{tr}(AF - \tfrac{1}{3}A^3)$$

$$= \mathrm{tr}\big(A(dA + A^2) - \tfrac{1}{3}A^3\big)$$

$$= \mathrm{tr}\,(A\,dA + \tfrac{2}{3}A^3)$$

So we have verified

$$\mathrm{tr}\, F^2 = d\omega_3^0 \tag{13.43}$$

and given the explicit form for the 3-form ω_3^0:

$$\omega_3^0 = tr\left(AdA + \frac{2}{3}A^3\right) \tag{13.44}$$

which brings us back to equation (13.27).

Similarly, in D = 2n dimensions, the $U_A(1)$ anomaly is given by the 2-n form

$$\begin{aligned}\Omega_{2n}(A) = tr\,F^n &= d\omega_{2n-1}^0 \\ &= d\; n\int_0^1 dt\, t^{n-1}\, tr\, A\{dA + tA^2\}^{n-1}\end{aligned} \tag{13.45}$$

Explicitly for n = 3, D = 6:

$$\begin{aligned}\omega_5^0 &= 3\,tr\int_0^1 dt\; A\,(tdA + t^2A^2)^2 \\ &= tr\left(A\,(dA)^2 + \frac{3}{5}A^5 + \frac{3}{2}A^3 dA\right) \\ tr\,F^3 &= d\omega_5^0\end{aligned} \tag{13.46}$$

Also note that with

$$\partial_\mu J^\mu \sim \varepsilon^{\mu_1\mu_2\cdots\mu_D}\, tr\,(F_{\mu_1\mu_2}\cdots F_{\mu_{D-1}\mu_D})$$

we obtain from (13.25)

$$d{*}J \sim tr\,F^n = \Omega_{2n}(A) \tag{13.47}$$

To find the non-abelian anomalies in D = 2n dimensions, we have to find the functional $G_i[A]$ which solves the Wess-Zumino conditions (cf. chapter 9)

$$[X_i(x), X_j(y)] = f_{ijk}\, X_k(x)\, \delta(x-y) \tag{13.48}$$

$$X_i(x)\, W[A] = G_i(A(x)) \tag{13.49}$$

Recall that the gauge transformations are represented by the functional differential operators

$$X_i(x) = -\partial_\mu \frac{\delta}{\delta A^i_\mu(x)} - \left(A_\mu \times \frac{\delta}{\delta A_\mu(x)}\right)_i \tag{13.50}$$

and that their action on the functional of one-particle-irreducible Green's functions $W[A]$ just yields the current (non-) conservation equation (13.49).

Acting with (13.48) on W and using (13.49) yields

$$X_i(x)\, G_j(y) - X_j(y)\, G_i(x) = f_{ijk}\, G_k(x)\, \delta(x-y) \tag{13.51}$$

Now we introduce an infinitesimal 0-form $v(x)$, taking values in the Lie algebra; $v(x) = -i v_i(x)\, \lambda^i$. Then we integrate (13.51) over $\int d^4x\, d^4y\, u_i(x)\, v_j(y)$:

$$\left(\int dx\, u_i(x)\, X_i(x)\right)\left(\int dy\, v_j(y)\, G_j(y)\right) - \left(\int dy\, v_j(y)\, X_j(y)\right)\left(\int dx\, u_i(x)\, G_i(x)\right)$$

$$= \int dx\, \underbrace{f_{ijk}\, u_i(x)\, v_j(x)}_{=(u\times v)_k(x)}\, G_k(x) \tag{13.52}$$

Let us introduce the notation

$$G(v) := \int_{S_D} dx\, v_i(x)\, G_i(x) =: v\cdot G \tag{13.53}$$

$$\delta_v := \int_{S_D} dx\, v_i(x)\, X_i(x) =: v\cdot X \tag{13.54}$$

δ_v generates the gauge transformation with $v(x) = -i v_i(x)\lambda^i$ as the gauge function. The integration is meant to be over the compactified R^D, i.e., the D-sphere S_D, if we limit ourselves to those gauge fields whose field strengths vanish at infinity.

So far, the W.-Z. consistency conditions are given by

$$\delta_u G(v) - \delta_v G(u) = G(u \times v) \tag{13.55}$$

Next, we need to prove that $\Omega_{2n} = \mathrm{tr}F^n$ is gauge invariant. To do so, we go over to forms

$$A = -i A_\mu^k \lambda_k dx^\mu \in \Lambda^1 \tag{13.56}$$

$$v = -i v^k \lambda_k \quad \in \Lambda^0$$

The gauge transformation on A_μ,

$$\delta_v A_\mu = v \cdot X A_\mu$$

$$= -\partial_\mu v - [A_\mu, v]$$

can then be reexpressed in the form $(d = dx^\mu \partial_\mu)$

$$\delta_v A = -dv - [A, v] \tag{13.57}$$

Likewise,

$$\delta_v F = -[F, v] \tag{13.58}$$

After these preparations, we are able to prove that $\mathrm{tr}F^n$ is indeed gauge invariant

$$\begin{aligned}
\delta_v \Omega_{2n} &= \delta_v \mathrm{tr} F^n \\
&= \mathrm{tr}\left(\delta_v F F^{n-1} + F \delta_v F^{n-2} + \dots + F^{n-1} \delta_v F\right) \\
&= n\, \mathrm{tr}\left[(\delta_v F) F^{n-1}\right] \\
&= -n\, \mathrm{tr}\left[[F, v] F^{n-1}\right] \\
&= -n\, \mathrm{tr}\left(F v F^{n-1} - v F F^{n-1}\right)
\end{aligned}$$

$$\delta_v \Omega_{2n} = 0 \tag{13.59}$$

Furthermore, $\Omega_{2n} = \mathrm{tr}F^n$ is closed:

$$\begin{aligned} d\,\mathrm{tr}\,F^n &= \mathrm{tr}\left[dF\,F^{n-1} + F dF\,F^{n-2} + \dots + F^{n-1} dF\right] \\ &= n\,\mathrm{tr}\left(dF\,F^{n-1}\right) \quad ; \quad DF \equiv dF + [A,F] = 0 \\ &= -n\,\mathrm{tr}\left[(AF - FA)F^{n-1}\right] \\ &= 0 \end{aligned}$$

$$d\Omega_{2n} = 0 \quad \text{or with } n \longrightarrow n+1 : d\Omega_{2n+2} = 0 \tag{13.60}$$

(Of course, for Ω_{2n+2} = (2n+2)-form in (2n+2)-dimensional space, the result $d\Omega_{2n+2} = 0$ is trivial.)

From (13.60) we learn that there exists a (2n+1)-form (locally) such that

$$\Omega_{2n+2} = d\omega^0_{2n+1} \tag{13.61}$$

with ω^0_{2n+1} given by (cf. (13.45)

$$\omega^0_{2n+1} = (n+1)\int_0^1 dt\; t^n\, \mathrm{tr}\, A\{dA + tA^2\}^n \tag{13.62}$$

From the gauge invariance of Ω, equation (13.59), we know

$$0 = \delta_v \Omega_{2n+2} \underset{(61)}{=} \delta_v (d\omega^0_{2n+1}) = d(\delta_v \omega^0_{2n+1}) \tag{13.63}$$

so that we conclude that $\delta_v \omega^0_{2n+1}$ is closed, too. Hence, there exists (locally) a v-dependent 2n-form such that

$$\delta_v \omega^0_{2n+1} = d\,\omega^1_{2n}(v) \tag{13.64}$$

Here, the superscript indicates that $\omega^1_{2n}(v)$ is of first order in v.

Now we define

$$G(v) := \int dx\, v_i(x)\, G_i(x) := \int_{S_D} \omega^1_{2n}(v, A) \tag{13.65}$$

and show that the so constructed anomaly $G_i[A]$ satisfies the W.-Z. conditions (13.55). To this end, we assume that we can extend the gauge fields from our D-dimensional space, S_D, to a (D+1)-dimensional ball, B_{D+1}, which has S_D as its boundary, $S_D = \partial B_{D+1}$. Next, consider the following functional

$$U[A] := \int_{B_{D+1}} \omega^0_{2n+1}(A) \tag{13.66}$$

which responds under gauge variation as follows:

$$\begin{aligned} \delta_v U[A] &= \int_{B_{D+1}} \delta_v \omega^0_{2n+1}(A) \\ &\underset{(64)}{=} \int_{B_{D+1}} d\omega^1_{2n}(v, A) \\ &= \int_{\partial B_{D+1} = S_D} \omega^1_{2n}(v, A) = G(v, A) \end{aligned} \tag{13.67}$$

This is to be substituted in the W.-Z. conditions:

$$\begin{aligned} \delta_u G(v) - \delta_v G(u) &= \delta_u \delta_v U - \delta_v \delta_u U \\ &= [\delta_u, \delta_v] U \end{aligned}$$

$$= \int dx\,dy\; u_i(x)\, v_j(y) \underbrace{\left[X_i(x), X_j(y)\right]}_{= f_{ijk} X_k(x)\, \delta(x-y)} U$$

$$= \int dx\; \underbrace{f_{ijk}\, u_i(x)\, v_j(x)}_{=(u\times v)_k(x)} X_k(x)\, U$$

$$= \delta_{u\times v}\, U$$

$$= G(u\times v)\;,\; \text{qed.}$$

So we want to know ω^1_{2n} with $d\omega^1_{2n}(v) = \delta_v \omega^0_{2n+1}$.
The general formula for computing $\omega^1_{2n}(v,A)$ is given by

$$-\omega^1_{2n}(v,A) = (n+1)\int_0^1 dt \left\{ \mathrm{Str}(v\, F_t^{\,n}) - t(t-1)\, n\, \mathrm{Str}(v[v,A]F_t^{\,n-1}) \right\} \tag{13.68}$$

where $\quad F_t = t\,dA + t^2 A^2$

For D = 4, n = 2 we find

$$\omega^1_4(v,A) = -\,\mathrm{tr}\left\{ v\, d\left(A\,dA + \tfrac{1}{2} A^3\right)\right\} \tag{13.69}$$

This leads to the anomaly for the chiral SU(N) gauge theory, cf. equation (13.29). Note that the non-abelian chiral anomaly $G^i[A]$ in both equations (13.29) and (13.69) is a total divergence-- like the $U_A(1)$ anomaly.

According to equation (13.60), $\mathrm{tr}F^n$ is closed. Hence, there exists locally a (2n-1)-form ω^0_{2n-1} so that

$$\mathrm{tr}\, F^n = d\omega^0_{2n+1} \tag{13.70}$$

with the solution

$$\omega^0_{2n-1} = n\int_0^1 dt\; t^{n-1}\, \mathrm{tr}\, A(dA + tA^2)^{n-1} \; . \tag{13.71}$$

Again: topological quantization always involves a closed form which is not globally exact. To further analyze the topological significance of ω^{0}_{2n-1}, let us assume that A(x) becomes a pure gauge asymptotically (eucl. E^{2n}-gauge theory)

$$A \longrightarrow v = g^{-1} dg \,, \quad v \in \Lambda^1, \, g \in \Lambda^0 \tag{13.72}$$

$$\left(A_\mu(x) \longrightarrow v_\mu(x) = g^{-1}(x)\, \partial_\mu g(x) \right)$$

and $g(x) \in G$ = simple group.

The exterior derivative of (13.72) is then

$$dA \rightarrow dv = dg^{-1} dg = - g^{-1} dg\, g^{-1} dg = -A^2 \rightarrow -v^2 \tag{13.73}$$

since

$$d1 = d(g^{-1} g) = dg^{-1} g + g^{-1} dg = 0$$

$$\Rightarrow \quad dg^{-1} = - g^{-1} dg\, g^{-1}$$

In equation (13.71) we need

$$\begin{aligned} \mathrm{tr}\, A \,(dA + t A^2)^{n-1} \\ &\longrightarrow \mathrm{tr}\, v(-v^2 + t v^2)^{n-1} \\ &= (-)^{n-1}\, \mathrm{tr}\, v\, v^{2n-2}\, (1-t)^{n-1} \\ &= (-)^{n-1}\, \mathrm{tr}\, v^{2n-1}\, (1-t)^{n-1} \end{aligned} \tag{13.74}$$

So we obtain

$$\omega^0_{2n-1} \longrightarrow tr(v^{2n-1})\, n(-1)^{n-1} \underbrace{\int_0^\infty dt\, t^{n-1}(1-t)^{n-1}}_{=\frac{(n-1)!\,(n-1)!}{(2n-1)!}}$$

$$= \frac{(-1)^{n-1}\, n!\,(n-1)!}{(2n-1)!}\, tr(v^{2n-1}) \tag{13.75}$$

Consider the integral

$$\int_{E^{2n}} tr\, F^n = \int_{E^{2n}} d\omega^0_{2n-1} = \int_{\partial E^{2n} \sim S^{2n-1}} \omega^0_{2n-1}$$

$$\sim \int_{S^{2n-1}} tr(v^{2n-1}) \tag{13.76}$$

This expression classifies the mapping $S_{2n-1} \rightarrow G$. Thus, the integral of the abelian anomaly in D =2n dimensions, trF^n, is associated with the homotopy group $\pi_{2n-1}(G)$.

Let us investigate a particular example, namely, D = 4, n = 2, and G = SU(2), which corresponds to the instanton. ∂E^4 is taken to be a cylinder C.

$$\int_{E^4} d^4x\, F_{\mu\nu}\, {}^*F^{\mu\nu} \sim \int_C tr\, v^3$$

$$= \int_C tr(g^{-1}dg\, g^{-1}dg\, g^{-1}dg)$$

$$= \int_C \underbrace{tr(g^{-1}\partial_\alpha g\; g^{-1}\partial_\beta g\; g^{-1}\partial_\gamma g)}_{=:\frac{1}{3!}t_{\alpha\beta\gamma}}\, dx^\alpha \wedge dx^\beta \wedge dx^\gamma$$

$$= \int_c \frac{1}{3!} t_{[\alpha\beta\gamma]} dx^\alpha \wedge dx^\beta \wedge dx^\gamma$$

$$= \sum \int du^1 du^2 du^3 \, t_{[\alpha\beta\gamma]}(x(u)) \frac{\partial x^\alpha}{\partial u^1} \frac{\partial x^\beta}{\partial u^2} \frac{\partial x^\gamma}{\partial u^3}$$

over faces at $t = \pm\infty$ — param. of cylinder

With the choice of parametrization

$$x^1 = u^1$$
$$x^2 = u^2$$
$$x^3 = u^3$$

and $x^0 = \pm\infty$, so that

$$\frac{\partial x^\alpha}{\partial u^i} = \begin{cases} 0 \,, & \alpha = 0 \\ \delta_i^\alpha \,, & \alpha = 1,2,3 \end{cases}$$

we can continue to write

$$\int d^4x \, F_{\mu\nu} {}^*F^{\mu\nu} \sim \sum \iiint d^3u \; t_{[1,2,3]}$$

$$\sim \sum \iiint d^3u \; \varepsilon_{ijk} t_{ijk}$$

$$\sim \sum \int d^3u \; \varepsilon_{ijk} \, \mathrm{tr}\left[g^{-1}\partial_i g \, g^{-1}\partial_j g^{-1} \partial_k g\right]$$

$$\sim \{Q_T(+\infty) - Q_T(-\infty)\}$$

$$= q[A] \quad \text{cf. chapter 4.} \qquad (13.77)$$

($q[A]$ ← Pontryagin index)

where the integrals are taken over large 3-spheres, one at $t = -\infty$, the other at $t = +\infty$.

If $\mathrm{tr}\,v^{2n-1}$ were globally exact, i.e., $\mathrm{tr}\,v^{2n-1} = d\gamma$, $\gamma \in \Lambda^{2n-2}$, then, by Stokes' theorem (n = 2),

$$\int_{E^4} d^4x\, F_{\mu\nu}{}^*F^{\mu\nu} \sim \int_{\partial E^4} \mathrm{tr}\, v^{2n-1} = \int_{\partial E^4} d\gamma = \int_{\partial\partial E^4} \gamma = 0 \qquad (13.78)$$

since $\partial\partial E^4 = 0$. Consequently, there could be no instantons in the theory.

In general, then, let $g(x) \in G$, $x \in S_D$,

$$v \equiv g^{-1} dg \quad \text{, i.e.,} \quad v_\mu(x) = g^{-1}(x)\, \partial_\mu g(x),\ \mu = 1 \ldots D$$

For G = SU(N), .. the quantity

$$Q \equiv \int_{S_D} \mathrm{tr}\, v^D \qquad (13.79)$$

classifies the mapping

$$g : S_D \to G, \quad x \mapsto g(x)$$

i.e., of S_D into the group G. Whenever $\mathrm{tr}\,v^D$ is globally exact, the homotopy group $\Pi_D(G)$ is trivial:

$$\mathrm{tr}\, v^D = d\gamma$$

$$Q = \int_{S_D} \mathrm{tr}\, v^D = \int_{S_D} d\gamma = \int_{\partial S_D} \gamma = 0 \qquad (13.80)$$

Here are three examples:

(1) $G = U(1) = \{e^{i\Theta} \mid \Theta \in \mathbb{R}\}$, $D = 1$

$$S_D = S_1 = \left\{ \begin{pmatrix} \cos\alpha \\ \sin\alpha \end{pmatrix} \middle| \alpha \in [0, 2\pi[\right\}$$

$$g : [0, 2\pi[\to U(1) : \alpha \mapsto e^{i\Theta(\alpha)}$$

Then we have $\vartheta = g^{-1} d g$ or explicitly

$$\vartheta(\alpha) = e^{-i\Theta(\alpha)} \frac{d}{d\alpha} e^{i\Theta(\alpha)} = i \frac{d\Theta}{d\alpha}$$

so that ϑ is exact locally but not globally, because

$$Q = \int_{S_1} \vartheta = i \int_0^{2\pi} d\alpha \frac{d\Theta}{d\alpha} = i\left(\Theta(2\pi) - \Theta(0)\right) \tag{13.81}$$

and so ϑ fails to be exact at $\Theta = 0$ if $\Theta(0) \neq \Theta(2\pi)$.

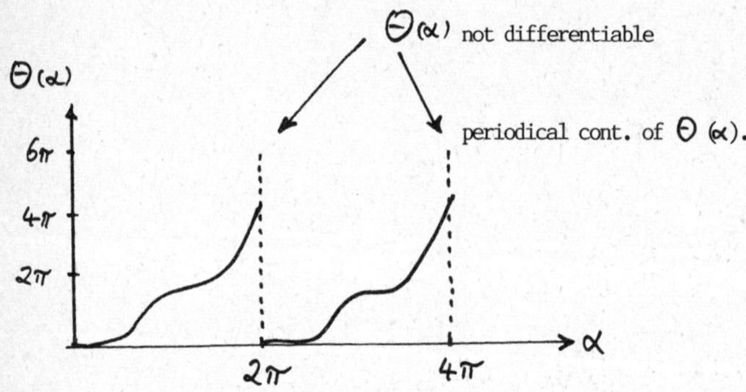

This fact is associated with flux quantization.

(2) For D = 3, G = SU(2), Q is essentially the baryon number B of the Skyrme model (g = U)

$$B \sim \int_{S_3} tr\, \vartheta^3 = \int_{S_3} \underbrace{tr[g^{-1}\partial_i g\, g^{-1}\partial_j g\, g^{-1}\partial_k g]}_{=: \frac{1}{3!} t_{ijk}} dx^i \wedge dx^j \wedge dx^k$$

$$= \int_{S_3} \frac{1}{3!} t_{[ijk]}\, dx^i \wedge dx^j \wedge dx^k$$

$$= \int d^3u \; t_{[ijk]} \underbrace{\frac{\partial x^i}{\partial u^1} \frac{\partial x^j}{\partial u^2} \frac{\partial x^k}{\partial u^3}}_{x^i = u^i \,:\; \delta^i_1 \delta^j_2 \delta^k_3}$$

$$= \int d^3u \; t_{[123]}$$

$$\sim \int d^3u \; \varepsilon_{ijk} t_{ijk} \tag{13.82}$$

If tr $v^3 = d\gamma$ globally, the result would be a zero baryon number.

(3) The action of the non-linear σ-model contains the term

$$\int_{M_5} \mathrm{tr}\, v^5 \tag{13.83}$$

where M_5 is a five-dimensional manifold whose boundary ∂M_5 is 4-dimensional space-time. We have shown that tr v^5 is not globally exact; hence the coefficient of this term (Wess-Zumino effective action) is indeed topologically quantized.

Additional References to
Chiral Anomalies and Differential Geometry

B. Zumino, Wu Yong-Shi, A. Zee, Nucl. Phys. B239, 477 (1984)

B. Zumino in: Relativity, Groups and Topology II, B.S. De Witt, R. Stora (Eds.), North-Holland (1984)

Differential Forms:

H. Flanders, "Differential Forms", Academic Press (1963)

T. Eguchi, P. Gilkey, A. Hanson, Phys. Rep. 66, 213 (1980)

C. Nash, S. Sen, "Topology and Geometry for Physicists", Academic Press, London (1983)

14. Effective Gauge Theory Actions in (2n + 1) Dimensions

In this chapter we want to explore the relationship between fermion number fractionization and the topological properties of odd space-time dimensional gauge theories. Roughly speaking, charge fractionization means the appearance of a fractional charge number whenever a fermi field interacts with a topologically nontrivial gauge field configuration. In the following we will show that the fermionic vacuum becomes charged. In particular, we shall compute the fermion number induced by static classical background fields. By way of illustration, we shall first analyze (2 + 1)-dimensional QED. Our goal is to calculate the vacuum current density in presence of a static prescribed external field and then functionally integrate this quantity to obtain the effective action.

We use the coordinate and metric convention (euclidean space) $x = (x^o, x^1, x^2) = (x^o, \vec{x})$, $g = (-,-,-)$.

Then we introduce the euclidean version of the generating functional for the gauge field by integrating out the fermi fields:

$$Z[A] = \int [d\psi d\bar{\psi}]\, e^{\int d^3x\, \bar{\psi}(i\not{D} - m)\psi}$$

$$\equiv \int [d\psi d\bar{\psi}]\, e^{\int d^3x\, \bar{\psi}(i\not{\partial} - e\gamma^\mu A_\mu - m)\psi}$$

$$= \det[i\not{D} - m] \tag{14.1}$$

Next, consider the quantity

$$e\langle 0|\bar{\psi}(w)\gamma^\mu\psi(w)|0\rangle^A = \int [d\psi d\bar{\psi}]\, e(\bar{\psi}\gamma^\mu\psi)(w)\, e^{\int d^3x\, \bar{\psi}(i\not{\partial} - e\not{A} - m)\psi}$$

$$= -\frac{\delta}{\delta A_\mu(w)} \int [d\psi d\bar{\psi}]\, e^{\int d^3x\, \bar{\psi}(i\not{\partial} - e\not{A} - m)\psi}$$

$$= -\frac{\delta Z[A]}{\delta A_\mu(w)}$$

which yields for the vacuum current induced by $A_\mu(x)$:

$$\langle j^\mu(w)\rangle^A = e\,\frac{\langle 0|\bar\psi(w)\gamma^\mu\psi(w)|0\rangle^A}{\langle 0|0\rangle^A} = -\frac{1}{Z[A]}\,\frac{\delta Z[A]}{\delta A_\mu(w)}$$

$$= -\frac{\delta}{\delta A_\mu(w)}\ln Z[A]$$

$$= -\frac{\delta}{\delta A_\mu(w)}\ln\det[i\not{D}-m] \tag{14.2}$$

We assume that $A^0 = 0$; $\vec{A} = \vec{A}(\vec{x})$ is the external prescribed static field. However, in the following, it is convenient to keep $A^0 = A^0(x^0,\vec{x})$ as infinitesimal source (to generate $\langle j^0\rangle$) which has to be set equal to zero after the functional differentiation with respect to A_0 has been performed.

$$\langle j^0(w)\rangle = -\frac{\delta}{\delta A_0(w)}\ln\det\left[\gamma^0(i\partial_0 - eA_0) + i\gamma^k D_k - m\right]\Big|_{A_0=0} \tag{14.3}$$

or

$$\langle j^0(w)\rangle = -\frac{\delta}{\delta A_0(w)}\ln\det\Big[-\gamma^0 eA_0 + \underbrace{\{\gamma^0 iD_0 + \gamma^k iD_k\}}_{=\,i\not{D}[A^0=0]\,=:\,i\tilde{\not{D}}} - m\Big]\Big|_{A_0=0}$$

$$\tilde{\not{D}} \equiv \gamma^0\partial_0 + \not{D}_2,\quad \not{D}_2 = \gamma^k D_k$$

$$= -\frac{\delta}{\delta A_0(w)}\ln\det[-\gamma^0 eA_0 + i\tilde{\not{D}} - m]\Big|_{A^0=0}$$

$$= -\frac{\delta}{\delta A_0(w)}\Big\{\ln\det[-\gamma^0 eA_0 + i\tilde{\not{D}} - m] + \ln\det[-i\tilde{\not{D}} - m]\Big\}\Big|_{A^0=0}$$

$$= -\frac{\delta}{\delta A_0(w)}\ln\det\Big\{[-\gamma^0 eA_0 + i\tilde{\not{D}} - m][-i\tilde{\not{D}} - m]\Big\}\Big|_{A^0=0}$$

$$= - \frac{\delta}{\delta A_0(w)} \ln \det \{ \underbrace{\gamma^0 e A_0 (i\tilde{\not{D}} + m)}_{\text{inf. perturbation}} + \underbrace{\tilde{\not{D}}^2 + m^2}_{\text{positive, hermitean}} \} \Big|_{A^0=0}$$

At this point we use the ζ -function representation of the determinant of an operator Ω:
$\det \Omega = e^{-\zeta'(\Omega|0)}$, which gives

$$\langle j^0(w) \rangle = \frac{d}{ds}\Big|_0 \frac{\delta \zeta(\gamma^0 e A_0(i\tilde{\not{D}}+m) + \tilde{\not{D}}^2 + m^2 | s)}{\delta A_0(w)} \Big|_{A^0=0}$$

$$= \frac{d}{ds}\Big|_0 (-s) \sum_n \lambda_n^{-(1+s)} \int d^3_z \, \varphi_n^+(z) \frac{\delta(\gamma^0 e A_0 (i\tilde{\not{D}}+m) + \cdots)}{\delta A_0(w)} \varphi_n(z) \Big|_{A^0=0} \tag{14.4}$$

with $$(\tilde{\not{D}}^2 + m^2)\varphi_n = \lambda_n \varphi_n \,, \quad \varphi_n = \varphi_n(z^0, z^1, z^2) \tag{14.5}$$

Performing the functional differentiation in (14.4), and setting A^0 equal to zero produces

$$\langle j^0(w) \rangle = \frac{d}{ds}\Big|_0 (-s) \sum_n \lambda_n^{-(1+s)} \int d^3_z \, \varphi_n^+(z) \, e\gamma^0 \delta^3(z-w) (i\tilde{\not{D}} - m)_z \varphi_n(z)$$

$$= -e \frac{d}{ds}\Big|_0 s \sum_n \lambda_n^{-(1+s)} \varphi_n^+(w) \gamma^0 (i\tilde{\not{D}} + m)_w \varphi_n(w)$$

$$= -e \frac{d}{ds}\Big|_0 s \sum_n \varphi_n^+(w) \gamma^0 (i\tilde{\not{D}} + m)(\tilde{\not{D}}^2 + m^2)^{-(1+s)} \varphi_n(w) ,$$

$$\langle j^0(w) \rangle = -e \frac{d}{ds}\Big|_0 s \lim_{x' \to x} \operatorname{tr}_\gamma \left[\gamma^0 (i\tilde{\not{D}} + m)(\tilde{\not{D}}^2 + m^2)^{-(1+s)} \right]_x \delta(x - x')$$

With the aid of the representation

$$A^{-s} = \frac{1}{\Gamma(s)} \int_0^\infty dt\, t^{s-1} e^{-At} ,$$

$$A^{-(1+s)} = \frac{1}{\Gamma(s+1)} \int_0^\infty dt\, t^{s} e^{-At} , \quad A > 0$$

we obtain

$$\langle j^0(x) \rangle = -e \frac{d}{ds}\Big|_0 \frac{s}{\Gamma(1+s)} \int_0^\infty dt\, t^s \lim_{x' \to x} tr_\gamma [\gamma^0 (i\tilde{\not{D}} + m) e^{-(\tilde{\not{D}}^2 + m^2)t}]_x \delta^3(x-x') \tag{14.6}$$

The trace operation can be further reduced to

$$tr_\gamma [\gamma^0 (i\gamma^0 \partial_0 + i\not{D}_2 + m) e^{-(\tilde{\not{D}}^2 + m^2)t}]$$

$$\overset{tr AB = tr BA}{=} tr_\gamma [(i\gamma^0 \partial_0 + i\not{D}_2 + m) \underbrace{e^{-(\tilde{\not{D}}^2 + m^2)t}}_{\text{even number of } \gamma\text{'s}} \gamma^0]$$

$$= tr_\gamma [(i\gamma^0 \partial_0 + i\not{D}_2 + m) \gamma^0 e^{-(\tilde{\not{D}}^2 + m^2)t}]$$

$$= tr_\gamma [\gamma^0 (i\gamma^0 \partial_0 \underbrace{- i\not{D}_2 + m) e^{-(\tilde{\not{D}}^2 + m^2)t}}_{\to 0}]$$

So far, we have obtained

$$\langle j^0(x) \rangle = -e \frac{d}{ds}\Big|_0 \frac{s}{\Gamma(1+s)} \int_0^\infty dt\, t^s \lim_{x' \to x} tr_\gamma [\gamma^0 (i\gamma^0 \partial_0 + m) e^{-(\tilde{\not{D}}^2 + m^2)t}]_x \delta^3(x-x') \tag{14.7}$$

Substituting

$$\delta(x-x') = \int \frac{d^3k}{(2\pi)^3} e^{ik(x-x')}$$

into the right-hand side of (14.7) gives

$$\langle j^0(x) \rangle = -e \frac{d}{ds}\Big|_0 \frac{s}{\Gamma(1+s)} \int_0^\infty dt\, t^s \int \frac{d^3k}{(2\pi)^3} e^{-ikx} tr_\gamma [(-i\partial_0 + m\gamma^0) e^{-(\not{D}^2[A^0=0] + m^2)t}] e^{ikx}$$

Now recall the definition of $\not{D}^2$:

$$\tilde{\not{D}}^2 \equiv \not{D}^2[A^0=0] = \left(D_\mu D^\mu - \tfrac{i}{2} e \gamma_\mu \gamma_\nu F^{\mu\nu}\right)_{\substack{A^0=0 \\ \vec{A}=\vec{A}(\vec{x})}}$$

$$= \partial_0 \partial^0 + D_i D^i - \tfrac{i}{2} e \gamma_i \gamma_j F^{ij}$$

$$= -\partial_0^2 + \not{D}_2^2$$

since

$$F^{0j} = \partial^0 A^j - \partial^j A^0 = 0 \qquad (f_{abc} A_b^0 A_c^j = 0)$$

The vanishing commutator, $[\partial_0^2, \not{D}_2^2] = 0$, then allows us to write

$$e^{-\tilde{\not{D}}t} = e^{\partial_0^2 t} e^{-\not{D}_2^2 t}$$

Therefore,

$$\langle j^0(x)\rangle = -e \frac{d}{ds}\Big|_0 \frac{s}{\Gamma(1+s)} \int_0^\infty dt\, t^s e^{-m^2 t} \int \frac{d^3k}{(2\pi)^3} e^{-ikx} \cdot$$

$$\cdot\, tr_\gamma\left[(-i\partial_0 + m\gamma^0) e^{\partial_0^2 t} e^{-\not{D}_2^2 t}\right] e^{ikx}$$

$$= -e \frac{d}{ds}\Big|_0 \frac{s}{\Gamma(1+s)} \int_0^\infty dt\, t^s e^{-m^2 t} \int \frac{d^3k}{(2\pi)^3}$$

$$tr_\gamma \Big\{ \underbrace{e^{-ik_0x^0} (-i\partial_0) e^{\partial_0^2 t} e^{ik_0x^0}}_{\to 0} e^{-ik_ix^i} e^{-\not{D}_2^2 t} e^{ik_jx^j}$$

$$+ m\gamma^0 \underbrace{e^{-ik_0x^0} e^{\partial_0^2 t} e^{ik_0x^0}}_{= e^{-k_0^2 t}} e^{-ik_ix^i} e^{-\not{D}_2^2 t} e^{ik_jx^j} \Big\}$$

which yields our final formula (with possible color degrees of freedom)

$$< j^0_{[a]}(x) > = -em \int_{-\infty}^{\infty} \frac{dk_0}{(2\pi)} \frac{d}{ds}\Big|_0 \frac{s}{\Gamma(1+s)} \int_0^{\infty} dt\, t^s e^{-(m^2+k_0^2)t}\,.$$

$$\cdot\, tr_{\gamma[c]}[T^a]\gamma^0 \int \frac{d^2k}{(2\pi)^2} e^{-ik_i x^i} e^{-\not{D}_2^2 t} e^{ik_j x^j} \tag{14.8}$$

The right-hand side can be evaluated in a twofold way:

(1) $$<j^0(x)> = -em \int \frac{dk_0}{(2\pi)} \frac{d}{ds}\Big|_0 \frac{s}{\Gamma(1+s)} \int_0^{\infty} dt\, t^s e^{-(m^2+k_0^2)t}$$

$$\cdot\, tr\,\gamma^0 \int \frac{d^2k}{(2\pi)^2} e^{-[-k^2+2ik\cdot D+D^2-\frac{ie}{2}\gamma_i\gamma_j F^{ij}]t}$$

$$k^2 \equiv k_j k^j\,, \quad k\cdot D \equiv k_j D^j$$

where use has been made of ($D_2 \equiv D$)

$$e^{-ikx} e^{-\not{D}_2^2 t} e^{ikx} = \exp\{-e^{-ikx} \not{D}_2^2 e^{ikx}\}$$

and $$[e^{-ikx} \not{D}_2^2 e^{ikx}] f(x) = [-k^2+2ikD_2+D_2^2-\frac{ie}{2}\gamma_i\gamma_j F^{ij}] f(x)$$

Substituting $t=\tau s^2$ yields

$$<j^0(x)> = -em \int \frac{dk_0}{(2\pi)} \frac{d}{ds}\Big|_0 \frac{s}{\Gamma(1+s)} (s^2)^{1+s} \int_0^{\infty} d\tau\, \tau^s e^{-(m^2+k_0^2)\tau s^2}$$

$$\cdot\, tr_\gamma\, \gamma^0 \int \frac{d^2k}{(2\pi)^2} e^{-[-k^2+2ikD+D^2-\frac{ie}{2}\gamma_i\gamma_j F^{ij}]\tau s^2}$$

Introducing the variable

$$k_i' = k_i s\sqrt{\tau}\,, \quad k'^2 = k^2 s^2 \tau\,, \quad d^2k' = d^2k\, s^2 \tau$$

we can rewrite our last expression for $\langle j^0 \rangle$ as

$$\langle j^0(x) \rangle = -em \int \frac{dk_0}{2\pi} \frac{d}{ds}\Big|_0 \frac{s}{\Gamma(1+s)} \frac{(s^2)^{1+s}}{s^2} \int_0^\infty d\tau \frac{\tau^s}{\tau} e^{-(m^2+k_0^2)\tau s^2}$$

$$\underbrace{\cdot \, tr_\gamma \gamma^0 \int \frac{d^2K'}{(2\pi)^2} e^{-[-K'^2 + 2iK'Ds\sqrt{\tau} + (D^2 + \frac{ie}{2}\gamma_i\gamma_j F^{ij})s^2\tau]}}_{S}$$

The quantity S has already been computed in the Schwinger model ($\gamma^0 \hat{=} \gamma_5$) with the result

$$S = -\frac{e}{4\pi} \varepsilon_{ij} F^{ij} s^2\tau + O(s^3\tau^{3/2}) \tag{14.9}$$

so that

$$\langle j^0(x) \rangle = \frac{e^2 m}{4\pi} \int \frac{dk_0}{2\pi} \frac{d}{ds}\Big|_0 \frac{s}{\Gamma(1+s)} (s^2)^{1+s} \underbrace{\int_0^\infty d\tau \, \tau^s e^{-(m^2+k_0^2)\tau s^2}}_{= \Gamma(1+s)(M^2 s^2)^{-(1+s)}} \varepsilon_{ij} F^{ij} + \dots \quad , \; M^2 := m^2 + K_0^2$$

$$= \frac{e^2 m}{4\pi} \varepsilon_{ij} F^{ij} \int \frac{dk_0}{2\pi} \frac{d}{ds}\Big|_0 s(s^2)^{1+s} (M^2)^{-(1+s)} (s^2)^{-(1+s)} + \dots$$

$$= \frac{e^2 m}{4\pi} \varepsilon_{ij} F^{ij} \int \frac{dk_0}{2\pi} \underbrace{\frac{d}{ds}\Big|_0 s \, e^{-(1+s)\ln M^2}}_{= \frac{1}{M^2}} + \dots$$

$$= \frac{e^2 m}{4\pi} \varepsilon_{ij} F^{ij} \underbrace{\int_{-\infty}^{\infty} \frac{dk_0}{2\pi} \frac{1}{m^2 + k_0^2}}_{= \frac{1}{2|m|}} + \dots$$

$$= \frac{e^2}{8\pi} \frac{m}{|m|} \varepsilon_{ij} F^{ij} + \dots$$

The $O(s^3\tau^{3/2})$ terms of (14.9) yield a non-vanishing contribution of order $\frac{1}{m}$. Here, then , is our final result

$$\langle j^0(x)\rangle = \frac{e^2}{8\pi}\,\frac{m}{|m|}\,\varepsilon_{ij}F^{ij} + O\left(\frac{1}{m}\right) \qquad (14.10)$$

(2) The second method also starts from (14.8), which we rewrite in a slightly different form:

$$\langle j^0(x)\rangle = -em\int\frac{dk_0}{2\pi}\,\frac{d}{ds}\Big|_0\frac{s}{\Gamma(1+s)}\int_0^\infty dt\,t^s e^{-(m^2+k_0^2)t}\,\cdot$$

$$\cdot\lim_{\vec{x}'\to\vec{x}}\mathrm{tr}_\gamma\left[\gamma^0 e^{-\not{D}_2^2 t}\right]_x\delta^{(2)}(\vec{x}-\vec{x}')$$

$$= e\int_{-\infty}^{\infty}\frac{dk_0}{2\pi}\,\frac{m}{M}\,(-M)\frac{d}{ds}\Big|_0\frac{s}{\Gamma(1+s)}\int_0^\infty dt\,t^s\lim_{\vec{x}'\to\vec{x}}\mathrm{tr}_\gamma\left[\gamma^0 e^{-(\not{D}_2^2+M^2)t}\right]_x\delta^{(2)}(\vec{x}-\vec{x}')$$

$$= e\int_{-\infty}^{\infty}\frac{dk_0}{2\pi}\,\frac{m}{M}\,\frac{\langle 0|\bar{\Psi}(\vec{x})\gamma^0\Psi(\vec{x})|0\rangle^A}{\langle 0|0\rangle^A}\Bigg|_{M=(m^2+k_0^2)^{1/2}} \qquad (14.11)$$

To prove the last line, equation (14.11), let us consider a two-dimensional euclidean theory, g = (-,-), and introduce an infinitesimal source function a(x) according to $(\not{D}\equiv\not{D}_2)$

$$\frac{\langle 0|\bar{\Psi}(\vec{w})\gamma_5\Psi(\vec{w})|0\rangle^A}{\langle 0|0\rangle^A} = \frac{1}{Z[A,a]}\,\frac{\delta}{\delta a(\vec{w})}\underbrace{\int[d\Psi d\bar{\Psi}]\,e^{\int\bar{\Psi}(i\not{D}-m+a\gamma_5)\Psi}}_{=Z[A,a]}\Bigg|_{a=0}$$

$$= \frac{1}{Z[A,a]}\,\frac{\delta}{\delta a(\vec{w})}\,Z[A,a]\Big|_{a=0}$$

$$= \frac{\delta}{\delta a(\vec{w})}\ln Z[A,a]\Big|_{a=0}$$

$$= \frac{\delta}{\delta a(\vec{w})} \ln \det [\, i \not{D} - m + a\gamma_5 \,] \Big|_{a=0}$$

$$= \frac{\delta}{\delta a(\vec{w})} \left\{ \ln \det [\, i\not{D} - m + a\gamma_5 \,] + \ln \det [\, -i\not{D} - m \,] \right\} \Big|_{a=0}$$

$$= \frac{\delta}{\delta a(\vec{w})} \ln \det \{ \underbrace{-a\gamma_5 (i\not{D} + m)}_{\text{inf. perturb.}} + \underbrace{\not{D}^2 + m^2}_{\text{positive, hermitean}} \} \Big|_{a=0}$$

$$= - \frac{d}{ds}\Big|_0 \frac{\delta \zeta(-a\gamma_5 (i\not{D}+m) + \not{D}^2 + m^2 | s)}{\delta a(\vec{w})} \Big|_{a=0}$$

$$= - \frac{d}{ds}\Big|_0 s \sum_n \lambda_n^{-(s+1)} \varphi_n^+(\vec{w}) \gamma_5 (i\not{D} + m) \varphi_n(\vec{w})$$

The complete orthonormal system φ_n satisfies the equation

$$(\not{D}^2 + m^2) \varphi_n = \lambda_n \varphi_n$$

so that we can continue to write

$$= - \frac{d}{ds}\Big|_0 s \sum_n \varphi_n^+(\vec{w}) \gamma_5 (i\not{D} + m) (\not{D}^2 + m^2)^{-(1+s)} \varphi_n(\vec{w})$$

$$\frac{\langle 0 | \bar{\Psi}(\vec{x}) \gamma_5 \Psi(\vec{x}) | 0 \rangle^A}{\langle 0 | 0 \rangle^A} = - \frac{d}{ds}\Big|_0 s \lim_{\vec{x}' \to \vec{x}} tr_\gamma \left[\gamma_5 (i\not{D} + m) (\not{D}^2 + m^2)^{-(s+1)} \right]_x \delta^{(2)}(\vec{x} - \vec{x}')$$

$$= - \frac{d}{ds}\Big|_0 \frac{s}{\Gamma(1+s)} \int_0^\infty dt\, t^s \lim_{\vec{x}' \to \vec{x}} tr_\gamma \left[\gamma_5 (i\underbrace{\not{D}}_{\substack{\text{does not} \\ \text{contr. when taking tr}}} + m) e^{-(\not{D}^2 + m^2)t} \right]_x \delta^{(2)}(\vec{x} - \vec{x}')$$

This, then, proves our statement (14.11)

$$\frac{\langle 0 | \bar{\Psi}(\vec{x}) \gamma_5 [T^a] \Psi(\vec{x}) | 0 \rangle^A}{\langle 0 | 0 \rangle^A}\Big|_M = -M \frac{d}{ds}\Big|_0 \frac{s}{\Gamma(1+s)} \int_0^\infty dt\, t^s \lim_{\vec{x}' \to \vec{x}} tr_\gamma \Big[\gamma_5 [T^a] \cdot e^{-(\not{D}^2 + M^2)t} \Big]_x \delta^{(2)}(\vec{x} - \vec{x}') \tag{14.12}$$

In two dimensions, the axial vector anomaly is given by

$$\frac{\partial^i \langle 0|\bar{\Psi}\gamma_i\gamma_5\Psi|0\rangle}{\langle 0|0\rangle} = 2iM\frac{\langle 0|\bar{\Psi}\gamma_5\Psi|0\rangle}{\langle 0|0\rangle} - \frac{ie}{2\pi}\varepsilon_{ij}F^{ij} \tag{14.13}$$

which can be used in (14.11) to produce ($\gamma_5 \hat{=} \gamma^0$)

$$\langle j^0(x)\rangle = e\int_{-\infty}^{\infty}\frac{dk_0}{2\pi}\,\frac{m}{M}\,\frac{1}{2iM}\left\{\frac{\partial_j\langle 0|\bar{\Psi}\gamma^j\gamma^0\Psi|0\rangle^A}{\langle 0|0\rangle^A} + \frac{ie}{2\pi}\varepsilon_{ij}F^{ij}\right\} \tag{14.14}$$

The contribution of the anomaly is then contained in

$$\langle j^0(x)\rangle = e\,\frac{1}{2i}\,\frac{ie}{2\pi}\,\varepsilon_{ij}F^{ij}\underbrace{\int_{-\infty}^{\infty}\frac{dk_0}{2\pi}\,\frac{m}{M^2}}_{=\frac{1}{2}\frac{m}{|m|}} + \cdots$$

$$= \frac{e^2}{8\pi}\,\frac{m}{|m|}\,\varepsilon_{ij}F^{ij} + \cdots \tag{14.15}$$

The non-anomalous contributions, $\partial_j\langle 0|\bar{\Psi}\gamma^j\gamma_0\Psi|0\rangle$, vanish for
(1) constant fields,
(2) for $m \to \infty$, cf. eq. (14.10).
They never contribute to $\int d^2x \langle j^0(x)\rangle$, assuming that surface terms can be neglected.

Integrating (14.15) over two-space, we obtain

$$Q = \int d^2x\,\langle j^0(x)\rangle = \frac{m}{|m|}\,\frac{e}{2}\,\frac{e}{4\pi}\int d^2x\,\varepsilon_{ij}F^{ij}$$

or

$$Q = \frac{m}{|m|}\,\frac{e^2}{2}\,\Phi[A] = eN \tag{14.16}$$

which relates the fractional fermion number N and the magnetic flux:

$$\Phi[A] := \frac{1}{4\pi}\int d^2x\, \varepsilon_{ij} F^{ij} = q[A] \tag{14.17}$$

where $q[A]$ is the two-dimensional Pontryagin index.

Note that equations (14.16) and (14.17) hold for arbitrary stationary magnetic fields

$$F^{\mu\nu} = \left(\begin{array}{c|cc} 0 & 0 & 0 \\ \hline 0 & 0 & B \\ 0 & -B & 0 \end{array}\right), \quad F^{0j} = 0$$

By Lorentz covariance we conclude that

$$\langle j^{\mu}(x)\rangle = \frac{e^2}{8\pi}\frac{m}{|m|}\,\varepsilon^{\mu\alpha\beta} F_{\alpha\beta} + \cdots \tag{14.18}$$

which is gauge invariant and conserved.

Given $\langle j^{\mu}(x)\rangle^{A}$, we now want to determine the effective action $\Gamma[A]$ which is related to the vacuum current by

$$\langle j^{\mu}(x)\rangle^{A} = -\frac{\delta\Gamma[A]}{\delta A_{\mu}(x)}, \quad \Gamma[0] = 0 \tag{14.19}$$

First, let us rewrite $\Gamma[A]$ in the form

$$\Gamma[A] = \Gamma[A] - \Gamma[0] = \ln e^{\Gamma[A]} - \ln e^{\Gamma[0]}$$

$$= \ln\frac{e^{\Gamma[A]}}{e^{\Gamma[0]}} \equiv \ln\frac{F[A]}{F[0]}, \quad F[A] := e^{\Gamma[A]} \tag{14.20}$$

Choosing a path $\{A_t, t\in[0,1]\}$ with $A_o = 0$ and $A_1 = A$, we can continue to write

$$\Gamma[A] = \lim_{N\to\infty} \ln \prod_{n=0}^{N-1} \frac{F[A_{\frac{n+1}{N}}]}{F[A_{n/N}]}$$

$$= \lim_{N\to\infty} \sum_{n=0}^{N-1} \ln \frac{F[A_{\frac{n+1}{N}}]}{F[A_{n/N}]}$$

$$= \lim_{N\to\infty} N \int_0^1 dt \left(\ln F[A_{t+\frac{1}{N}}] - \ln F[A_t]\right)$$

$$= \lim_{N\to\infty} N \int_0^1 dt \left(\Gamma[A_{t+\frac{1}{N}}] - \Gamma[A_t]\right)$$

$$= \lim_{N\to\infty} N \int_0^1 dt \left(\Gamma[A_t + \tfrac{1}{N}\dot{A}_t] - \Gamma[A_t]\right)$$

$$= \lim_{N\to\infty} N \int_0^1 dt \int d^3x \, \frac{1}{N} \, \dot{A}_t^\mu \frac{\delta \Gamma[A_t]}{\delta A_t^\mu(x)}$$

Here, then, is the formula for computing the effective action

$$\Gamma[A] = \int d^3x \int_0^1 dt \;\; \dot{A}_t^\mu(x) \frac{\delta \Gamma[A_t]}{\delta A_t^\mu(x)} \tag{14.21}$$

In the non-abelian case, we will choose $A_t^\mu = tA^\mu$, so that

$$\Gamma[A] = \int d^3x \int_0^1 dt \;\; A_a^\mu(x) \frac{\delta \Gamma[tA]}{\delta (tA_a^\mu(x))} \tag{14.22}$$

$$\underset{(19)}{=} - \int d^3x \int_0^1 dt \;\; A_a^\mu(x) \left\langle j_\mu^a(tA(x)) \right\rangle \tag{14.23}$$

Now let us return to our example QED_3.

$$\Gamma[A] = -\frac{m}{|m|}\frac{e^2}{8\pi}\varepsilon_{\mu\alpha\beta}\int d^3x \int_0^1 dt\ \dot{A}_t^\mu(\partial^\alpha A_t^\beta - (\alpha \leftrightarrow \beta))$$

$$= -\frac{m}{|m|}\frac{e^2}{8\pi} 2\varepsilon_{\mu\alpha\beta}\int d^3x \int_0^1 dt\ \dot{A}_t^\mu\ \partial^\alpha A_t^\beta$$

Here we make use of

$$\frac{d}{dt}\varepsilon_{\mu\alpha\beta}\int d^3x\ A_t^\mu\partial^\alpha A_t^\beta = \varepsilon_{\mu\alpha\beta}\int d^3x\,(\dot{A}_t^\mu\partial^\alpha A_t^\beta + A_t^\mu\partial^\alpha\dot{A}_t^\beta)$$

$$= \varepsilon_{\mu\alpha\beta}\int d^3x\ (\dot{A}_t^\mu\partial^\alpha A_t^\beta - \dot{A}_t^\beta\partial^\alpha A_t^\mu)$$

$$= 2\ \varepsilon_{\mu\alpha\beta}\int d^3x\ \dot{A}_t^\mu\partial^\alpha A_t^\beta$$

which means for $\Gamma[A]$:

$$\Gamma[A] = -\frac{m}{|m|}\frac{e^2}{8\pi}\int_0^1 dt\,\frac{d}{dt}\varepsilon_{\mu\alpha\beta}\int d^3x\ A_t^\mu\partial^\alpha A_t^\beta$$

$$= -\frac{m}{|m|}\frac{e^2}{8\pi}\varepsilon_{\mu\alpha\beta}\int d^3x\ A^\mu\partial^\alpha A^\beta$$

$$= -\frac{m}{|m|}\frac{e^2}{16\pi}\varepsilon_{\mu\alpha\beta}\int d^3x\ A^\mu F^{\alpha\beta}$$

$$= -\frac{m}{|m|}\frac{e^2}{8\pi}\int d^3x\ A^\mu\,{}^*F_\mu\ ,\quad {}^*F_\mu := \frac{1}{2}\varepsilon_{\mu\alpha\beta}F^{\alpha\beta} \tag{14.24}$$

or

$$\Gamma[A] = \frac{m}{|m|}\,2\pi\,W[A] \tag{14.25}$$

with

$$W[A] = -\frac{e^2}{16\pi^2}\int d^3x \, A^\mu \, {}^*F_\mu$$

$$= -\frac{e^2}{32\pi^2}\int d^3x \, \varepsilon_{\mu\alpha\beta} \, A^\mu F^{\alpha\beta}$$

$$= -\frac{e^2}{16\pi^2}\int d^3x \, \varepsilon_{\mu\alpha\beta} \, A^\mu \partial^\alpha A^\beta \tag{14.26}$$

Here we recognize the Chern-Simons form.

Incidentally, W[A] is gauge invariant if surface terms are assumed to vanish.

Under $A^\mu \to A^\mu + \partial^\mu \Lambda$, we have $F^{\mu\nu} \to F^{\mu\nu}$ and

$$W[A] \longrightarrow W[A] - \frac{e^2}{32\pi^2}\underbrace{\int d^3x \, \varepsilon_{\mu\alpha\beta} \, \partial^\mu \Lambda \, F^{\alpha\beta}}$$

$$= 2\int d^3x \, \varepsilon_{\mu\alpha\beta} \, \partial^\mu \Lambda \, \partial^\alpha A^\beta$$

$$= -2\int d^3x \, \varepsilon_{\mu\alpha\beta} \, \Lambda \, \partial^\mu \partial^\alpha A^\beta$$

$$= 0$$

$$\longrightarrow W[A]$$

Let us now extend our previous results of three dimensional space-time to non-abelian background fields, i.e., SU(N) gauge theory in 3 dimensions. According to equation (14.11), we are interested in

$$\langle j^0_a(x) \rangle = g \int_{-\infty}^{\infty} \frac{dk_0}{2\pi} \, \frac{m}{M} \, \frac{\langle 0|\bar\Psi \gamma^0 T^a \Psi|0\rangle^A}{\langle 0|0\rangle^A}\Bigg|_{M=(m^2+k_0^2)^{1/2}} \tag{14.27}$$

where we need the two-dimensional anomalous divergence of the $SU(N)_A$ current. In chapter 6 , we found for the Jacobian in two dimensions

$$J[\alpha] = \exp\left\{\frac{ig}{2\pi}\int d^2x \, \varepsilon_{\mu\nu} \, \alpha \, tr_c[T^a F^{\mu\nu}]\right\}$$

which is needed in $\left(\tilde{\Psi} = e^{i\alpha T^a\gamma_5}\psi,\ \tilde{\bar{\Psi}} = \bar{\psi} e^{i\alpha T^a\gamma_5}\right)$

$$I := \int [d\bar{\Psi} d\tilde{\Psi}]\, e^{\int d^2x\, \tilde{\bar{\Psi}}(i\partial\!\!\!/ - g A\!\!\!/ - m)\tilde{\Psi}}$$

$$= J[\alpha] \int [d\psi d\bar{\Psi}]\, e^{\int d^2x\, \bar{\Psi} e^{i\alpha T^a\gamma_5}(i\partial\!\!\!/ - g A\!\!\!/ - m) e^{i\alpha T^a\gamma_5}\psi} \tag{14.27$'$}$$

Taking $\alpha(x)$ infinitesimal, one can easily see that $(14.27)'$ takes the form

$$I = \int [d\psi d\bar{\Psi}]\ e^{\frac{ig}{2\pi}\int d^2x\ \varepsilon_{\mu\nu}\ \alpha\, tr_c(T^a F^{\mu\nu})}$$

$$\cdot \exp\left\{\int d^2x\left(\mathcal{L} + \alpha\, \partial_\mu(\bar{\Psi}\gamma^\mu\gamma_5 T^a\psi) - 2im\alpha(\bar{\Psi}\gamma_5 T^a\psi) + ig\alpha\, \bar{\Psi}[T^a, A\!\!\!/]\gamma_5\psi\right)\right\} \tag{14.28}$$

The divergence of the axial vector current which follows from (14.28) is then given by

$$\partial_\mu(\bar{\Psi}\gamma^\mu\gamma_5 T^a\psi) + ig\,\bar{\Psi}[T^a, A\!\!\!/]\gamma_5\psi = 2im(\bar{\Psi}\gamma_5 T^a\psi) - \frac{ig}{2\pi} tr_c(T^a F^{\mu\nu})\varepsilon_{\mu\nu} \tag{14.29}$$

(In four dimensions the last term has to be replaced by

$$+\frac{ig^2}{8\pi^2}\ tr_c\left(T^a\, F_{\mu\nu}\, {}^*F^{\mu\nu}\right).\ )$$

The left-hand side of (14.29) is indeed the covariant derivative, since

$$ig\ \bar{\Psi}\,[T^a, T^b]\,\gamma_\mu\gamma_5\,\psi A^\mu_b = -g f^{abc}\,\bar{\Psi}\gamma_\mu\gamma_5 T^c\psi\, A^\mu_b$$

so that

$$\begin{aligned}(\partial_\mu\delta^{ac} - g f^{abc} A^b_\mu)\, j_5^{\mu c} &= (\partial_\mu\delta^{ac} + g f^{acb} A^b_\mu)\, j_5^{\mu c} \\ &\equiv D^{ac}_\mu\, j_5^{\mu c}\end{aligned} \tag{14.30}$$

In order to apply equation (14.29) further, we must restrict the background gauge field so that for fixed a:

$$\left[T^a, A_\mu\right] = 0 \tag{14.31}$$

In equation (14.27), we need

$$\left.\frac{\langle 0|\bar\Psi \gamma_5 T^a \Psi|0\rangle^A}{\langle 0|0\rangle^A}\right|_M \overset{(29)}{=} \frac{\partial_\mu \langle 0|\bar\Psi\gamma^\mu\gamma_5 T^a \Psi|0\rangle}{2iM\langle 0|0\rangle^A} + \frac{ig}{2\pi(2iM)} tr_c\left(T^a \varepsilon_{\mu\nu} F^{\mu\nu}\right) \tag{14.32}$$

$$\text{in 4 dim.:} \quad -\frac{ig^2}{8\pi^2}\frac{1}{2iM} tr_c\left(T^a F_{\mu\nu} {}^*F^{\mu\nu}\right) \tag{14.33}$$

Returning to our (2+1)-dimensional vacuum charge density, equation (14.27), we obtain

$$\langle j_a^0(x)\rangle = g\int_{-\infty}^{\infty}\frac{dk_0}{2\pi}\,\frac{m}{M}\frac{1}{2iM}\left\{\frac{\partial_j \langle 0|\bar\Psi\gamma^j\gamma_0 T^a\Psi|0\rangle}{\langle 0|0\rangle} + \frac{ig}{2\pi} tr_c\left(T^a \varepsilon_{ij} F^{ij}\right)\right\} \tag{14.34}$$

Neglecting again the non- anomalous contribution (cf. equation (14.14) ff.) which vanishes for constant fields and in the limiting case $m\to\infty$, we find

$$\langle j_a^0(x)\rangle = g\,\frac{ig}{2\pi}\,\frac{1}{2i}\,\underbrace{tr_c\left(T^a\varepsilon_{ij}F^{ij}\right)}_{=\varepsilon_{ij}F_b^{ij}\,\underbrace{tr(T^aT^b)}_{=\frac{1}{2}\delta^{ab}}\;=\;\frac{1}{2}\varepsilon_{ij}F_a^{ij}}\;\underbrace{\int_{-\infty}^{\infty}\frac{dk_0}{2\pi}\,\frac{m}{m^2+k_0^2}}_{=\frac{1}{2}\frac{m}{|m|}}$$

$$= \frac{g^2}{16\pi}\,\frac{m}{|m|}\,\varepsilon_{ij}F_a^{ij} + \cdots \tag{14.35}$$

The general result is then obtained by requiring both Lorentz and gauge covariance

$$< j_a^{\mu}(x) > = \frac{m}{|m|} \frac{g^2}{16\pi} \varepsilon^{\mu\alpha\beta} F_{a\,\alpha\beta} + \cdots \tag{14.36}$$

Repeating the former steps in (4+1) dimensions, we find analogously

$$< j_a^0(x) > = g \int_{-\infty}^{\infty} \frac{dk_0}{2\pi} \frac{m}{M} \frac{1}{2iM} (-i) \frac{g^2}{8\pi^2} tr_c (T^a F_{ij}\, {}^*F^{ij}) + \cdots$$

$$i,j = 1,2,3,4$$

$$= -\frac{g^3}{16\pi^2} tr_c (T^a F_{ij}\, {}^*F^{ij}) \underbrace{\int_{-\infty}^{\infty} \frac{dk_0}{2\pi} \frac{m}{m^2 + k_0^2}}_{= \frac{1}{2}\frac{m}{|m|}} + \cdots$$

$$= -\frac{m}{|m|} \frac{g^3}{32\pi^2} \frac{1}{2} \varepsilon^{ijkl} tr_c (T^a F_{ij} F_{kl}) + \cdots \tag{14.37}$$

The induced non-abelian charge is

$$Q^a = \int d^4x < j_a^0(x) > = -\frac{m}{|m|} \frac{g^3}{32\pi^2} \int d^4x \, tr_c (T^a F_{ij}\, {}^*F^{ij}) \tag{14.38}$$

and the covariant version of (14.37) is given by

$$< j_a^{\mu}(x) > = -\frac{m}{|m|} \frac{g^3}{64\pi^2} \varepsilon^{\mu\alpha\beta\gamma\delta} tr_c (T^a F_{\alpha\beta} F_{\gamma\delta}) \tag{14.39}$$

Having obtained the vacuum current in three and five dimensions, we now turn to the calculation of the associated effective actions. The formula to be used is stated in (14.23).

In three dimensions we have

$$\Gamma[A] = -\int d^3x \int_0^1 dt \; A_a^{\mu}(x) \langle j_\mu^a(tA(x)) \rangle$$

$$\underset{(34)}{=} C \int d^3x \int_0^1 dt \; \varepsilon^{\mu\alpha\beta} \, tr(A_\mu F_{\alpha\beta}(tA)) \tag{14.40}$$

where we introduced

$$-C = 2 \frac{m}{|m|} \frac{g^2}{16\pi} = \frac{m}{|m|} \frac{g^2}{8\pi} \tag{14.41}$$

and

$$F_{\alpha\beta} = \partial_\alpha A_\beta - \partial_\beta A_\alpha + ig[A_\alpha, A_\beta] \tag{14.42}$$

so that the effective action reads

$$\Gamma[A] = C \int d^3x \int_0^1 dt \; [t \, tr(A_\mu \varepsilon^{\mu\alpha\beta} \{\partial_\alpha A_\beta - \partial_\beta A_\alpha\})$$

$$+ ig \, t^2 \, tr(A_\mu \varepsilon^{\mu\alpha\beta} [A_\alpha, A_\beta])]$$

$$= C \int d^3x \, [\tfrac{1}{2} tr(A_\mu \varepsilon^{\mu\alpha\beta} \{\partial_\alpha A_\beta - \partial_\beta A_\alpha\} + \tfrac{ig}{3} tr(A_\mu \varepsilon^{\mu\alpha\beta} [A_\alpha, A_\beta])]$$

$$= C \int d^3x \, [\tfrac{1}{2} tr(A_\mu \varepsilon^{\mu\alpha\beta} \{\partial_\alpha A_\beta - \partial_\beta A_\alpha + ig[A_\alpha, A_\beta]\})$$

$$- \tfrac{ig}{2} \, tr(A_\mu \varepsilon^{\mu\alpha\beta} [A_\alpha, A_\beta]) + \tfrac{ig}{3} tr(A_\mu \varepsilon^{\mu\alpha\beta} [A_\alpha A_\beta])]$$

$$= C \int d^3x \, [\tfrac{1}{2} tr(A_\mu \varepsilon^{\mu\alpha\beta} F_{\alpha\beta} - \tfrac{ig}{6} tr(A_\mu \varepsilon^{\mu\alpha\beta} A_\alpha A_\beta - \underbrace{A_\mu \varepsilon^{\mu\alpha\beta} A_\beta A_\alpha}_{\to - A_\alpha A_\mu A_\beta \varepsilon^{\alpha\mu\beta}})$$

$$= -\frac{m}{|m|} \frac{g^2}{8\pi} tr \int d^3x [\underbrace{\tfrac{1}{2} A_\mu \varepsilon^{\mu\alpha\beta} F_{\alpha\beta}}_{= A_\mu {}^*F^\mu} - \tfrac{ig}{3} \varepsilon^{\mu\alpha\beta} A_\mu A_\alpha A_\beta] \tag{14.43}$$

$$\Gamma[A] = \frac{m}{|m|}\,\pi W[A] \tag{14.44}$$

The final result for $\Gamma[A]$ in (2+1) dimensions is then contained in

$$\Gamma[A] = \frac{m}{|m|}\,\pi W[A]$$

with
$$W[A] := -\frac{g^2}{8\pi^2}\, tr \int d^3x \left(A_\mu {}^*F^\mu - \frac{ig}{3}\varepsilon^{\mu\alpha\beta} A_\mu A_\alpha A_\beta \right) \tag{14.45}$$

This is the Chern-Simons topological invariant in three dimensions. It is known that $W[A]$ is not gauge invariant against homotopically non-trivial gauge transformations. Furthermore, the "anomaly" in odd--here, (2+1)--dimensions appears as a parity-violating topological term in the vacuum current $\langle j_\mu^a \rangle$, rather than as a topological term ($\sim$ *FF) in the divergence of the axial vector current, $\partial_\mu \langle j_5^\mu \rangle$.

Let us now find the effective action in (4+1) dimensions. We begin with equation (14.39):

$$\langle j_\mu^a \rangle = -C\, \varepsilon_{\mu\alpha\beta\gamma\delta}\, tr_c (T^a F^{\alpha\beta} F^{\gamma\delta}) \tag{14.46}$$

where
$$C = \frac{m}{|m|}\frac{g^3}{64\pi^2}\,, \quad F_{\alpha\beta} = \partial_\alpha A_\beta + ig A_\alpha A_\beta - (\alpha \leftrightarrow \beta) \tag{14.47}$$

Following our previous steps starting with (14.40), we have for the effective action

$$\begin{aligned}
\Gamma[A] &= -\int d^5x \int_0^1 dt\, A_a^\mu(x) \langle j_\mu^a(tA) \rangle \\
&= C \int d^5x \int_0^1 dt\, A_\mu^a(x)\, \varepsilon^{\mu\alpha\beta\gamma\delta}\, tr(T^a F_{t\alpha\beta} F_{t\gamma\delta})\,, \quad F_{t\mu\nu} \equiv F_{\mu\nu}(tA) \\
&= C \int d^5x \int_0^1 dt\, \varepsilon^{\mu\alpha\beta\gamma\delta}\, tr(A_\mu F_{t\alpha\beta} F_{t\gamma\delta}) \\
&= 4C \int d^5x \int_0^1 dt\, \varepsilon^{\mu\alpha\beta\gamma\delta}\, tr[A_\mu (\partial_\alpha A_\beta t + ig t^2 A_\alpha A_\beta) \\
&\qquad \cdot (\partial_\gamma A_\delta t + ig t^2 A_\gamma A_\delta)]
\end{aligned}$$

$$= 4C \int d^5x \int_0^1 dt\, \varepsilon^{\mu\alpha\beta\gamma\delta}\, tr\, A_\mu \left(\partial_\alpha A_\beta \partial_\gamma A_\delta t^2 + \right.$$

$$\left. + igt^3 \partial_\alpha A_\beta A_\gamma A_\delta + igt^3 A_\alpha A_\beta \partial_\gamma A_\delta - g^2 t^4 A_\alpha A_\beta A_\gamma A_\delta \right)$$

The third term can be shown to be equal to the second one:

$$\varepsilon^{\mu\alpha\beta\gamma\delta}\, tr\, A_\mu A_\alpha A_\beta \partial_\gamma A_\delta = \varepsilon^{\mu\alpha\beta\gamma\delta}\, tr\, \underset{\mu}{A_\beta}\, \underset{\alpha}{\partial_\gamma}\, \underset{\beta}{A_\delta}\, \underset{\gamma}{A_\mu}\, \underset{\delta}{A_\alpha}$$

$$= \varepsilon^{\gamma\delta\mu\alpha\beta}\, tr\, A_\mu \partial_\alpha A_\beta A_\gamma A_\delta = \varepsilon^{\mu\alpha\beta\gamma\delta}\, tr\, A_\mu \partial_\alpha A_\beta A_\gamma A_\delta$$

Therefore, $$\Gamma[A] = 4C \int d^5x \int_0^1 dt\, \varepsilon^{\mu\alpha\beta\gamma\delta}\, tr \left(t^2 A_\mu \partial_\alpha A_\beta \partial_\gamma A_\delta \right.$$

$$\left. + 2igt^3 A_\mu A_\alpha A_\beta \partial_\gamma A_\delta - g^2 t^4 A_\mu A_\alpha A_\beta A_\gamma A_\delta \right)$$

Performing the trivial integration over t yields

$$\Gamma[A] = 4C \int d^5x\, \varepsilon^{\mu\alpha\beta\gamma\delta}\, tr \left(\tfrac{1}{3} A_\mu \partial_\alpha A_\beta \partial_\gamma A_\delta \right.$$

$$\left. + \frac{ig}{2} A_\mu A_\alpha A_\beta \partial_\gamma A_\delta - \frac{g^2}{5} A_\mu A_\alpha A_\beta A_\gamma A_\delta \right)$$

Introducing

$$\frac{4C}{3} = \frac{m}{|m|} \frac{g^3}{48\pi^2}$$

yields the final result

$$\Gamma[A] = \frac{m}{|m|}\frac{g^3}{48\pi^2}\int d^5x\, \varepsilon^{\mu\alpha\beta\gamma\delta}\, tr\Big(A_\mu \partial_\alpha A_\beta \partial_\gamma A_\delta$$

$$+ \frac{3}{2} ig\, A_\mu A_\alpha A_\beta \partial_\gamma A_\delta - \frac{3}{5} g^2 A_\mu A_\alpha A_\beta A_\gamma A_\delta\Big)$$

(14.48)

Now it is easy to compute the induced fermion number. If we limit ourselves to the abelian U(1) current, $T^a \rightarrow 1$, and introduce $g J^\mu := \langle j^\mu \rangle$, we obtain in (4+1) dimensions

$$J^\mu \underset{(39)}{=} - \frac{m}{|m|}\frac{g^2}{64\pi^2}\varepsilon^{\mu\alpha\beta\gamma\delta}\, tr(F_{\alpha\beta} F_{\gamma\delta}) \qquad (14.49)$$

The integral over 4-space is then

$$N = \int d^4x\, J^0(x)$$

$$= - \frac{m}{|m|}\frac{g^2}{64\pi^2}\varepsilon^{ijkl}\int d^4x\, tr(F_{ij} F_{kl})$$

$$= - \frac{m}{|m|}\frac{1}{2}\frac{g^2}{16\pi^2}\int d^4x\, tr(F_{ij}\, {}^*F^{ij})$$

or $$N = - \frac{m}{|m|}\frac{1}{2}\, q[A] \qquad (14.50)$$

with the four dimensional Pontryagin index

$$q[A] = \frac{g^2}{16\pi^2} \int d^4x \; tr\,(F_{ij} {}^*F^{ij}) \qquad i,j = 1,\dots,4 \tag{14.51}$$

At last we come to the general (2n + 1) dimensional theory. With

$$A_\mu = A^a_\mu T^a \tag{14.52}$$

we have in D = 2n + 1 dimensions ($x = (x^o,\vec{x}) = (x^o,x^i) = x^\mu$)

$$\langle j^\mu(x)\rangle = g\frac{\langle 0|\bar\psi(x)\gamma^\mu\psi(x)|0\rangle^A}{\langle 0|0\rangle^A} \, , \quad \langle j^\mu_a(x)\rangle = g\frac{\langle 0|\bar\psi(x)\gamma^\mu T^a\psi(x)|0\rangle^A}{\langle 0|0\rangle^A} \tag{14.53}$$

$$\langle j^o_a(x)\rangle = g\int_{-\infty}^{\infty}\frac{dk_o}{2\pi}\,\frac{m}{M}\,\frac{\langle 0|\bar\psi(\vec{x})\gamma^o T^a\psi(\vec{x})|0\rangle^A}{\langle 0|0\rangle^A}\Bigg|_{M=(m^2+k_o^2)^{1/2}} \tag{14.54}$$

Again, we assumed a stationary background field $A^o \equiv 0$, $\vec{A} = \vec{A}(\vec{x})$.

The integrand in (14.54) is given by the (non-) abelian anomaly in 2n dimensions. The divergence of the U(1) current reads (in the following, we set g = 1):

$$\partial^\mu(\bar\psi\gamma_\mu\gamma_5\psi) = 2i\,M(\bar\psi\gamma_5\psi) + i\,K_n\,\varepsilon_{\mu_1\dots\mu_{2n}}\,tr\left(F^{\mu_1\mu_2}\dots F^{\mu_{2n-1}\mu_{2n}}\right) \tag{14.55}$$

where

$$K_n = \frac{(-1)^n}{2^{2n-1}\pi^n n!} \tag{14.56}$$

which contains the specific examples studied so far:

$n = 1 \;:\; K_1 = -\frac{1}{2\pi} \qquad (D = 2)$

$n = 2 \;:\; K_2 = \frac{1}{16\pi^2} \qquad (D = 4).$

The non-abelian anomaly has the same form. Restricting the background gauge field again so that for fixed a $[T^a, A_\mu] = 0$, equation (14.55) is modified according to

$$\partial^\mu(\bar{\Psi}\gamma_\mu\gamma_5 T^a\Psi) = 2iM(\bar{\Psi}\gamma_5 T^a\Psi) + iK_n \varepsilon_{\mu_1\dots\mu_{2n}} tr(T^a F^{\mu_1\mu_2} \dots \dots F^{\mu_{2n-1}\mu_{2n}}), \mu_i = 1\dots 2n \quad (14.57)$$

As many times before, γ_5 of the 2n-dimensional theory corresponds to γ^0 of the associated (2n+1)-dimensional theory.

The induced non- abelian charge density is then given by

$$< j_a^0(x) > = \int_{-\infty}^{\infty} \frac{dk_0}{2\pi} \frac{m}{M} \frac{1}{2iM} \left\{ \frac{\partial_i \langle 0|\bar{\Psi}\gamma^i\gamma_0 T^a \Psi|0\rangle^A}{\langle 0|0\rangle^A}\Bigg|_M - iK_n \varepsilon_{\mu_1\dots\mu_{2n}} tr(T^a F^{\mu_1\mu_2}\dots F^{\mu_{2n-1}\mu_{2n}}) \right\} \quad (14.58)$$

whereby the contribution of the anomaly is contained in

$$< j_a^0(x) > = -g K_n \frac{1}{2} \varepsilon_{\mu_1\dots\mu_{2n}} tr(T^a F^{\mu_1\mu_2}\dots F^{\mu_{2n-1}\mu_{2n}}) \underbrace{\int_{-\infty}^{\infty} \frac{dk_0}{2\pi} \frac{m}{m^2+k_0^2}}_{=\frac{1}{2}\frac{m}{|m|}}$$

$$= -\frac{m}{|m|}\frac{K_n}{4} \varepsilon_{\mu_1\dots\mu_{2n}} tr(T^a F^{\mu_1\mu_2}\dots F^{\mu_{2n-1}\mu_{2n}}), \quad (14.59)$$

$$\mu_i = 1\dots 2n$$

By implementing both Lorentz and gauge covariance, we arrive at

$$< j_a^\mu(x) > = C_n \varepsilon^{\mu\mu_1\dots\mu_{2n}} tr(T^a F_{\mu_1\mu_2}\dots F_{\mu_{2n-1}\mu_{2n}}) \quad (14.60)$$

$$C_n = -\frac{m}{|m|}\frac{K_n}{4}, \quad \mu_i = 0, 1, 2, \dots, 2n$$

A slightly more elegant form of this important result arises from introducing differential forms

$$j_a(x) := \langle j_{a\mu}(x) \rangle dx^\mu$$

$$= C_n \underbrace{\varepsilon_\mu{}^{\mu_1 \cdots \mu_{2n}} dx^\mu}_{= * dx^{\mu_1} \wedge .. \wedge dx^{\mu_{2n}}} tr(T^a F_{\mu_1 \mu_2} \cdots F_{\mu_{2n-1} \mu_{2n}})$$

$$= 2^n C_n * tr(T^a \tfrac{1}{2} F_{\mu_1\mu_2} dx^{\mu_1} \wedge dx^{\mu_2} \ldots \tfrac{1}{2} F_{\mu_{2n-1}} dx^{\mu_{2n-1}} \wedge dx^{\mu_{2n}})$$

$$= (-i)^n 2^n C_n * tr(T^a F^n), \qquad \begin{array}{l} A \equiv i A_\mu dx^\mu, \ A_\mu = A_\mu^a T^a \\ F = dA + A^2 = \frac{i}{2} F_{\mu\nu} dx^\mu \wedge dx^\nu \\ F_{\mu\nu} \equiv F_{\mu\nu}^a T^a \end{array}$$

leaving us with the compact expression ($g \equiv 1$).

$$j^a(x) = (-i)^n 2^n C_n * tr(T^a F^n) \tag{14.61}$$

$$= -\frac{m}{|m|} \frac{1}{2} \left(\frac{i}{2\pi}\right)^n \frac{1}{n!} tr(T^a * F^n)$$

Specific examples are contained in

$D = 3$, $n = 1$, $g \equiv 1$, $m > 0$:

$$j^a = -\frac{1}{2} \frac{i}{2\pi} tr(T^a * F) = -\frac{i}{4\pi} tr(T^a * \tfrac{i}{2} F_{\alpha\beta} dx^\alpha dx^\beta)$$

$$= \frac{1}{8\pi} \underbrace{tr(T^a T^b)}_{= \frac{1}{2}\delta^{ab}} F^b_{\alpha\beta} \underbrace{*(dx^\alpha dx^\beta)}_{= \varepsilon_\mu{}^{\alpha\beta} dx^\mu}$$

$$= \frac{1}{16\pi} \varepsilon_{\mu\alpha\beta} F^{a\,\alpha\beta} dx^\mu$$

so that $$\langle j_{a\mu}(x) \rangle = \frac{1}{16\pi} \varepsilon_{\mu\alpha\beta} F_a{}^{\alpha\beta}$$

which reproduces our result (14.36).

Likewise, for D = 5, n = 2, g ≡ 1, m > 0:

$$j^a = -\frac{1}{2}\left(\frac{i}{2\pi}\right)^2 \frac{1}{2}\, tr(T^a * \mathcal{F}^2)$$

$$= \frac{1}{16\pi^2}\, tr\left(T^a * i^2 \frac{1}{4} \mathcal{F}_{\alpha\beta}\mathcal{F}_{\gamma\delta}\, dx^\alpha dx^\beta dx^\gamma dx^\delta\right)$$

$$= -\frac{1}{64\pi^2}\, tr(T^a \mathcal{F}_{\alpha\beta}\mathcal{F}_{\gamma\delta})\, \underbrace{*dx^\alpha dx^\beta dx^\gamma dx^\delta}_{=\varepsilon_\mu{}^{\alpha\beta\gamma\delta}\, dx^\mu}$$

$$\Rightarrow \langle j_{a\mu}\rangle = -\frac{1}{64\pi^2}\, \varepsilon_{\mu\alpha\beta\gamma\delta}\, tr(T^a \mathcal{F}^{\alpha\beta}\mathcal{F}^{\gamma\delta})$$

which is the result of our explicit calculation (14.39).

Given the vacuum current, equation (14.60), we can easily construct the effective action:

$$\Gamma[A] = -\int d^{2n+1}x \int_0^1 dt\; A^a_\mu(x) \langle j^\mu_a(tA(x))\rangle$$

$$= -C_n \int d^{2n+1}x \int_0^1 dt\; A^a_\mu\, \varepsilon^{\mu\mu_1\ldots\mu_{2n}}\, tr(T^a \mathcal{F}^t_{\mu_1\mu_2}\cdots \mathcal{F}^t_{\mu_{2n-1}\mu_{2n}})$$

$$= -C_n \int d^{2n+1}x \int_0^1 dt\; \underbrace{\varepsilon^{\mu\mu_1\ldots\mu_{2n}}}_{=*(dx^\mu dx^{\mu_1}\ldots dx^{\mu_{2n}})}\; tr(A_\mu \mathcal{F}^t_{\mu_1\mu_2}\cdots \mathcal{F}^t_{\mu_{2n-1}\mu_{2n}})$$

The x-integration is meant to be over the compactified space-time, i.e., over the (2n + 1) sphere, S^{2n+1}. Note that $\int dt$ is not a differential form. So we have

$$\Gamma[A] = -(-i)^{n+1} 2^n C_n \int_{S^{2n+1}} \int_0^1 dt\; * tr\Big(A \frac{1}{2}\mathcal{F}^t_{\mu_1\mu_2}\, dx^{\mu_1} dx^{\mu_2} \cdots \cdots \frac{1}{2}\mathcal{F}^t_{\mu_{2n-1}\mu_{2n}}\, dx^{\mu_{2n-1}} dx^{\mu_{2n}}\Big)$$

$$= -(-i)^{n+1}\, 2^n C_n \int_{S^{2n+1}} * \int_0^1 dt\, tr(A F_t^n) \tag{14.62}$$

In this last formula for $\Gamma[A]$, we introduce the particular parametrization

$$F_{t\mu\nu} \equiv F^t_{\mu\nu} = F_{\mu\nu}(tA)\,, \quad F_t = t\,dA + tA^2 \tag{14.63}$$

which yields

$$\Gamma[A] = -(-i)^{n+1}\, 2^n C_n \int_{S^{2n+1}} * \int_0^1 dt\, tr\left[A(t\,dA + t^2A^2)^n\right] \tag{14.64}$$

Now consider the n-th Chern character

$$\Omega_{2n}(A) = tr\, F^n \tag{14.65}$$

where $F = dA + A^2$ is the field strength 2-form. The (2n - 1)-form $\omega_{2n-1}(A)$ with

$$\Omega_{2n}(A) = d\omega_{2n-1} \tag{14.66}$$

was already given in chapter 13

$$\omega_{2n-1} = n\, tr \int_0^1 dt\, A(t\,dA + t^2A^2)^{n-1} \tag{14.67}$$

Replacing $n \to n + 1$ then yields

$$\omega_{2n+1} = (n+1)\, tr \int_0^1 dt\, A(t\,dA + t^2A^2)^n \tag{14.68}$$

which turns equation (14.64) into the compact form

$$\Gamma[A] = -(-i)^{n+1} \frac{2^n C_n}{n+1} \int_{S^{2n+1}} * \omega_{2n+1}(A) \tag{14.69}$$

where the dual of the (2n+1)-form ω_{2n+1} in (2n+1) dimensional space is just a scalar function (0-form) which has to be integrated over the compactified manifold S^{2n+1}.

Using the definition of ω as an n-form on an n-dimensional manifold M_n = infinitely large n-sphere, det g = 1, we have

$$\begin{aligned}\int_{M_n}\omega &= \int_{M_n}\omega_{\mu_1\ldots\mu_n}(x)\,dx^{\mu_1}\wedge\ldots\wedge dx^{\mu_n}\\ &= \int_{M_n}\omega_{\mu_1\ldots\mu_n}\,\varepsilon^{\mu_1\ldots\mu_n}\,dx^1\wedge\ldots\wedge dx^n\\ &= \int d^n u\;\omega_{\mu_1\ldots\mu_n}(x(u))\,\varepsilon^{\mu_1\ldots\mu_n}\\ &= \int d^n u\;{*\omega}(x(u))\end{aligned} \qquad (14.70)$$

where we assumed that there is a coordinate system u^μ which covers the manifold and applied the formula

$$\int f\,dx^1\wedge\ldots\wedge dx^n = \int f(u^1,\ldots,u^n)\,\underbrace{du^1\ldots du^n}_{\equiv d^n u} \qquad (14.71)$$

On the other hand, the dual of ω is given by

$$\begin{aligned}*\omega &= \omega_{\mu_1\ldots\mu_n}\,\underbrace{*\,dx^{\mu_1}\wedge\ldots\wedge dx^{\mu_n}}_{=\,\varepsilon^{\mu_1\ldots\mu_n}}\\ &= \varepsilon^{\mu_1\ldots\mu_n}\,\omega_{\mu_1\ldots\mu_n}\end{aligned}$$

so that (14.72)

$$\int_{M_n}\omega = \int d^n u\;{*\omega}(x(u))$$

and therefore, (14.69) can also be written as

$$\Gamma[A] = -\frac{2^n C^n}{(n+1)} (-i)^{n+1} \int_{S^{2n+1}} \omega_{2n+1}(A)$$

$$= -\frac{m}{|m|} \frac{\pi}{(n+1)!} \left(\frac{i}{2\pi}\right)^{n+1} \int_{S^{2n+1}} \omega_{2n+1}(A) \tag{14.73}$$

Let us use this formula to reproduce our former results:

D = 3, n = 1, m > 0:

$$\Gamma[A] = -\frac{\pi}{2}\left(\frac{i}{2\pi}\right)^2 \int_{S^3} \omega_3(A)$$

$$= \frac{1}{8\pi} \int_{S^3} 2 \int_0^1 dt \; tr\left[A(t dA + t^2 A^2)\right]$$

$$= \frac{1}{4\pi} \int_{S^3} tr\left[\frac{1}{2} A dA + \frac{1}{3} A^2\right]$$

$$= \frac{1}{8\pi} \int_{S^3} tr\left[AF - \frac{1}{3} A^3\right]$$

or

$$\Gamma[A] = \frac{1}{8\pi} \int d^3x \; * tr\left(AF - \frac{1}{3} A^3\right)$$

$$= -\frac{1}{8\pi} \int d^3x \; tr\left(\frac{1}{2} A_\mu F_{\alpha\beta} - \frac{i}{3} A_\mu A_\alpha A_\beta\right) \underbrace{*dx^\mu dx^\alpha dx^\beta}_{= \varepsilon^{\mu\alpha\beta}}$$

$$= -\frac{1}{8\pi} tr \int d^3x \; \varepsilon^{\mu\alpha\beta} \left(\frac{1}{2} A_\mu F_{\alpha\beta} - \frac{i}{3} A_\mu A_\alpha A_\beta\right)$$

which is precisely our result (14.45).

Similarly, for n = 2, D = 2n+1 = 5, m > 0:

$$\Gamma[A] = -\frac{\pi}{3\cdot 2}\left(\frac{i}{2\pi}\right)^3 \int_{S^5} \omega_5 = \frac{i}{48\pi^2}\int d^5_x * tr\left(A(dA)^2 + \frac{3}{2}A^3 dA + \frac{3}{5}A^5\right)$$

$$= \frac{i}{48\pi^2}\int d^5_x\, tr\left(i^3 A_\mu \partial_\alpha A_\beta \partial_\gamma A_\delta + \frac{3}{2} i^4 A_\mu A_\alpha A_\beta \partial_\gamma A_\delta + \frac{3}{5} i^5 A_\mu A_\alpha A_\beta A_\gamma A_\delta\right)$$

$$\cdot * dx^\mu dx^\alpha dx^\beta dx^\gamma dx^\delta$$

$$= \frac{1}{48\pi^2} tr \int d^5_x\, \varepsilon^{\mu\alpha\beta\gamma\delta}\Big(A_\mu \partial_\alpha A_\beta \partial_\gamma A_\delta + \frac{3}{2} i A_\mu A_\alpha A_\beta \partial_\gamma A_\delta$$

$$- \frac{3}{5} A_\mu A_\alpha A_\beta A_\gamma A_\delta\Big)$$

Again, we have reproduced the result of an explicit calculation, namely, eq. (14.48).

The induced particle number in (2n+1) dimensions is also easily obtained. Limiting ourselves to the $U_A(1)$ current ($T^a \to I$, $g \equiv 1$), we obtain from (14.61), $j = j_\mu dx^\mu$),

$$j = -\frac{m}{|m|}\frac{1}{2}\left(\frac{i}{2\pi}\right)^n \frac{1}{n!} tr(*T^n) \tag{14.74}$$

and for the particle number, i.e., U(1) charge,

$$Q_{2n+1} \equiv N = \int d^{2n}_x\, j^0(x) = C_n\, \varepsilon^{0\mu_1 \ldots \mu_{2n}} \int d^{2n}_x\, tr(F_{\mu_1\mu_2} \ldots F_{\mu_{2n-1}\mu_{2n}})$$

$$= C_n\, \varepsilon^{k_1 k_2 \ldots k_{2n}} \int d^{2n}_x\, tr(F_{k_1 k_2} \ldots F_{k_{2n-1} k_{2n}}) \tag{14.75}$$

Inserting the constant C_n yields

$$Q_{2n+1} = \frac{m}{|m|} \frac{(-1)^{n+1}}{2^{2n+1} \pi^n n!} \varepsilon^{k_1 k_2 \dots k_{2n}} \int d^{2n}x \, tr(\mathcal{F}_{k_1 k_2} \dots \mathcal{F}_{k_{2n-1} k_{2n}}) \tag{14.76}$$

Exploiting this formula, we can easily derive our former results:

D = 3, n = 1, m>0:

$$Q_3 = \frac{1}{8\pi} \varepsilon^{k_1 k_2} \int d^2x \, \mathcal{F}_{k_1 k_2}$$

an expression which we have obtained explicitly in (14.16).

D = 5, n = 2, m>0:

$$Q_5 = -\frac{1}{64\pi^2} \varepsilon^{k_1 k_2 k_3 k_4} \int d^4x \, tr(\mathcal{F}_{k_1 k_2} \mathcal{F}_{k_3 k_4})$$

$$= -\frac{1}{32\pi^2} \int d^4x \, tr(\mathcal{F}_{k_1 k_2} {}^*\mathcal{F}^{k_1 k_2})$$

$$= -\frac{1}{2} q[A]$$

$\left(= -\frac{1}{2} \text{ index } \not{D}_4 \right.$, cf. end of this chapter $\left.\right)$

Here we rediscovered our former results (14.50, 14.51)

Let us return to formula (14.76), which we express as

$$Q_{2n+1} = \frac{m}{|m|} \frac{(-1)^{n+1}}{2^{2n+1} \pi^n n!} 2^n (-i)^n * (dx^{k_1} \dots dx^{k_{2n}})$$

$$\cdot \int d^{2n}x \, tr\left(\frac{i}{2} \mathcal{F}_{k_1 k_2} \cdots \frac{i}{2} \mathcal{F}_{k_{2n-1} k_{2n}}\right)$$

$$= -\frac{m}{|m|} \frac{1}{2} \left(\frac{i}{2\pi}\right)^n \frac{1}{n!} \int d^{2n}x \; * tr \, \mathcal{F}^n$$

or

$$Q_{2n+1} = -\frac{m}{|m|}\frac{1}{2}\left(\frac{i}{2\pi}\right)^n \frac{1}{n!}\int_{R^{2n}} tr\, \mathcal{F}^n$$

$$= -\frac{m}{|m|}\frac{1}{2}\int_{R^{2n}} tr \sum_{m=0}^{\infty}\frac{1}{m!}\left(\frac{i}{2\pi}\mathcal{F}\right)^m$$

so that $$Q_{2n+1} = -\frac{1}{2}\frac{m}{|m|}\int_{R^{2n}} tr\, e^{\frac{i}{2\pi}\mathcal{F}} = -\frac{1}{2}\frac{m}{|m|}\int_{R^{2n}} ch(\mathcal{F})$$

or (cf. end of this chapter)

$$Q_{2n+1} = -\frac{1}{2}\frac{m}{|m|}\,\text{index}\, \not{D}_{2n} = -\frac{1}{2}\frac{m}{|m|}(n_+ - n_-)$$

Let us summarize our findings: in (2n+1) dimensions, the vacuum vector current density $\langle 0|\bar{\psi}\gamma^\mu T^a\psi|0\rangle$ contains an anomalous contribution which is determined by the non-abelian anomaly in 2n dimensions: $j^a(x) \sim *tr(T^aF^n)$. Recalling $\frac{\delta\Gamma}{\delta A^\mu} = -\langle j^\mu\rangle$, we conclude that the effective action in (2n+1) dimensions contains a term $\int\omega_{2n+1}$, where we note that $d\omega_{2n+1} = \Omega_{2n+2} = trF^{n+1}$ is the abelian anomaly in (2n+2) dimensions.

As another useful application, let us look at the "quantum Hall-effect" in QED_3. From equation (14.18), we obtain

$$\langle j^\mu(x)\rangle = C\,\varepsilon^{\mu\alpha\beta}\mathcal{F}_{\alpha\beta}\,,\quad C = \frac{m}{|m|}\frac{e^2}{8\pi} \tag{14.77}$$

This relation is exact for constant background fields.

The zero-component of (14.77) shows a pure B-dependence:

$$j^0 = C\,\varepsilon^{0ij} F_{ij} = 2C\, F_{12} = 2C\, B$$

whereas the vector component of $\langle j^\mu \rangle$ is only dependent on E:

$$j^1 = 2C\,\varepsilon^{102} F_{02}\,, \quad j^2 = 2C\,\varepsilon^{201} F_{01}\,, \quad \vec{j} = \vec{j}(\vec{E})$$

or $$j^1 = 2C\,E^2 = \pm\,\sigma_{12} E^2 \tag{14.78}$$

which shows that the "Hall conductivity" of the vacuum is given by

$$\sigma_{12} = \frac{1}{2}\,\frac{e^2}{2\pi} \tag{14.79}$$

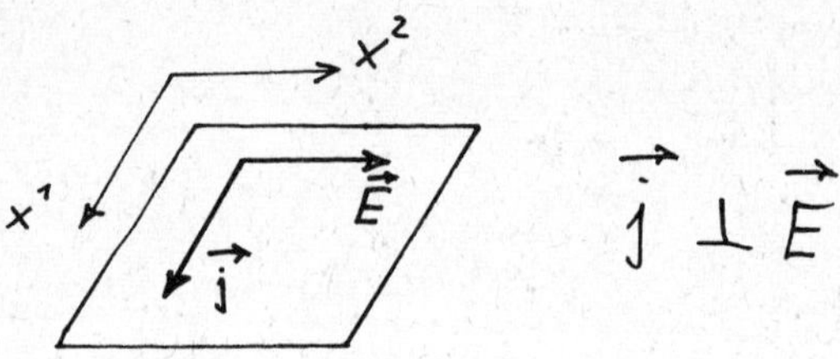

It has been noted that the occurrence of the topological mass term (for n = 3), e.g., (14.25), in the effective action is associated with the asymmetry in the spectrum of the Dirac operator between positive and negative eigenvalues. In this way, parity violation is introduced via quantum corrections. So let us have a closer look at the spectral asymmetry and its related "η-function". To this end, we return to equation (14.6) which contains the following expressions:

$$\gamma^0(i\tilde{\not{D}} + m) = \gamma^0(i\gamma^0\partial_0 + i\not{D}_2 + m) = -i\,\partial_0 + \gamma^0(i\not{D}_2 + m)$$

$$e^{-\tilde{\not{D}}^2 t} = e^{\partial_0^2 t}\, e^{-\not{D}_2^2 t}$$

Hence, we can also write $\langle j^0(x)\rangle$ as

$$\langle j^0(x)\rangle = -e\frac{d}{ds}\Big|_0 \frac{s}{\Gamma(1+s)}\int_0^\infty dt\, t^s \lim_{x'\to x} tr\Big[\{\underset{\substack{\downarrow\\ A}}{-i\partial_0} + \underset{\substack{\downarrow\\ B}}{\gamma^0(i\not{D}_2+m)}\}e^{\partial_0^2 t}e^{-(\not{D}_2^2+m^2)t}\Big]_x \delta^{(3)}(x-x')$$

To evaluate A and B, we follow the same steps that led us to equation (14.8), namely,

$$A = \lim_{x'\to x} tr\left[(-i\partial_0)e^{\partial_0^2 t}e^{-(\not{D}_2^2+m^2)t}\right]_x \int_{-\infty}^{\infty}\frac{dk_0}{2\pi}e^{ik_0(x-x')^0}\delta^{(2)}(\vec{x}-\vec{x}')$$

$$= \lim_{\vec{x}'\to\vec{x}} tr \underbrace{\int_{-\infty}^{\infty}\frac{dk_0}{2\pi}e^{-ik_0x^0}\underbrace{(-i\partial_0)e^{\partial_0^2 t}e^{ik_0x^0}}_{=k_0 e^{-k_0^2}}}_{=0} e^{-(\not{D}_2^2+m^2)t}\delta^{(2)}(\vec{x}-\vec{x}')$$

$$= 0$$

$$B = \lim_{\vec{x}'\to\vec{x}} tr \underbrace{\int_{-\infty}^{\infty}\frac{dk_0}{2\pi}e^{-ik_0x^0}\underbrace{e^{\partial_0^2 t}e^{ik_0x^0}}_{=e^{-k_0^2 t}}}_{=\frac{1}{2\pi}\sqrt{\pi/t}}\gamma^0(i\not{D}_2+m)e^{-(\not{D}_2^2+m^2)t}\delta^{(2)}(\vec{x}-\vec{x}')$$

$$= \frac{1}{2\sqrt{\pi}}t^{-\frac{1}{2}}\lim_{\vec{x}'\to\vec{x}} tr_\gamma\left[\gamma^0(i\not{D}_2+m)e^{-(\not{D}_2^2+m^2)t}\right]_x\delta^{(2)}(\vec{x}-\vec{x}')$$

The vacuum current density is therefore reduced to

$$\langle j^0(x)\rangle = -e\frac{1}{2\sqrt{\pi}}\frac{d}{ds}\Big|_0\frac{s}{\Gamma(1+s)}\int_0^\infty dt\, t^{s-\frac{1}{2}}\lim_{\vec{x}'\to\vec{x}} tr_\gamma\left[\gamma^0(i\not{D}_2+m)e^{-(\not{D}_2^2+m^2)t}\right]_x\delta^{(2)}(\vec{x}-\vec{x}')$$

and $Q = \int d^2x \langle j^0(x)\rangle$

$$= -\frac{e}{2\sqrt{\pi}}\frac{d}{ds}\Big|_0\frac{s}{\Gamma(1+s)}\int_0^\infty dt\, t^{s-\frac{1}{2}} tr_{\gamma x}\left[\gamma^0(i\not{D}_2+m)e^{-(\not{D}_2^2+m^2)t}\right] \qquad (14.80)$$

At this stage, we return to Minkowski space $\gamma^{0+} = \gamma^0$, $(\gamma^0)^2 = +1$. The Dirac operator which is important in the following analysis is stated in

$$H := \gamma^0(i\not{D}_2 + m) = i\gamma^0\gamma^k D_K + m\gamma^0$$
$$= \alpha^k iD_K + m\beta$$

$$\begin{pmatrix} \gamma^0\gamma^k =: \alpha^k \\ \gamma^0 =: \beta \end{pmatrix}$$

H is hermitean, since $\not{D}^+ = \not{D}$ $(+ = +_{\gamma X})$:

$$H^+ = [\gamma^0(i\not{D}_2 + m)]^+ = (-i\not{D}_2 + m)\gamma^0 = \gamma^0(i\not{D}_2 + m) = H \tag{14.81}$$

We also need the square of H:

$$H^2 = \gamma^0(i\not{D}_2 + m)\gamma^0(i\not{D}_2 + m) = \underbrace{(\gamma^0)^2}_{=1}(-i\not{D}_2 + m)(i\not{D}_2 + m)$$
$$= \not{D}_2^2 + m^2 \tag{14.82}$$

Following from (14.81), there exists a complete orthonormal system of eigenfunctions:

$$H\psi_n = \lambda_n \psi_n \ , \ \lambda_n \in \mathbb{R} \tag{14.83}$$

so that the induced charge is given by

$$Q = -\frac{e}{2\sqrt{\pi}} \frac{d}{ds}\Big|_0 \frac{s}{\Gamma(1+s)} \int_0^\infty dt\, t^{s-\frac{1}{2}}\, tr_{\gamma X}[He^{-H^2 t}]$$
$$= -\frac{e}{2\sqrt{\pi}} \sum_n \frac{d}{ds}\Big|_0 \frac{s}{\Gamma(1+s)} \int_0^\infty dt\, t^{s-\frac{1}{2}} \lambda_n e^{-\lambda_n^2 t} \tag{14.84}$$

Here we need the integral ($\tau = \lambda_n^2 t$)

$$\int_0^\infty dt\, t^{s-\frac{1}{2}} e^{-\lambda_n^2 t} = (\lambda_n^2)^{-(s+\frac{1}{2})} \int_0^\infty d\tau\, \tau^{(s+\frac{1}{2})-1} e^{-\tau}$$

$$= (\lambda_n^2)^{-(s+\frac{1}{2})} \Gamma(s+\tfrac{1}{2})$$

which we substitute into (14.84):

$$Q = -\frac{e}{2\sqrt{\pi}} \sum_n \lambda_n \frac{d}{ds}\Big|_0 s \frac{\Gamma(s+\frac{1}{2})}{\Gamma(s+1)} \underbrace{(\lambda_n^2)^{-(s+\frac{1}{2})}}_{= |\lambda_n|^{-2s} |\lambda_n|^{-1}}$$

$$= -\frac{e}{2\sqrt{\pi}} \frac{d}{ds}\Big|_0 s \frac{\Gamma(s+\frac{1}{2})}{\Gamma(s+1)} \sum_n \underbrace{\frac{\lambda_n}{|\lambda_n|}}_{= \mathrm{sign}(\lambda_n)} |\lambda_n|^{-2s} \tag{14.85}$$

At this point we introduce the spectral function

$$\eta(H,s) = \sum_n |\lambda_n|^{-s} \mathrm{sign}(\lambda_n) \tag{14.86}$$

which converges for $\mathrm{Re}(s)$ sufficiently large (s) and has a meromorphic continuation to the entire complex s-plane. Moreover, s = 0 is not a pole, and $\eta(0)$, the eta invariant, is finite. These facts will be used in (14.85):

$$Q = -\frac{e}{2\sqrt{\pi}} \Big\{ \frac{\Gamma(s+\frac{1}{2})}{\Gamma(s+1)} \eta(H,2s) + \underbrace{s \frac{d}{ds} \frac{\Gamma(s+\frac{1}{2})}{\Gamma(s+1)} \eta(H,2s)}_{\text{continuous at } s = 0} \Big\} \Big|_{s=0}$$

$$= -\frac{e}{2\sqrt{\pi}} \frac{\Gamma(\frac{1}{2})}{\Gamma(1)} \eta(H,0)$$

$$= -\frac{e}{2} \eta(H,0)$$

The final result is then given by

$$Q = -\frac{e}{2} \lim_{s \to 0^+} \eta(H,s) \tag{14.87}$$

where $$\eta(H,s) = \sum_n |\lambda_n|^{-s} \, sign(\lambda_n)$$

These results are extended to (2n+1) dimensions. In the (2 +1) dimensional formula (14.14),

$$\langle j^0(x) \rangle = e \int \frac{dk_0}{2\pi} \frac{m}{2iM^2} \left\{ \frac{\partial_j \langle 0|\bar{\psi}\gamma^j\gamma^0\psi|0\rangle^A}{\langle 0|0\rangle^A} + \frac{ie}{2\pi} \varepsilon_{ij} F^{ij} \right\}$$

we merely need to replace the two-dimensional Pontryagin density $\frac{1}{2\pi}{}^*F$ by the (2n)-dimensional one, or by the (2n)-dimensional Pontryagin index $q_{2n}[A_j]$ in Q:

$$Q = e \int_{-\infty}^{\infty} \frac{dk_0}{2\pi} \frac{m}{2iM^2} \left\{ \oint df_j \frac{\langle 0|\bar{\psi}\gamma^j\gamma^0\psi|0\rangle^A}{\langle 0|0\rangle^A} + const \underbrace{q_{2n}[A_j]}_{\text{topol. invariant}} \right\} \tag{14.88}$$

Although the first term can lead to nonlocal functions of $A_\mu(x)$, we want to limit ourselves to a subclass of models where only local terms survive for $R \to \infty$. Then $\oint df_j \, \%$ is invariant under local deformations of $A_\mu(x)$. Hence, Q and $\eta(H,0)$ are topological invariants with respect to local deformations of $A_\mu(x)$.

At last we come to the index theorem for the Dirac operator

$$H = \alpha^k iD_k + \beta m \tag{14.89}$$

In contrast to chapter 12 where we were interested in the difference $\nu = n_+ - n_-$ between the number of eigenstates of the Dirac operator $\not{D}(A^a_\mu)$ with vanishing eigenvalues of positive and negative chirality, $\gamma_5' = \pm 1$, we are now investigating the parity or conjugation matrix $\gamma^0 = \beta$ with eigenvalues $\gamma_0' = \pm 1$. Let us first discuss the massless case, m = 0 (conjugation symmetrical case). Then, to every eigenvalue $\lambda_n > 0$, there also exists an eigenvalue $-\lambda_n < 0$. This follows from the anti-commutator (cf. 14.3) (14.90)

$$\{\beta, H(m=0)\} = 0 \tag{14.91}$$

$$H(\beta\psi_n) = -\beta H\psi_n = -\lambda_n(\beta\psi_n)$$

where (14.92)

$$H\psi_n = \lambda_n \psi_n$$

The spectrum lies symmetrically around $\lambda_n = 0$. Hence, only the zero modes of H contribute to the η -invariant.

Earlier in this chapter we found, eq. (14.16),

$$Q = \frac{m}{|m|}\frac{e^2}{2}\Phi = e\,N = \frac{m}{|m|}\frac{e}{2}q[A] \tag{14.93}$$

flux fermion number Pontryagin index

Therefore, $Q = \frac{m}{|m|}\frac{e}{2}q[A] \overset{(87)}{=} -\frac{e}{2}\eta(H,0)$ (14.94)

or $\eta(H,0) = \mp q[A] = \pm\,\mathrm{index}(\not{D}_2|S_+)$ (14.95)

(Our convention: $\mathrm{index}(\not{D}_2/S_+) = -q[A] = n_+ - n_-$)

From now on, we follow the procedure outlined in chapter 12 where we discussed chirality instead of parity.

$$\eta(H,0) = \pm\{\dim\mathrm{Ker}\,\not{D}_2|S_+ - \dim\mathrm{Ker}\,\not{D}_2|S_-\} \tag{14.96}$$

Using the relation $H(m = 0) = \gamma^0 i\not{D}_2$ we see that $H(m = 0)$ and $H^{+}(m = 0)$ have the same zero modes as $\not{D}_2$, so that index $\not{D}_2/S_+$ = index $H(m = 0)/S_+$.

Our result is then summarized in the formulae

$$\eta(H,0) = \pm \text{ index } H(m=0)|S_+ \tag{14.97}$$

$$Q = -\frac{e}{2}\frac{m}{|m|} \text{ index } H(m=0)|S_+ \tag{14.98}$$

It is important to note that the spectral asymmetry, i.e., the induced charge of the vacuum, is solely determined by the zero-modes of $H(m = 0)$; although (14.98) is also true for $m \neq 0$, only zero modes contribute.

Going back to equation (14.8), we can write it as

$$\langle j^0(x)\rangle = -em\frac{d}{ds}\Big|_0 \frac{s}{\Gamma(1+s)}\int_0^\infty dt\, t^s e^{-m^2 t}\underbrace{\int\frac{dk_0}{2\pi}e^{-k_0^2 t}}_{=\frac{1}{2\sqrt{\pi}}t^{-\frac{1}{2}}}\lim_{\vec{x}'\to\vec{x}} tr_\gamma\left[\gamma^0 e^{-\not{D}_2^2 t}\right]_{\vec{x}}\delta^{(2)}(\vec{x}-\vec{x}')$$

or $$Q = \int d^2x\langle j^0(x)\rangle = -\frac{em}{2\sqrt{\pi}}\frac{d}{ds}\Big|_0\frac{s}{\Gamma(1+s)}\int_0^\infty dt\, t^{s-\frac{1}{2}}e^{-m^2 t}\, tr_{\gamma x}\left[\gamma^0 e^{-\not{D}_2^2 t}\right] \tag{14.99}$$

Our statement is that $tr_{\gamma x}\left[\gamma^0 e^{-\not{D}_2^2 t}\right]$ is determined alone by the zero-modes. With $P_\pm = \frac{1}{2}(1\pm\gamma^0)$ and $\gamma^0 = P_+ + P_-$, $S = S_+ \cup S_-$, $\mathrm{Tr} \equiv tr_{\gamma x}$, we can write

$$\begin{aligned} Tr_S\left[\gamma^0 e^{-\not{D}_2^2 t}\right] &= Tr_S\left[P_+ e^{-\not{D}_2^2 t} - P_- e^{-\not{D}_2^2 t}\right] \\ &= \text{index }(\not{D}_2/S_+), \qquad \forall\, t = 0 \\ &= \underset{\gamma^{0\prime}=1}{n_+} - \underset{\gamma^{0\prime}=-1}{n_-} \end{aligned} \tag{14.100}$$

We use this information in (14.99) and end up with

$$Q = -\frac{em}{2\sqrt{\pi}}\ \text{index}\,(\not{D}_2|S_+)\ \frac{d}{ds}\Big|_0 s\,\frac{1}{\Gamma(s)}\underbrace{\int_0^\infty dt\, t^{s-\frac{1}{2}}\, e^{-m^2 t}}_{=(m^2)^{-(s+\frac{1}{2})}\,\Gamma(s+\frac{1}{2})}$$

$$= -\frac{em}{2\sqrt{\pi}}\ \text{index}\,(\not{D}_2|S_+)\ \underbrace{\frac{d}{ds}\Big|_0 s\,\frac{\Gamma(s+\frac{1}{2})}{\Gamma(s+1)}\,(m^2)^{-(s+\frac{1}{2})}}_{\frac{\Gamma(\frac{1}{2})}{\Gamma(1)}(m^2)^{-\frac{1}{2}} = \sqrt{\pi}\,\frac{1}{|m|}}$$

$$= -\frac{e}{2}\,\frac{m}{|m|}\ \text{index}\,(\not{D}_2|S_+) \tag{14.101}$$

$$= \frac{e}{2}\,\frac{m}{|m|}\, q[A] \quad \text{, as before.} \tag{14.102}$$

In particular for the configuration with index 1, we obtain for the vacuum charge in three dimensions $Q_3 = \pm\frac{1}{2}e$.

Additional References to
Effective Gauge Theory Actions in (2n+1) Dimensions

A.J. Niemi, G.W. Semenoff, Phys. Rev. Lett. 51, 2077 (1983)

A.N. Redlich, Phys. Rev. Lett. 52, 18 (1984)

R.J. Hughes, Phys. Lett. 148B, 215 (1984)

K. Ishikawa, Phys. Rev. Lett. 53, 1615 (1984)

K. Ishikawa, Phys. Rev. D31, 1432 (1985)

R. Jackiw, C. Rebbi, Phys. Rev. D13, 3398 (1976)

J. Goldstone, F. Wilczek, Phys. Rev. Lett. 49, 986 (1981)

M.B. Paranjape, G.W. Semenoff, Phys. Lett. 132B, 369 (1983)

A.J. Niemi, G.W. Semenoff, Phys. Lett. 135B, 121 (1984)

S. Midorikawa, Phys. Rev. D31, 1499 (1985)

A.J. Niemi, G.W. Semenoff, Phys. Rev. D30, 809 (1984)

A.J. Niemi, G.W. Semenoff, Nucl. Phys. B251 FS 13 , 155 (1985)

A.J. Niemi, G.W. Semenoff, Nucl. Phys. B253, 14 (1985)

R. MacKenzie, F. Wilczek, Phys. Rev. D30, 2194 and 2260 (1984)

R. MacKenzie, "Vacuum Polarization by Solitons", Santa Barbara Preprint NSF-ITP-84-135 (1984)

R. Jackiw, J.R. Schrieffer, Nucl. Phys. B190 FS 3 , 253 (1981)

(continued next page)

Additional References to

Effective Gauge Theory Actions in (2n+1) Dimensions, continued

R. Jackiw, Comm. Nucl. Part. Phys. 13, 15 (1984)

R. Jackiw, Phys. Rev. D29, 2375 (1984)

S. Deser, R. Jackiw, S. Templeton, Phys. Rev. Lett. 48, 975 (1982); Ann. Phys. 140, 372 (1982)

R. Jackiw in: Asymptotic Realms of Physics, edited by A. Guth, K. Wan, R. Jaffe, MIT-Press, Cambridge (1983)

Differential Forms:

H. Flanders, "Differential Forms", Academic Press (1963)

T. Eguchi, P. Gilkey, A. Hanson, Phys. Rep. 66, 213 (1980)

C. Nash, S. Sen, "Topology and Geometry for Physicists", Academic Press, London (1983)

15. Chiral Anomaly with Gravitational Background Field

So far, we formulated QED as a theory with local gauge invariance, i.e., we changed the phase of any charge-carrying field at each space-time point while simultaneously changing the electromagnetic vector potential by the derivative of the corresponding scalar function, $\partial_\mu \Lambda(x)$. Then we developed a similar idea in connection with non-abelian Yang-Mills fields. The equations governing the theory of strong interaction, QCD, were based on the invariance with respect to an arbitrary rotation at each space-time point in 8-dimensional color space. This invariance implies the "existence" of the eight color carrying vector fields $A^a_\mu(x)$.

Now, following an old idea of Weyl's, we want to introduce the gravitational field in a similar way, namely, by demanding invariance under an arbitrary change (gauge transformation) in the orientation at each point with a simultaneous change in the gravitational potential.

The fields to be used to set up our gauge covariant equations for a spin-carrying field in presence of an external gravitational field are the 4×4 vierbein fields $h_a{}^\mu(x)$ and the 4×6 vierbein connections $(A_\mu)_{ab}(x) = -(A_\mu)_{ba}(x)$. These are vector fields with respect to general coordinate transformations

$$h'_a{}^\mu(x') = \frac{\partial x'^\mu}{\partial x^\nu} h_a{}^\nu(x) \tag{15.1}$$

$$(A'_\mu)_{ab}(x') = \frac{\partial x^\nu}{\partial x'^\mu} (A_\nu)_{ab}(x) \tag{15.2}$$

with $\mu, \nu, \ldots, a, b, \ldots = 0, 1, 2, 3$.

The response to a local Lorentz transformation is

$$h'_a{}^\mu(x) = \Lambda_a{}^b(x)\, h_b{}^\mu(x) \tag{15.3}$$

$$(A'_\mu)_{cb}(x) = \Lambda_a{}^c(x)\Lambda_b{}^d(x)(A_\mu)_{cd}(x) + \Lambda_b{}^c(x)\,\partial_\mu \Lambda_{ac}(x) \tag{15.4}$$

where $\Lambda_a{}^c(x)\,\eta_{cd}\,\Lambda_b{}^d(x) = \eta_{ab} := diag(-1,+1,+1,+1)$ (15.5)

The fermionic Lagrangian is given by

$$\mathcal{L} = -\bar{\psi}\left(\frac{1}{i}\gamma^\mu D_\mu + m\right)\psi \tag{15.6}$$

where the covariant derivative of a Dirac spinor is defined by

$$D_\mu := \partial_\mu - \frac{1}{2}\Sigma_{ab}\,A_\mu^{ab} \tag{15.7}$$

with $\Sigma_{ab} := \frac{1}{4}[\gamma^a, \gamma^b]\,,\quad \{\gamma^a, \gamma^b\} = -2\eta^{ab}$ (15.8)

$$\gamma^\mu(x) = h_a{}^\mu(x)\,\gamma^a\,,\quad A_{\mu ab}(x) = h_a{}^\rho(x)\,h_{\rho b;\mu}(x) \tag{15.9}$$

and

$$\begin{aligned}\{\gamma^\mu(x), \gamma^\nu(x)\} &= h_a{}^\mu(x)\,h_b{}^\nu(x)\,\{\gamma^a, \gamma^b\} = -2\,h_a{}^\mu(x)\,h^{a\nu}(x)\\ &= -2\,g^{\mu\nu}(x)\end{aligned} \tag{15.10}$$

In the future, we will always understand a semicolon as a covariant derivative that contains only Γ-affine connections as in

$$T_{\dots\nu\dots;\mu}(x) = T_{\dots\nu\dots,\mu}(x) - \Gamma^\lambda_{\nu\mu}\,T_{\dots\lambda\dots}(x) + \dots$$

Now we perform a Wick rotation in the local Lorentz frame ξ^a:

$$\xi^0 = i\xi^4\,,\quad \gamma^0 = -i\gamma^4\,,\quad h_0{}^\mu = i\,h_4{}^\mu\,,\quad h^0{}_\mu = -i\,h^4{}_\mu \tag{15.11}$$

which follows from

$$h_a{}^\mu = \frac{\partial x^\mu}{\partial \xi^a} \quad : \quad h_0{}^\mu = \frac{\partial x^\mu}{\partial \xi^0} = \frac{\partial x^\mu}{\partial(-i\xi^4)} = i\,\frac{\partial x^\mu}{\partial \xi^4} \equiv i\,h_4{}^\mu \tag{15.12}$$

$$h^a{}_\mu = \frac{\partial \xi^a}{\partial x^\mu} \; : \; h^0{}_\mu = \frac{\partial \xi^0}{\partial x^\mu} = \frac{\partial(-i\xi^4)}{\partial x^\mu} = -i\,h^4{}_\mu \tag{15.13}$$

The metric for the local Lorentz frame is then altered according to

$$\eta = (-+++) \longrightarrow \eta = (++++) \tag{15.14}$$

and

$$\gamma_5 = i\gamma^0\gamma^1\gamma^2\gamma^3 = \gamma^4\gamma^1\gamma^2\gamma^3 \tag{15.15}$$

remains unchanged.

Since

$$g_{\mu\nu} = h^a{}_\mu\, h^b{}_\nu\, \eta_{ab} \tag{15.16}$$

and

$$g \equiv \det(g_{\mu\nu}) = \left(\det(h^a{}_\mu)\right)^2 \det(\eta_{ab}) \tag{15.17}$$

we have

in Minkowski space: $\det\eta = -1 \quad : \quad g = -\left(\det(h^a{}_\mu)\right)^2$ (15.18)

in Euclidean space: $\det\eta = +1 \quad : \quad g = \left(\det(h^a{}_\mu)\right)^2 \equiv h^2$ (15.19)

so that

$$h(x) = \det(h^a{}_\mu(x)) = \sqrt{g(x)} \tag{15.20}$$

This means that

$$h^M(x) = -i\,h^E(x) \tag{15.21}$$

The action is then changed according to

$$i\int d^4x\, h^M(x)\, \mathcal{L}^M(x) = i\int d^4x\,(-i\, h^E(x))\, \mathcal{L}^E(x)$$

$$= +\int d^4x\, h^E\, \mathcal{L}^E(x)$$

(15.22)

Notice that $$\{\gamma^a, \gamma^b\} = -2\delta^{ab} \tag{15.23}$$

and all the γ matrices are anti-hermitean:

$$\gamma^{a+} = -\gamma^a \tag{15.24}$$

so that

$$\Sigma_{ab}^{+} = \frac{1}{4}(\gamma_a\gamma_b - \gamma_b\gamma_a)^{+}$$

$$= \frac{1}{4}(\gamma_b^{+}\gamma_a^{+} - \gamma_a^{+}\gamma_b^{+})$$

$$\underset{(24)}{=} \frac{1}{4}(\gamma_b\gamma_a - \gamma_a\gamma_b)$$

$$= -\Sigma_{ab} \tag{15.25}$$

One can show that the (euclidean) operator $\not{D} = \gamma^\mu(\partial_\mu - \frac{1}{2}A_\mu^{ab}\Sigma_{ab})$ is hermitean with regard to the scalar product

$$(\varphi_1, \varphi_2) = \int d^4x\, h(x)\, \varphi^*_{1\alpha}(x)\, \varphi_{2\alpha}(x)$$

(α is a spinor index), i.e., it holds that

$$(\not{D}\varphi_1, \varphi_2) = (\varphi_1, \not{D}\varphi_2)$$

(15.26)

From now on, the discussion follows the previous chapters on chiral anomalies very closely.

So, first, we study the response of the functional integration measure under chiral transformation

$$\psi(x) \rightarrow \tilde{\psi}(x) = e^{i\alpha(x)\gamma_5}\psi(x)$$

$$\bar{\psi}(x) \rightarrow \tilde{\bar{\psi}}(x) = \bar{\psi}(x)e^{i\alpha(x)\gamma_5} \tag{15.27}$$

with the result

$$[d\psi d\bar{\psi}] \longrightarrow [d\tilde{\psi}d\tilde{\bar{\psi}}] = J[\alpha]\,[d\psi d\bar{\psi}] \tag{15.28}$$

Recall that

$$Z = \int [d\psi d\bar{\psi}]\, e^{-i\int d^4x\, h^M(x)\,\bar{\psi}(-i\not{D}+m)_M\psi}$$

$$= \int [d\psi d\bar{\psi}]\, e^{-\int d^4x\, h^E(x)\,\bar{\psi}(-i\not{D}+m)_E\psi} \tag{15.29}$$

So, the euclidean version of Z changes as follows:

$$\int [d\tilde{\psi}d\tilde{\bar{\psi}}]\, e^{-\int d^4x\, h\,\tilde{\bar{\psi}}(-i\not{D}+m)\tilde{\psi}} = \det[i\not{D}+m]$$

$$= J[\alpha]\int [d\psi d\bar{\psi}]\, e^{-\int d^4x\, h\,\bar{\psi}e^{i\alpha\gamma_5}(-i\not{D}+m)e^{i\alpha\gamma_5}\psi}$$

$$= J[\alpha]\,\det\left[e^{i\alpha\gamma_5}(-i\not{D}+m)e^{i\alpha\gamma_5}\right]$$

or

$$J[\alpha] = \frac{\det[-i\not{D}+m]}{\det\left[e^{i\alpha\gamma_5}(-i\not{D}+m)e^{i\alpha\gamma_5}\right]}$$

As many times before, we can evaluate

$$e^{i\alpha\gamma_5}(-i\not{D}+m)\,e^{i\alpha\gamma_5}\psi =$$

$$= \left[-i\not{D}+m+\partial_\mu\alpha\,\gamma^\mu\gamma_5+2im\alpha\gamma_5+O(\alpha^2)\right]\psi \tag{15.30}$$

Then we write

$$J[\alpha] = \frac{\det[-i\not{D}+m]}{\det[-i\not{D}+m+\partial_\mu\alpha\,\gamma^\mu\gamma_5+2im\alpha\gamma_5]}\cdot\frac{\det[i\not{D}+m]}{\det[i\not{D}+m]}$$

$$= \frac{\det[\not{D}^2+m^2]}{\det[\Omega]} \tag{15.31}$$

with $$\Omega := \not{D}^2+m^2+\partial_\mu\alpha\,\gamma^\mu\gamma_5(i\not{D}+m)+2im\alpha\gamma_5(i\not{D}+m) \tag{15.32}$$

which is hermitean and positive for $\alpha = 0$.

Taking the logarithm of (15.31) yields

$$\ln J[\alpha] = -\left\{\ln\det\Omega-\ln\det[\not{D}^2+m^2]\right\}+O(\alpha^2)$$

$$= -\int d^4w\, h(w)\,\alpha(w)\frac{\delta}{\delta\alpha(w)}\ln\det\Omega\Big|_{\alpha=0}+O(\alpha^2)$$

$$= \int d^4w\, h(w)\,\alpha(w)\frac{d}{ds}\Big|_0\frac{\delta\,\zeta(\Omega|s)}{\delta\alpha(w)}\Big|_{\alpha=0}+O(\alpha^2) \tag{15.33}$$

Now we consider a complete set of eigenfunctions belonging to Ω :

$$\Omega_{x\alpha\beta}\,\Phi_{n\beta}(x) = \Lambda_n\,\Phi_{n\alpha}(x) \tag{15.34}$$

with

$$\int d^4x\; h(x)\,\Phi^*_{n\alpha}(x)\,\Phi_{n'\alpha}(x) = \delta_{nn'}$$

and

$$\sum_n \Phi^*_{n\alpha}(x)\,\Phi_{n\beta}(x') = \delta_{\alpha\beta}\,\frac{\delta(x-x')}{h(x)}$$

Then we obtain

$$\frac{\delta\zeta(\Omega|s)}{\delta\alpha(w)} = -s\sum_n \Lambda_n^{-(1+s)}\int d^4z\; h(z)\,\Phi_n^+(z)\,\frac{\delta\Omega_z}{\delta\alpha(w)}\,\Phi_n(z) \tag{15.35}$$

Using (15.36)

$$\Omega(\alpha=0) = \not{D}^2+m^2,\quad (\not{D}^2+m^2)\varphi_n = \lambda_n\varphi_n$$

we have (15.37)

$$\frac{\delta\zeta(\Omega|s)}{\delta\alpha(w)}\bigg|_{\alpha=0} = -s\sum_n \lambda_n^{-(1+s)}\int d^4z\; h(z)\,\varphi_n^+(z)\,\frac{\delta\Omega_z}{\delta\alpha(w)}\bigg|_{\alpha=0}\varphi_n(z)$$

and so we end up with

$$\ln J[\alpha] = -\int d^4w\; h(w)\,\alpha(w)\,\frac{d}{ds}\bigg|_0 s\sum_n \lambda_n^{-(1+s)}\cdot \tag{15.38}$$

$$\cdot\int d^4z\; h(z)\,\varphi_n^+(z)\,\frac{\delta\Omega_z}{\delta\alpha(w)}\bigg|_{\alpha=0}\varphi_n(z)$$

$$=: -i\int d^4w\;\alpha(w)\,A(w)$$

where $A(w) = \frac{i}{2}\,h(w)\,\frac{d}{ds}\Big|_0 s\sum_n \lambda_n^{-(1+s)}\int d^4z\; h(z)\,\varphi_n^+(z)\,\frac{\delta\Omega_z}{\delta\alpha(w)}\Big|_{\alpha=0}\varphi_n(z)$ (15.39)

The quantity A(x) has been evaluated by various people with the result

$$A(x) = -\frac{1}{384\pi^2}\,\varepsilon^{\rho\sigma\alpha\beta}\,R_{\mu\nu\rho\sigma}\,R^{\mu\nu}{}_{\alpha\beta} \tag{15.40}$$

$$\varepsilon^{\rho\sigma\alpha\beta}(x) := h(x)\,\varepsilon^{abcd}\left(h_a{}^\rho h_b{}^\sigma h_c{}^\alpha h_d{}^\beta\right),\quad \varepsilon^{1234} = +1$$

and use has been made of

$$\begin{aligned}[D_\mu, D_\nu]\psi &= \tfrac{1}{2} R_{\mu\nu}{}^{ab}\,\Sigma_{ab}\psi \\ &= \tfrac{1}{4} R_{\mu\nu}{}^{ab}\,\gamma_a\gamma_b\psi \\ &= \tfrac{1}{4} R_{\mu\nu\rho\sigma}\,\gamma^\rho\gamma^\sigma\psi\end{aligned} \tag{15.41}$$

with $$R_{\mu\nu}{}^{ab} = -A^{ab}_{\mu,\nu} + A^{ab}_{\nu,\mu} - A^{ac}_\mu A_{\nu c}{}^b + A^a_{\nu c} A_\mu{}^{cb} \tag{15.42}$$

With the aid of (15.40), it is then easy to obtain the divergence of the axial vector current. We merely need to combine A(x) with the change of the Lagrangian under chiral transformation (cf. eq. (15.30)). Here are the simple steps:

$$[d\tilde\psi\, d\tilde{\bar\psi}] = J[\alpha]\,[d\psi\, d\bar\psi] = e^{-i\int d^4x\,\alpha(x)\,A(x)}\,[d\psi\, d\bar\psi]$$

$$\Rightarrow \int [d\tilde\psi\, d\tilde{\bar\psi}]\, e^{-\int d^4x\, h(x)\,\tilde{\bar\psi}(-i\not{D}+m)\tilde\psi}$$

$$= \int [d\psi\, d\bar\psi]\, e^{-\int d^4x\,\alpha(x)\, i A(x)}\; e^{-\int d^4x\, h(x)\,\bar\psi\left[-i\not{D}+m+\partial_\mu\alpha\,\gamma^\mu\gamma_5 + 2im\alpha\gamma_5\right]\psi}$$

$$= \int [d\psi\, d\bar\psi]\, \underbrace{e^{-\int d^4x\,\alpha(x)\left\{iA(x) - \partial_\mu\left(h\bar\psi\gamma^\mu\gamma_5\psi\right) + 2imh\bar\psi\gamma_5\psi\right\}}}_{=1-\int d^4x\,\{\ldots\}\alpha(x)+O(\alpha^2)}\; e^{-\int d^4x\, h\bar\psi(-i\not{D}+m)\psi}$$

integr. b. parts

So we obtain the relation

$$\partial_\mu h(x) \langle \bar{\psi} \gamma^\mu \gamma_5 \psi \rangle = 2im\, h(x) \langle \bar{\psi} \gamma_5 \psi \rangle + i\, A(x) \tag{15.43}$$

or $$\partial_\mu h(x) \langle \bar{\psi} \gamma^\mu \gamma_5 \psi \rangle = 2im\, h \langle \bar{\psi} \gamma_5 \psi \rangle - \frac{i}{384\pi^2} \varepsilon^{\rho\sigma\alpha\beta} R_{\mu\nu\rho\sigma} R^{\mu\nu}{}_{\alpha\beta}$$

Employing the relation

$$\frac{1}{\sqrt{g}} \partial_\mu \left(\sqrt{g} \langle \bar{\psi} \gamma^\mu \gamma_5 \psi \rangle \right) = \langle \bar{\psi} \gamma^\mu \gamma_5 \psi \rangle_{;\mu}$$

we obtain for the derivative of the axial-vector current in a gravitational background field

$$\langle \bar{\psi} \gamma^\mu \gamma_5 \psi \rangle_{;\mu} = 2im \langle \bar{\psi} \gamma_5 \psi \rangle - \frac{i}{384\pi^2} \frac{\varepsilon^{\rho\sigma\alpha\beta}}{h(x)} R_{\mu\nu\rho\sigma} R^{\mu\nu}{}_{\alpha\beta} \tag{15.44}$$

Finally, we want to apply the index theorem to gravity. In close analogy to Yang-Mills theory, this gives the difference between the number of positive chirality and negative chirality zero-frequency solutions to the Dirac equation.

The quickest way to derive the index theorem is to investigate the limit $m \rightarrow 0$ in (15.43) and to notice that the expression of the divergence of the axial vector current (on the way to (15.43)) was given by

$$\partial_\mu h \langle \bar{\psi} \gamma^\mu \gamma_5 \psi \rangle = 2im\, h \langle \bar{\psi} \gamma_5 \psi \rangle + 2ih \left. \frac{d}{ds} \right|_0 s \sum_n \frac{\varphi_n^+ \gamma_5 \varphi_n}{\lambda_n^s} \tag{15.45}$$

$$\langle \bar{\psi} \gamma_5 \psi \rangle = -m \left. \frac{d}{ds} \right|_0 s \sum_n \frac{\varphi_n^+ \gamma_5 \varphi_n}{\lambda_n^{1+s}}$$

Now,

$$\lim_{m \to 0} m \langle \bar{\psi} \gamma_5 \psi \rangle = \lim_{m \to 0} (-m^2) \left. \frac{d}{ds} \right|_0 s \sum_n \frac{\varphi_n^+ \gamma_5 \varphi_n}{\lambda_n^{1+s}}$$

$$= \lim_{m\to 0} (-m^2) \frac{d}{ds}\Big|_0 \, s \left(\sum_n{}' \frac{\varphi_n^+ \gamma_5 \varphi_n}{(m^2)^{1+s}} + \sum{}'' \frac{\varphi_n^+ \gamma_5 \varphi_n}{\lambda_n^{1+s}} \right)$$

$$= \lim_{m\to 0} \Big(\underbrace{\frac{-m^2}{m^2} \frac{d}{ds}\Big|_0 \, s(m^2)^{-s}}_{=1} \sum_n{}' \varphi_n^+ \gamma_5 \varphi_n - \underbrace{m^2 \frac{d}{ds}\Big|_0 \, s \sum{}'' \frac{\varphi_n^+ \gamma_5 \varphi_n}{\lambda_n^{1+s}}}_{\text{finite} \xrightarrow[m\to 0]{} 0} \Big)$$

$$= - \sum_n{}' \varphi_n^+ \gamma_5 \varphi_n \tag{15.46}$$

where the sum only runs over zero-modes.

Hence, the integrated version of eq. (15.43) in the limit m$\to$0 yields

($\int_M d^4x \, \partial_\mu (\ldots) = \int_{\partial M = 0} (\ldots) = 0$, indicating that the index theorem holds true only for M with $\partial M = 0$.)

$$-2i \int d^4x \, h(x) \sum_n{}' \varphi_n^+(x) \underbrace{\gamma_5 \varphi_n(x)}_{= \chi_n \varphi_n(x)} + i \int d^4x \, A(x) = 0$$

or

$$2 \underbrace{\sum_n{}' \chi_n}_{n_+ - n_-} \cdot \int d^4x \, h(x) \, \varphi_n^+(x) \varphi_n(x) = \int d^4x \, A(x)$$

or

$$n_+ - n_- = \frac{1}{2} \int d^4x \, A(x) \tag{15.47}$$

Our final result is then given by

$$n_+ - n_- = - \frac{1}{768\pi^2} \int d^4x \, \varepsilon^{\rho\sigma\alpha\beta} R_{\mu\nu\rho\sigma} R^{\mu\nu}{}_{\alpha\beta} \in \mathbb{Z} \tag{15.48}$$

It is perhaps not uninteresting to study chiral anomalies coupled to both gauge and gravitational fields in even space-time dimensions. We begin with the Lagrangian

$$\mathcal{L} = - \bar{\psi} \left(\frac{1}{i} \gamma^\mu D_\mu + m \right) \psi \tag{15.49}$$

where the covariant derivative in presence of gauge fields is given by

$$D_\mu = D_\mu^{\text{grav.}} + iV_\mu^a T^a \tag{15.50}$$

and $D_\mu^{\text{grav.}}$ is defined in eq. (15.7).

The change of the fermionic integration measure under the generalized chiral transformation

$$\psi(x) \rightarrow e^{i\alpha(x)\gamma_{2n+1}}\psi(x) \tag{15.51}$$

then produces a Jacobian

$$J[\alpha] = e^{-i\int d^{2n}x\, \alpha(x)\mathcal{A}(x)} \tag{15.52}$$

where

$$\mathcal{A}(x) = \frac{2h(x)}{(4\pi)^n}\lim_{t\to 0}\frac{1}{t^n}\operatorname{tr}_{\gamma c}\left[\gamma_{2n+1}e^{-F_{\mu\nu}\Sigma^{\mu\nu}t}\left\{\det\frac{(\hat{R}/2t)}{\sinh(\hat{R}/2t)}\right\}^{\frac{1}{2}}\right] \tag{15.53}$$

and

$$\begin{aligned}
\Sigma^{\mu\nu} &= \tfrac{1}{4}[\gamma^\mu,\gamma^\nu]\\
\hat{R} &= (\hat{R}_{\mu\nu}) = (R_{\mu\nu\rho\sigma}\Sigma^{\rho\sigma}) = (\tfrac{1}{2}R_{\mu\nu\rho\sigma}\gamma^\rho\gamma^\sigma)\\
F_{\mu\nu} &= F^a_{\mu\nu}T^a\ , \quad T^a = \tfrac{1}{2}\lambda^a
\end{aligned} \tag{15.54}$$

In analogy with γ_5 in four dimensions, one can show (dim M = 2n)

$$tr[\gamma_{2n+1}\gamma^{a_1}\dots\gamma^{a_j}] = \begin{cases} 0 & \text{for } 1 \leq j < 2n \\ 2^n \varepsilon^{a_1 \dots a_{2n}} & \text{for } j = 2n \end{cases}$$

and

$$tr[\gamma_{2n+1}\gamma^{\mu_1}\dots\gamma^{\mu_j}] = h_{a_1}{}^{\mu_1}\dots h_{a_j}{}^{\mu_j}\, tr[\gamma_{2n+1}\gamma^{a_1}\dots\gamma^{a_j}]$$

$$= \begin{cases} 0 & \text{for } 1 \leq j < 2n \\ \underline{2^n h_{a_1}{}^{\mu_1}\dots h_{a_{2n}}{}^{\mu_{2n}}\,\varepsilon^{a_1\dots a_{2n}}\,h(x)}\,\frac{1}{h(x)} & \text{for } j = 2n \end{cases}$$

$$= \begin{cases} 0 & \text{for } 1 \leq j < 2n \\ \frac{2^n}{h(x)}\,\varepsilon^{\mu_1\dots\mu_{2n}} & \text{for } j = 2n \end{cases} \tag{15.55}$$

Observe that the terms contributing to $\lim_{t\to 0} t^{-n}$ in (15.53) are proportional to t^k, $k \leq n$; moreover, terms contributing to $tr[\gamma_{2n+1}\dots]$ must be proportional to t^k, $k = n$ giving exactly 2n γ-matrices as required for a non-vanishing result in (15.55). Thus, we conclude that contributions to $\lim_{t\to 0} t^{-n}\, tr[\gamma_{2n+1}\dots]$ originate from terms $\sim t^n$ in $\exp\{-F\cdot\Sigma t\}[\det \cdot/\cdot]^{1/2}$ of eq. (15.53).

Now let us try to rederive some previously obtained results.

In the pure Yang-Mills limit (flat space: $h = 1$), we have immediately ($\exp\{-\Sigma\cdot F t\} = \sum_{k=0}^{\infty} \frac{(-)^k}{k!} t^k (\Sigma\cdot F)^k$)

$$\mathcal{A}(x) = 2(4\pi)^{-n} t^{-n}\, tr\Big[\gamma_{2n+1} \frac{(-)^n}{n!} t^n \underbrace{(\Sigma_{\mu\nu} F^{\mu\nu})}_{-\frac{1}{2}F_{\mu\nu}\gamma^\mu\gamma^\nu}{}^n\Big]$$

$$= \frac{(-1)^n}{n!} \frac{2}{(4\pi)^n} \frac{1}{2^n} tr_c(F_{\mu_1\nu_1} \cdots F_{\mu_n\nu_n}) \underbrace{tr[\gamma_{2n+1} \gamma^{\mu_1} \cdots \gamma^{\mu_n}]}_{= 2^n \varepsilon^{\mu_1\nu_1 \cdots \mu_n\nu_n}}$$

$$= (-1)^n \frac{1}{2^{2n-1}\pi^n n!} \varepsilon^{\mu_1\nu_1 \cdots \mu_n\nu_n} \; tr_c(F_{\mu_1\nu_1} \cdots F_{\mu_n\nu_n}) \tag{15.56}$$

This result should be compared with eq. (15.55/.56) of chapter 13 which we can now write as $\partial^\mu j^5_\mu = 2i M j^5 + i \mathcal{A}(x)$.

Assuming manifolds without boundary, $\partial M_{2n} = 0$, we are allowed to write

$$n_+ - n_- = \frac{1}{2} \int d^{2n}x \, \mathcal{A}(x)$$

$$= \frac{(-1)^n}{2^{2n}\pi^n n!} \int d^{2n}x \, \varepsilon^{\mu_1\nu_1 \cdots \mu_n\nu_n} \, tr_c(F_{\mu_1\nu_1} \cdots F_{\mu_n\nu_n}) \tag{15.57}$$

In the literature, one finds the more sophistocated expression

$$n_+ - n_- = \int_M ch(F) = \int_M tr\, e^{i\frac{F}{2\pi}} \tag{15.58}$$

where the integrand comes from

$$ch(F) = tr\, e^{\frac{i}{2\pi}F} \tag{15.59}$$

We then have $(F = \frac{i}{2} F_{\mu\nu} dx^\mu dx^\nu)$

$$n_+ - n_- = \int_{M_{2n}} tr\, e^{iF/2\pi} = \sum_{k=0}^{\infty} \frac{i^k}{(2\pi)^k k!} \int_{M_{2n}} tr\, F^k \tag{15.60}$$

$$= \frac{1}{(2\pi)^n} \frac{i^{2n}}{n!\,2^n} \int tr\, F_{\mu_1\nu_1 \dots \mu_n\nu_n} \underbrace{dx^{\mu_1} dx^{\nu_1} \dots dx^{\mu_n} dx^{\nu_n}}_{= \varepsilon^{\mu_1\nu_1 \dots \mu_n\nu_n} dx^1 dx^2 \dots dx^{2n}}$$

$$= \int dx^1 \dots dx^{2n} \frac{(-1)^n}{(4\pi)^n n!} \varepsilon^{\mu_1\nu_1 \dots \mu_n\nu_n} tr(F_{\mu_1\nu_1} \dots F_{\mu_n\nu_n}) \tag{15.61}$$

as before.

The special case of gravitation is contained in eq. (15.53), through the determinant

$$\det \left(\frac{\hat{R}t/2}{\sinh(\hat{R}t/2)} \right)^{\frac{1}{2}} \tag{15.62}$$

Observe that this is only a determinant with regard to the space-time (or bein) indices of $\hat{R} = (\hat{R}_{\mu\nu}) = (\frac{1}{2} R_{\mu\nu\rho\sigma} \gamma^\rho \gamma^\sigma)$, i.e., det $(\cdots)^{1/2}$ is still a spin matrix.

Expanding the determinant (15.62)

$$\det \left(\frac{\hat{R}t/2}{\sinh \hat{R}t/2} \right)^{\frac{1}{2}} =: \sum_{k=0}^{\infty} t^k F_k \left(\tfrac{1}{2} R\gamma\gamma \right)$$

$$=: \sum_{k=0}^{\infty} \frac{t^k}{2^k} M^{\nu_1 \dots \nu_k}_{\mu_1 \dots \mu_k} R^{\mu_1}{}_{\nu_1\rho_1\sigma_1} \gamma^{\rho_1} \gamma^{\sigma_1} \dots R^{\mu_k}{}_{\nu_k\rho_k\sigma_k} \gamma^{\rho_k} \gamma^{\sigma_k}$$

the first few terms are

$$= 1 - \frac{t^2}{12} \hat{R}_{\mu\nu} \hat{R}^{\mu\nu} + \frac{t^4}{6!} \left[5(\hat{R}_{\mu\nu} \hat{R}^{\mu\nu})^2 + 4 \hat{R}_{\mu\nu} \hat{R}^{\nu\rho} \hat{R}_{\rho\sigma} \hat{R}^{\sigma\mu} \right] + \dots \tag{15.63}$$

Note that F_k contains exactly k $\hat{R}$'s.

The gravitational anomaly is then given by

$$\mathcal{A}(x) = 2h(x)(4\pi)^{-n} \lim_{t\to 0} t^{-n} tr\left[\gamma_{2n+1} \det\left(\frac{\hat{R}t/2}{\sinh \hat{R}t/2}\right)^{1/2}\right]$$

$$= 2h(x)(4\pi)^{-n} \lim_{t\to 0} t^{-n} tr\left[\gamma_{2n+1} \sum_{k=0}^{\infty} t^k F_k\left(\tfrac{1}{2}R\gamma\gamma\right)\right]$$

$$= 2h(x)(4\pi)^{-n} \frac{1}{2^n} M^{\nu_1\ldots\nu_n}_{\mu_1\ldots\mu_n} R^{\mu_1}{}_{\nu_1\rho_1\sigma_1} \cdots R^{\mu_n}{}_{\nu_n\rho_n\sigma_n} \cdot$$

$$\cdot\, tr\left[\gamma_{2n+1}\gamma^{\rho_1}\gamma^{\sigma_1}\cdots\gamma^{\rho_n}\gamma^{\sigma_n}\right]$$

$$\underset{(55)}{=} 2(4\pi)^{-n} M^{\nu_1\ldots\nu_n}_{\mu_1\ldots\mu_n} \varepsilon^{\rho_1\sigma_1\ldots\rho_n\sigma_n} R^{\mu_1}{}_{\nu_1\rho_1\sigma_1} \cdots R^{\mu_n}{}_{\nu_n\rho_n\sigma_n} \tag{15.64}$$

Again, for a manifold M_{2n} without boundary, $\partial M_{2n} = 0$, we have the index theorem

$$n_+ - n_- = \frac{1}{2}\int d^{2n}x\, \mathcal{A}(x)$$

$$= \int d^{2n}x\, (4\pi)^{-n} M^{\nu_1\ldots\nu_n}_{\mu_1\ldots\mu_n} \varepsilon^{\rho_1\sigma_1\ldots\rho_n\sigma_n} R^{\mu_1}{}_{\nu_1\rho_1\sigma_1} \cdots R^{\mu_n}{}_{\nu_n\rho_n\sigma_n} \tag{15.65}$$

which is in agreement with $\left(R^{\mu}{}_{\nu} := \frac{1}{2} R^{\mu}{}_{\nu\rho\sigma}\, dx^\rho dx^\sigma\right)$

$$n_+ - n_- = \int_M \hat{A}(M) = \int_M \det\left(\frac{R/4\pi}{\sinh(R/4\pi)}\right)^{\frac{1}{2}} \tag{15.66}$$

where $\hat{A}(M)$ is the $\hat{A}$-genus of the manifold M.

Finally, mixed anomalies follow from eq. (15.53) by expanding the exponential as well as the determinant. The result is

$$\mathcal{A}(x) = 2(4\pi)^{-n} \sum_{m=0}^{\infty} \frac{(-1)^m}{m!} \varepsilon^{\alpha_1\beta_1 \dots \alpha_m\beta_m \rho_1\sigma_1 \dots \rho_{n-m}\sigma_{n-m}} \cdot$$

$$\cdot \, tr_c \left(F_{\alpha_1\beta_1} \dots F_{\alpha_m\beta_m}\right)$$

$$\cdot M^{\nu_1 \dots \nu_{n-m}}_{\mu_1 \dots \mu_{n-m}} R^{\mu_1}{}_{\nu_1\rho_1\sigma_1} \dots R^{\mu_{n-m}}{}_{\nu_{n-m}\rho_{n-m}\sigma_{n-m}} \tag{15.67}$$

and

$$n_+ - n_- = \int_{M_{2n}} \hat{A}(M_{2n}) \wedge ch(F) = \frac{1}{2} \int d^{2n}x \, \mathcal{A}(x) \tag{15.68}$$

$$= \int_{M_{2n}} \det\left(\frac{R/4\pi}{\sinh(R/4\pi)}\right)^{\frac{1}{2}} \wedge tr_c \, e^{iF/2\pi}$$

Additional References to

Chiral Anomaly with Gravitational Background Field

T. Kimura, Prog. Theor. Phys. 42, 1191 (1969)

R. Delbourgo, A. Salam, Phys. Lett. 40B, 381 (1972)

T. Eguchi, P.G.O. Freund, Phys. Rev. Lett. 37, 1251 (1976)

J. Gipson, Phys. Rev. D29, 2989 (1984)

R. Endo, M. Takao, Prog. Theor. Phys. 73, 803 (1985)

T. Eguchi, P. Gilkey, A. Hanson, Phys. Rep. 66, 213 (1980)

L. Alvarez-Gaumé, P. Ginsparg, Ann. Phys. 161, 423 (1985)

General Relativity:

S. Weinberg, "Gravitation and Cosmology", John Wiley, New York (1971)

R.U. Sexl, H.K. Urbantke, "Gravitation und Kosmoslogie", BI Mannheim (1974)

H. Stephani, "Allgemeine Relativitätstheorie", VEB Deutscher Verlag der Wissenschaften, Berlin (1980)

B. De Witt, "Dynamical Theory of Groups and Fields", Gordon and Breach, New York (1965)

16. Vacuum Charge in 2n + 1 Dimensions

In chapter 14 we investigated vacuum polarization by static background gauge fields in odd-dimensional space and observed that an integer or half-integer fermion number can be assigned to topological non-trivial field excitations.

We now want to consider the general case of fermions which, in addition to a SU(N) vector gauge field, interact as well with an external gravitational field. Such considerations are of interest with regard to Kaluza-Klein models; our results can be used, e.g., to determine the vacuum charge caused by the magnetic monopoles of the 5-dimensional Kaluza-Klein theory. It is interesting to see how our methods from chapter 14 can be applied in the more general case of a curved background metric.

The Lagrangian in 2n+ 1 dimensions now reads

$$\mathcal{L} = -\bar{\psi}\left[\frac{1}{i}\gamma^{\mu}\nabla_{\mu} + m\right]\psi \tag{16.1}$$

with
$$\nabla_{\mu} = \partial_{\mu} - \frac{1}{2}A_{\mu}^{ab}\Sigma_{ab} + iV_{\mu} \tag{16.2}$$

where $\mu = 0, 1, 2, \ldots, 2n$.

The fermion fields ψ transform according to the fundamental representation of the gauge group SU(N). We again assume the gauge field V_{μ} to be static and we choose the temporal gauge : $V^0 = 0$, $V^i = V^i(\vec{x})$, $\vec{x} \equiv (x^1, \ldots, x^{2n}) \equiv (x^i)$. The metric $g_{\mu\nu}(x)$ is similarly assumed to be static and can thus, without limiting its generality, be assumed in the form

$$g_{\mu\nu}(x) = \left(\begin{array}{c|c} 1 & 0 \\ \hline 0 & g_{ij}(\vec{x}) \end{array}\right) \tag{16.3}$$

Furthermore, it should be asymptotically (locally) flat. Besides, we imagine the Wick rotation in the time, i.e., the x^0-component, to have already been made so that we can work with a manifold of signature

(+, +, ..., +). Finally, $g_{\mu\nu}$ should be so composed that a well-defined spin connection A_μ^{ab} exists.

We now calculate the vacuum expectation value of the SU(N)-singlet charge operator in presence of background fields:

$$Q_{2n+1} = \int d^{2n}x \, h(x) \langle \bar{\psi}(x) \gamma^0 \psi(x) \rangle \tag{16.4}$$

Like in the earlier examples, an explicit normal ordering of the current operator is obsolete: the ζ -function regularization will automatically produce a vanishing vacuum charge for V = 0 and $g_{\mu\nu} = \eta_{\mu\nu}$. In the by now familiar way, the charge density in (16.4) can be represented as a derivation of the following generating functional with respect to $V_0(x)$:

$$\begin{aligned} Z[V_0] &= \int [d\psi d\bar{\psi}] e^{-\int d^{2n+1}x \, h(x) \bar{\psi}(-i\gamma_\mu \nabla^\mu - V_0\gamma^0 + m)\psi} \\ &= \det[-i\not{\nabla} + m - V_0\gamma^0] \end{aligned} \tag{16.5}$$

$$\begin{aligned} \langle \bar{\psi}(x)\gamma^0\psi(x) \rangle &\equiv \frac{\langle 0|\bar{\psi}(x)\gamma^0\psi(x)|0\rangle}{\langle 0|0\rangle} \\ &= \frac{\delta}{\delta V_0(x)} \ln \det[-i\not{\nabla} + m - V_0\gamma^0]\Big|_{V_0=0} \end{aligned} \tag{16.6}$$

The following steps are similar to their counterparts in chapter 14 so that we can sketch them briefly without further comment:

$$\begin{aligned} \langle \bar{\psi}(x)\gamma^0\psi(x) \rangle &= \frac{\delta}{\delta V_0(x)} \{ \ln \det[-i\not{\nabla} + m - V_0\gamma^0] \\ &\qquad + \ln \det[\,i\not{\nabla} + m] \}\Big|_{V_0=0} \\ &= -\frac{d}{ds}\Big|_0 \frac{\delta}{\delta V_0(x)} \zeta(\not{\nabla}^2 + m^2 - V_0\gamma^0(i\not{\nabla}+m)\,|\,s)\Big|_{V_0=0} \\ &= -\frac{d}{ds}\Big|_0 s \sum_n \lambda_n^{-(1+s)} \varphi_n^+(x)\gamma^0(i\not{\nabla}+m)\varphi_n(x) \end{aligned} \tag{16.7}$$

Here $$(\not{\nabla}^2 + m^2)\varphi_n = \lambda_n \varphi_n \tag{16.8}$$

The completeness of the φ's and the identity

$$A^{-s} = \frac{1}{\Gamma(s)} \int_0^\infty dt\, t^{s-1} e^{-At}, \quad A>0$$

produce

$$\langle \psi^+(x)\psi(x)\rangle = -\frac{1}{h(x)} \frac{d}{ds}\Big|_0 \frac{s}{\Gamma(1+s)} \int_0^\infty dt\, t^s \cdot$$

$$\cdot \lim_{x'\to x} tr_{\gamma c}\left[\gamma^0 (i\not{\nabla}_x + m) e^{-(\not{\nabla}_x^2 + m^2)t}\right] \delta^{(2n+1)}(x-x') \tag{16.9}$$

where the square of the Dirac operator reads

$$\not{\nabla}^2 = -g^{\mu\nu} \nabla_\mu \nabla_\nu + \frac{1}{4} R + i F_{\mu\nu} \Sigma^{\mu\nu} \tag{16.10}$$

Because of the special form (16.2) of the metric for which the Christoffel symbols as well as the components of the Riemann and Ricci tensors vanish as soon as one of the indices becomes zero, the simplification is

$$\not{\nabla}^2 = -\partial_0^2 + \vec{\not{\nabla}}^2_{2n} \tag{16.11}$$

where $\vec{\not{\nabla}}^2_{2n}$ has the same structure as (16.10); but it is only dependent on the 2n variables $\vec{x}$, i.e., the Greek indices in (16.10) are replaced by the Latin indices. In particular, the curvature scalar $R_{2n}(\vec{x})$ --as it occurs in $\vec{\not{\nabla}}^2_{2n}$ -- is constructed in exactly the same way from $g_{ij}(\vec{x})$ as $R(x)$ is formed from $g_{\mu\nu}(x)$. Following similar steps which led us to equation (16.11), we obtain in the present case

$$\langle \psi^+(x)\psi(x)\rangle = -m \frac{1}{h(x)} \int_{-\infty}^{\infty} \frac{dk_0}{2\pi} \frac{d}{ds}\Big|_0 \frac{s}{\Gamma(1+s)} \int_0^\infty dt\, t^s \cdot$$

$$\cdot e^{-(m^2+k_0^2)t} \lim_{\vec{x}'\to\vec{x}} tr_{\gamma c}\left[\gamma^0 e^{-\not{\nabla}^2_{2n} t}\right]_{\vec{x}} \delta^{(2n)}(\vec{x}-\vec{x}') \tag{16.12}$$

As before, the following fact is now very helpful: the right-hand side of (16.12) can be reproduced by calculating the vacuum expectation value $\langle \bar{\Psi}(\vec{x})\gamma^0 \Psi(\vec{x})\rangle$ in only 2n dimensions in a theory of fermions $\Psi(\vec{x})$ with mass $M = (m^2 + k_o^2)^{1/2}$ which interact with a classical gauge field $V^i(\vec{x})$ and a gravitational field $g_{ij}(\vec{x})$; more precisely,

$$\langle \psi(x)^+ \psi(x)\rangle = \int_{-\infty}^{\infty} \frac{dk_0}{2\pi} \frac{m}{M} \langle \bar{\Psi}(\vec{x})\gamma^0 \Psi(\vec{x})\rangle \Big|_{M=(m^2+k_0^2)^{1/2}} \tag{16.13}$$

The proof follows with our usual methods, by varying the generating functional,

$$Z[a] = \int [d\Psi d\bar{\Psi}]\, e^{-\int d^{2n}x\, h_{2n}(\vec{x}) \bar{\Psi}(\vec{x}) \left(\frac{1}{i}\nabla_{2n} + M - a(\vec{x})\gamma^0\right) \Psi(\vec{x})}$$

with respect to $a(\vec{x})$ followed by a ζ-function regularization.

Also note

$$h(x) = \left(\det(g_{\mu\nu}(x))\right)^{1/2} = \left(\det(g_{ij}(\vec{x}))\right)^{1/2} = h_{2n}(\vec{x}) \tag{16.14}$$

As has already been mentioned, the matrix γ^0 in the 2n-dimensional theory whose γ-matrices are γ^i, i = 1, 2, ..., 2n, plays the role of the generalized γ_5-matrix since it anti-commutes with all γ^i. Thus, it is possible to express the expectation value in (16.13) by the anomalous divergence in 2n dimensions

$$\partial_i h_{2n}(\vec{x}) \langle \bar{\Psi}(\vec{x})\gamma^i\gamma^0 \Psi(\vec{x})\rangle =$$

$$2iM h_{2n}(\vec{x}) \langle \bar{\Psi}(\vec{x})\gamma^0 \Psi\rangle + i\mathcal{A}_{2n}(\vec{x}) \tag{16.15}$$

where $\mathcal{A}_{2n}$ is given by (15.53) with the obvious replacements

$F_{\mu\nu}(x)\Sigma^{\mu\nu} \rightarrow F_{ij}(\vec{x})\Sigma^{ij}$, etc. Substituting this in (16.13) gives us for the vacuum charge (since the divergence term makes no contribution)

$$Q_{2n+1} = \int d^{2n}x \, h(x) \langle \psi^{+}(x)\psi(x)\rangle$$

$$= -\frac{1}{2} \int_{-\infty}^{\infty} \frac{dk_0}{2\pi} \frac{m}{m^2+k_0^2} \int d^{2n}x \, \mathcal{A}_{2n}(\vec{x})$$

from which follows

$$Q_{2n+1} = -\frac{1}{2}\frac{m}{|m|} \cdot \frac{1}{2}\int d^{2n}x \, \mathcal{A}_{2n}(\vec{x}) \qquad (16.16)$$

which, according to (15.68) can also be written as

$$Q_{2n+1} = -\frac{1}{2}\frac{m}{|m|} \int_{M_{2n}} \hat{A}(M_{2n}) \wedge ch(\mathcal{F}) \qquad (16.17)$$

with the 2n-dimensional manifold M_{2n} determined by $g_{ij}(\vec{x})$ and the field strength form

$$\mathcal{F} = i\,\frac{1}{2}\,\mathcal{F}_{ij}\, dx^i \wedge dx^j$$

Equation (16.17) is the most important result of this chapter; it shows that in a non-trivial metric background, the integral over the single Chern-character ch(F) present in the equation for Q_{2n+1} in chapter (14)[cf.(14.76)ff.] has to be replaced by the integral over $\hat{A}(M_{2n})\wedge ch(F)$.

From (16.17) we can see that a non-vanishing vacuum charge in a pure gravitational field only occurs in 4n + 1 dimensions, since the power expansion of the $\hat{A}$-genus contains only 4n-forms, i.e., because a chiral anomaly occurs only in 4n dimensions.

In order to elucidate (16.17), we write out the vacuum charge for n = 2 explicitly. We use the power expansion of $\hat{A}$ and ch(F) to obtain

$$Q_5 = -\frac{1}{2}\frac{m}{|m|}\left(N\frac{1}{24}\cdot\frac{1}{8\pi^2}\int \mathrm{tr}\, R^2 - \frac{1}{8\pi^2}\int \mathrm{tr}\, F^2\right)$$

$$= -\frac{1}{2}\frac{m}{|m|}\left(\frac{-N}{768\pi^2}\int d^4x\ \varepsilon^{\rho\sigma\alpha\beta} R_{\mu\nu\rho\sigma} R^{\mu\nu}{}_{\alpha\beta}\right.$$

$$\left. + \frac{1}{16\pi^2}\int d^4x \sqrt{g}\ \mathrm{tr}\, F_{\mu\nu}{}^{*}F^{\mu\nu}\right) \qquad (16.18)$$

$$= -\frac{1}{2}\frac{m}{|m|}\left(-\frac{N}{24}P_1 + q\right)$$

Here, P_1 is the first Pontryagin number and q is the Pontryagin index of the gauge field. Since the vacuum charge depends only on these topological invariants, it is clear that they do not change under local deformations of the background fields. The factor N arises from $\mathrm{tr}_c 1 = N$ and shows that the N components of the fermion multiplet make the same contribution to the gravitational anomaly, independently of each other. If we had done the above calculation for a U(1)-gauge theory we would have obtained (16.18) with N = 1. It is typical for n = 2 that Q splits into a sum of a gravitational and Yang-Mills term; in higher dimensions, mixed terms between field strength and the Riemann tensor appear.

Additional References to
Vacuum Charge in 2n+1 Dimensions

M.B. Paranjape, G.W. Semenoff, Phys. Rev. D31, 1324 (1985)

D.J. Gross, M.J. Perry, Nucl. Phys. B226, 29 (1983)

R.D. Sorkin, Phys. Rev. Lett. 51, 87 (1983)

S.W. Hawking, Phys. Lett. 60A, 81 (1977)

Kaluza-Klein Theory:

T. Kaluza, Sitzungsber. Preuss. Akad. Wiss. Phys. Math. Kl, 966 (1921); O. Klein, Z. Phys. 37, 895 (1926)

H.T. Flint, "The Quantum Equation and the Theory of Fields", Methuen, London (1966)

D.J. Toms in: An Introduction to Kaluza-Klein Theories, H.C. Lee (ed.), World Scientific, Singapore (1984)

Texts and Monographs in Physics

Editors: **W. Beiglböck, J. L. Birman, E. H. Lieb, T. Regge, W. Thirring**

M. Chaichian, N. F. Nelipa

Introduction to Gauge Field Theories

Translated from the Russian by J. Estrin

1984. 75 figures. XII, 332 pages. ISBN 3-540-13008-X

Contents: Introduction. - Invariant Lagrangians: Global Invariance. Local (Gauge) Invariance. Spontaneous Symmetry-Breaking. - Quantum Theory of Gauge Fields: Path Integrals and Transition Amplitudes. Covariant Perturbation Theory. - Gauge Theory of Electroweak Interactions: Lagrangians of the Electroweak Interactions. Quantum Electrodynamics. Weak Interactions. Higher Orders in Perturbation Theory. - Gauge Theory of Strong Interactions: Asymptotically Free Theories. Dynamical Structure of Hadrons. Quantum Chromodynamics; Perturbation Theory. Lattice Gauge Theories. Quantum Chromodynamics on a Lattice. Grand Unification. Topological Solitons and Instantons. - Conclusion. - Bibliography. - List of Symbols. - Subject Index.

F. J. Ynduráin

Quantum Chromodynamics

An Introduction to the Theory of Quarks and Gluons

1983. XI, 227 pages. ISBN 3-540-11752-0

Contents: Generalities. - QCD as a Field Theory. - Deep Inelastic Processes. - Quark Masses, PCAC, Chiral Dynamics, and the QCD Vacuum. - Functional Methods, Nonperturbative Solution. - References. - Index.

W. Greiner, B. Müller, J. Rafelski

Quantum Electrodynamics of Strong Fields

With an Introduction into Modern Relativistic Quantum Mechanics

1985. 258 figures. XI, 594 pages. ISBN 3-540-13404-2

Contents: Introduction. - The Wave Equation for Spin-1/2-Particles. - Dirac Particles in External Potentials. - The Hole-Theory. - The Klein Paradox. - Resonant States in Supercritical Fields. - Quantum Electrodynamics of Weak Fields. - The Classical Dirac Field Interacting with a Classical Electromagnetic Field - Formal Properties. - Second Quantization of the Dirac Field and Definition of the Vacuum. - Evolution of the Vacuum State in Supercritical Potentials. - Superheavy Quasimolecules. - The Dynamics of Heavy-Ion Collisions. - Experimental Test of Supercritical Fields in Heavy-Ion Collisions. - Vacuum Polarization. - Vacuum Polarization: Arbitrarily Strong External Potentials. - Many-Body Effects in QED of Strong Fields. - Bosons Bound in Strong Potentials. - Subcritical External Potentials. - Overcritical Potential for Bose Fields. - Strong Yang-Mills Fields. - Strong Fields in General Relativity. - References. - Subject Index.

Springer-Verlag
Berlin Heidelberg
New York Tokyo

Lecture Notes in Physics

Vol. 214: H. Moraal, Classical, Discrete Spin Models. VII, 251 pages. 1984.

Vol. 215: Computing in Accelerator Design and Operation. Proceedings, 1983. Edited by W. Busse and R. Zelazny. XII, 574 pages. 1984.

Vol. 216: Applications of Field Theory to Statistical Mechanics. Proceedings, 1984. Edited by L. Garrido. VIII, 352 pages. 1985.

Vol. 217: Charge Density Waves in Solids. Proceedings, 1984. Edited by Gy. Hutiray and J. Sólyom. XIV, 541 pages. 1985.

Vol. 218: Ninth International Conference on Numerical Methods in Fluid Dynamics. Edited by Soubbaramayer and J.P. Boujot. X, 612 pages. 1985.

Vol. 219: Fusion Reactions Below the Coulomb Barrier. Proceedings, 1984. Edited by S.G. Steadman. VII, 351 pages. 1985.

Vol. 220: W. Dittrich, M. Reuter, Effective Lagrangians in Quantum Electrodynamics. V, 244 pages. 1985.

Vol. 221: Quark Matter '84. Proceedings, 1984. Edited by K. Kajantie. VI, 305 pages. 1985.

Vol. 222: A. García, P. Kielanowski, The Beta Decay of Hyperons. Edited by A. Bohm. VIII, 173 pages. 1985.

Vol. 223: H. Saller, Vereinheitlichte Feldtheorien der Elementarteilchen. IX, 157 Seiten. 1985.

Vol. 224: Supernovae as Distance Indicators. Proceedings, 1984. Edited by N. Bartel. VI, 226 pages. 1985.

Vol. 225: B. Müller, The Physics of the Quark-Gluon Plasma. VII, 142 pages. 1985.

Vol. 226: Non-Linear Equations in Classical and Quantum Field Theory. Proceedings, 1983/84. Edited by N. Sanchez. VII, 400 pages. 1985.

Vol. 227: J.-P. Eckmann, P. Wittwer, Computer Methods and Borel Summability Applied to Feigenbaum's Equation. XIV, 297 pages. 1985.

Vol. 228: Thermodynamics and Constitutive Equations. Proceedings, 1982. Edited by G. Grioli. V, 257 pages. 1985.

Vol. 229: Fundamentals of Laser Interactions. Proceedings, 1985. Edited by F. Ehlotzky. IX, 314 pages. 1985.

Vol. 230: Macroscopic Modelling of Turbulent FLows. Proceedings, 1984. Edited by U. Frisch, J.B. Keller, G. Papanicolaou and O. Pironneau. X, 360 pages. 1985.

Vol. 231: Hadrons and Heavy Ions. Proceedings, 1984. Edited by W.D. Heiss. VII, 458 pages. 1985.

Vol. 232: New Aspects of Galaxy Photometry. Proceedings, 1984. Edited by J.-L. Nieto. XIII, 350 pages. 1985.

Vol. 233: High Resolution in Solar Physics. Proceedings, 1984. Edited by R. Muller. VII, 320 pages. 1985.

Vol. 234: Electron and Photon Interactions at Intermediate Energies. Proceedings, 1984. Edited by D. Menze, W. Pfeil and W. J. Schwille. VII, 481 pages. 1985.

Vol. 235: G.E.A. Meier, F. Obermeier (Eds.), Flow of Real Fluids. VIII, 348 pages. 1985.

Vol. 236: Advanced Methods in the Evaluation of Nuclear Scattering Data. Proceedings, 1985. Edited by H.J. Krappe and R. Lipperheide. VI, 364 pages. 1985.

Vol. 237: Nearby Molecular Clouds. Proceedings, 1984. Edited by G. Serra. IX, 242 pages. 1985.

Vol. 238: The Free-Lagrange Method. Proceedings, 1985. Edited by M.J. Fritts, W.P. Crowley and H. Trease. IX, 313 pages. 1985.

Vol. 239: Geometrics Aspects of the Einstein Equations and Integrable Systems. Proceedings, 1984. Edited by R. Martini. V, 344 pages. 1985.

Vol. 240: Monte-Carlo Methods and Applications in Neutronics, Photonics and Statistical Physics. Proceedings, 1985. Edited by R. Alcouffe, R. Dautray, A. Forster, G. Ledanois and B. Mercier. VIII, 483 pages. 1985.

Vol. 241: Numerical Simulation of Combustion Phenomena. Proceedings, 1985. Edited by R. Glowinski, B. Larrouturou and R. Temam. IX, 404 pages. 1985.

Vol. 242: Exactly Solvable Problems in Condensed Matter and Relativistic Field Theory. Proceedings, 1985. Edited by B.S. Shastry, S.S. Jha and V. Singh. V, 318 pages. 1985.

Vol. 243: Medium Energy Nucleon and Antinucleon Scattering. Proceedings, 1985. Edited by H.V. von Geramb. IX, 576 pages. 1985.

Vol. 244: W. Dittrich, M. Reuter, Selected Topics in Gauge Theories. V, 315 pages. 1986.